Lecture Notes in Mechanical Engineering

Lecture Notes in Mechanical Engineering (LNME) publishes the latest developments in Mechanical Engineering—quickly, informally and with high quality. Original research reported in proceedings and post-proceedings represents the core of LNME. Volumes published in LNME embrace all aspects, subfields and new challenges of mechanical engineering. Topics in the series include:

- Engineering Design
- Machinery and Machine Elements
- Mechanical Structures and Stress Analysis
- Automotive Engineering
- Engine Technology
- Aerospace Technology and Astronautics
- Nanotechnology and Microengineering
- Control, Robotics, Mechatronics
- MEMS
- Theoretical and Applied Mechanics
- Dynamical Systems, Control
- Fluid Mechanics
- Engineering Thermodynamics, Heat and Mass Transfer
- Manufacturing
- Precision Engineering, Instrumentation, Measurement
- Materials Engineering
- Tribology and Surface Technology

To submit a proposal or request further information, please contact the Springer Editor of your location:

China: Dr. Mengchu Huang at mengchu.huang@springer.com
India: Priya Vyas at priya.vyas@springer.com
Rest of Asia, Australia, New Zealand: Swati Meherishi
at swati.meherishi@springer.com
All other countries: Dr. Leontina Di Cecco at Leontina.dicecco@springer.com

To submit a proposal for a monograph, please check our Springer Tracts in Mechanical Engineering at http://www.springer.com/series/11693 or contact Leontina.dicecco@springer.com

Indexed by SCOPUS. All books published in the series are submitted for consideration in Web of Science.

More information about this series at http://www.springer.com/series/11236

Adolfo Crespo Márquez ·
Dragan Komljenovic · Joe Amadi-Echendu
Editors

14th WCEAM Proceedings

Springer

Editors
Adolfo Crespo Márquez
Department of Industrial Management
University of Seville
Seville, Spain

Dragan Komljenovic
Hydro-Quebec
Varennes, QC, Canada

Joe Amadi-Echendu
Graduate School of Technology
Management
University of Pretoria
Pretoria, South Africa

ISSN 2195-4356 ISSN 2195-4364 (electronic)
Lecture Notes in Mechanical Engineering
ISBN 978-3-030-64227-3 ISBN 978-3-030-64228-0 (eBook)
https://doi.org/10.1007/978-3-030-64228-0

This Springer imprint is published by the registered company Springer Nature Switzerland AG
The registered company address is: Gewerbestrasse 11, 6330 Cham, Switzerland

Preface

As the title implies, 14th WCEAM Proceedings arises from events and activities marking the fourteenth edition of the International Society of Engineering Asset Management (ISEAM) flagship World Congress on Engineering Asset Management (WCEAM) series. The content of 14th WCEAM Proceedings is divided into two parts. The first part includes sections 1 to 4 and comprises extended versions of the asset management track papers presented during 2019 World Congress on Reliability, Resilience, and Asset Management (WCRRAM). The Congress took place in Singapore from 28–31 July 2019, several months before the first cases of coronavirus infections were reported in December, 2019. The second part and section 5 of 14th WCEAM Proceedings includes papers presented during the 1st WCEAMOnline event held on 19 August 2020. During 1st WCEAMOnline, there were brief discussions on the ramifications of the ongoing COVID-19 pandemic on electricity and health infrastructure systems. The 1st WCEAMOnline event also featured presentations on academic curricula and programmes in Engineering Asset Management given that COVID-19 pandemic is imposing transformations towards online platforms for education and training in general.

A brief synopsis of 14th WCEAM follows. Section 1 of this e-book contains seven chapters that discuss business management issues ranging from investment and quality of serve challenges to servitization business models; the section also includes case studies of electrical utilities and urban commuter infrastructure. Section 2 contains six chapters addressing risk-based decision models for replacement of equipment and renovation of public infrastructure. Section 3 contains three chapters that discuss asset data and information including applications of industrial Internet of Things especially in manufacturing. Section 4 contains five chapters that examine asset management frameworks and systems from the viewpoint of resilience.

Section 5 contains three chapters with discourse on the management of resilience of asset systems in the context of major, large-scale instabilities and disruptions as exemplified by the COVID-19 pandemic. Since various global shocks and large-scale disruptions cannot be excluded in the future, it has become imperative to gain more in-depth knowledge on those phenomena not only to cope with them, but also to identify and seize opportunities they could create. It is worth remarking that COVID-19 pandemic is ongoing; hence, the intertwined topics of resilience and vulnerability of asset systems will continue to generate discussions in multifarious conferences, seminars, and workshops, as well as a plethora of discourse from the multi-disciplines that encompass knowledge and practice of asset management.

We humbly thank ISEAM for nominating and entrusting us with the responsibility to edit this 14th edition of WCEAM Proceedings. We are also grateful for the support provided to us through the established relationship between ISEAM and Springer.

October 2020

Adolfo Crespo Márquez
Dragan Komljenovic
Joe Amadi-Echendu

Contents

Business Management Issues

Servitization and the Management of Engineered Assets

Joe Amadi-Echendu(✉), Rayand Ramlal, and Floris Englebrecht

Department of Engineering and Technology Management, Graduate School of Technology Management, University of Pretoria, Pretoria, South Africa
joe.amadi-echendu@up.ac.za

Abstract. Managers of engineered assets have to make decisions to acquire, utilise, and retire equipment, machinery, and such man-made facilities. On the one hand, manufacturers, suppliers and vendors of machinery and equipment continuously seek to sustain their businesses. On the other hand, asset owners and users seek to achieve better performance of assets. Given the advent of the fourth industrial revolution technologies and globalization 4.0 business models, the question arises as to what level of product-to-service transformation is appropriate in a given user-manufacturer/supplier/vendor setting, and what should be the extent of the corresponding product-service system? This paper discusses an investigation into product-to-service transformation in terms of strategies adopted by equipment and machinery manufacturers, suppliers and vendors on the one hand, and also of asset owners and users on the other hand. Empirical data and information from the respective sides are summarily presented with regard to the effect of product-to-service transformation on the user-side management of engineered assets.

Keywords: Servitization · Product-service systems · Asset management

1 Introduction

Management revolves around decision making, and so, the management of engineered assets involves decisions as to how to acquire, utilise, (i.e., operate and maintain), and retire equipment, machinery, and such man-made facilities. With the advent of the fourth industrial revolution (4IR) technologies and globalization 4.0 business models, decisions about acquisition, utilisation (that is, operations and maintenance), and retirement of equipment, machinery, and such man-made facilities present fascinating challenges on the user side, that is, for users (i.e., managers, operators and maintainers), and owners (including custodians and/or stewards) of technology-laden engineered assets such as aircraft, earth-moving machines, power generation and water treatment facilities.

On the supply side, the 4IR technologies and globalization 4.0 business models also represent new and unprecedented challenges that are confronting original equipment manufacturers (OEMs), suppliers and vendors as they innovatively embed these technologies in the components, spare parts, equipment, machinery and systems that are

A. Crespo Márquez et al. (Eds.): WCEAM 2019, LNME, pp. 3–9, 2021.
https://doi.org/10.1007/978-3-030-64228-0_1

provided to asset users [1]. An asset such as a machine is a value item, and this value ethos implies that each stakeholder, whether from the side of the asset user or supplier, has a perception of the value of, say, a steam turbine or an earth moving machine.

Invariably, the interests of the stakeholders intersect where such perceptions of value intersect, thus influencing the interrelationships between the various stakeholders. An implication is that decisions to acquire, utilise, and retire an engineered asset would be influenced by the value perceptions of the pertinent stakeholders from both the user and supplier sides.

This paper discusses a longitudinal sequence of investigations into product-to-service transformation in the context of strategies adopted by manufacturers, suppliers and vendors of the product (engineered asset) on the one hand, and also of owners and users of the asset (product) on the other hand. Section 2 contains a brief discourse on the concepts of servitization, product-to-service transformation and product-service systems. Empirical data and information obtained from both the user- and supplier sides are summarized in Sect. 3. Section 4 contains some concluding remarks on the effect of servitization and product-to-service transformation on the user-side management of engineered assets.

2 Servitization and Product-Service Transformation

The concepts of servitization [2], deservitization [3], product-to-service transformation [4], and product-service systems [5] highlight the growing phenomenon whereby designers, OEMs, vendors, value-adding resellers, suppliers and agents supplement existing and new products (assets) with service offerings. The service(s) offered may be in the form tangible or intangible technology. Servitization involves bundling tangible and intangible services with tangible artefacts (i.e., products or engineered assets). It requires a two-way strategy. From the service provider side, it involves the delivery of a tangible or intangible service component as an embedded added value to providing or supplying an artefact or product. From the customer or user side, it involves the acceptance of a tangible or intangible service component as an integral part of the tangible artefact or product that is provided or supplied. With regard to asset acquisition, utilisation, or retirement, an interesting conundrum arises where, for example, a user of the asset may not subscribe to servitization strategy but, for technological and commercial reasons, the manufacturer, vendor, or supplier strategically prefers to bundle, for example, maintenance, repair and operations (MRO) as a service integral with the supply of an asset.

The picture in Fig. 1 captures several of the terms commonly found in literature about productization and servitization, such as 'technology as a product/service', within product-, use-, and result-oriented business models.

As a closely related concept, product-to-service transformation implies that the manufacturer or vendor supplies the asset (e.g., gas turbine or excavator) to the user as a service, say, on a pay-per-use basis, and on the presumption that the service bundle would be acceptable to the asset user. From the supply side, product-to-service transformation typically focuses on "bundling contracted services, equipment maintenance, consumable supplies, and financing" [7] to achieve recurring revenue and cash generation, that is, an annuity-income based business model. Obviously, evolutions in technology and business models greatly influence the product-to-service transformation processes and trajectory.

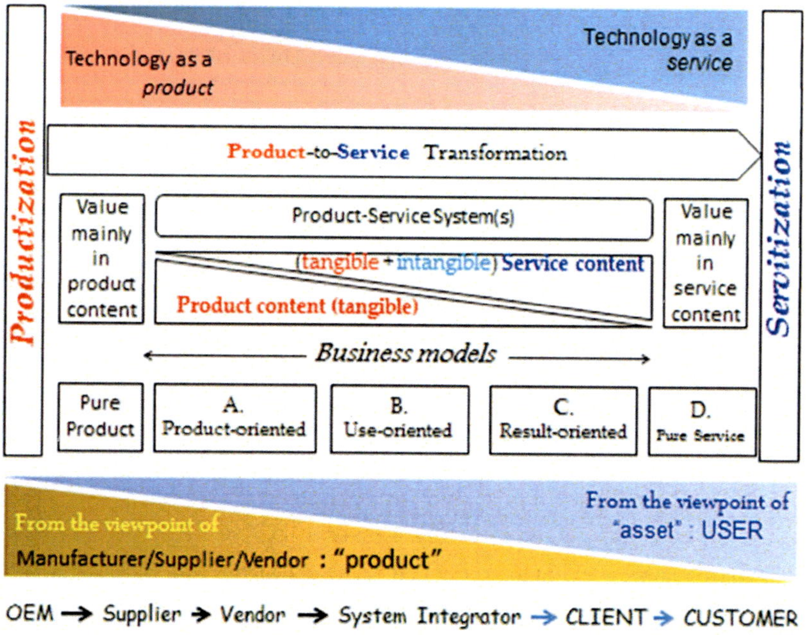

Fig. 1. Servitization and product-to-service transformation, (adapted from [6])

From a technology viewpoint, the spectrum ranges between commercialising technology as a product, as product-service or service-product combinations, and commercialising technology as a service. Theoretically, the product-to-service transformation journey offers an infinite range of possibilities to define the scope of product-service systems. However, in practice, the scope of a product-service system will be defined by the relationship between, say, the manufacturer, vendor, or supplier of a product (asset), and the user of the asset (product). Conversely, the scope of a product-service system establishes the relationships between the manufacturer, and vendor/supplier of the asset, the provider of services related to the asset, and the asset user.

As illustrated in Fig. 1, every servitization strategy inherently includes a certain degree of product-to-service transformation which is concomitant with the implementation of a product-service system [6]. As elucidated in [8], a servitization strategy imputes a product-service system with a number of features that encompass:

- the asset (or components of the asset) as the core product(s) (e.g., the gas turbine) or its constituent parts (e.g., blade, compressor, shaft, etc.);
- sustainability in terms of ability to re-use and/or re-cycle the asset or its constituent parts;
- capabilities to remotely monitor the asset (e.g., via application of sensors and IoT), and to exchange information as pertinent to the respective stakeholders to the asset;
- the range and scope of supplementary tangible (e.g., MRO) and intangible services possible within the product-service bundle;
- customizability of the product-service system; which reinforces

- value as perceived by the respective user-side and supply-side stakeholders to the asset.

These features are more or less qualitative interpretations of the intrinsic value ethos embodied in servitization and product-to-service transformation. The eventual scope of the product-service system will invariably influence decisions with regard to acquisition, utilisation and retirement of the asset or parts of the asset.

The following section briefly discusses empirical evidence obtained via semi-structured interviews of respondents associated with mining draglines and refinery turbo-machinery. The interview questions were designed to elucidate respondents' understanding of servitization, product-to-service transformation and product-service offerings. The respondent feedback provided some insight as to the extent of the deployment and use of product-service systems towards the management of engineered assets.

3 Empirical Information and Data

The first empirical evidence presented here arises from semi-structured interviews conducted in July 2014 with five representatives of four firms that provided maintenance services to companies operating draglines in coal mines. In each case, the dragline was owned by the operating company, whereas the firm providing maintenance services also represented the dragline manufacturer. Although the duration of the maintenance services agreements were typically more than five years, however, for reasons of confidentiality, the interviewer was not privy to the contracts of engagement between the respective maintenance services providers and the dragline operators in the three mines concerned.

Without mentioning 'servitization', an important point made by three of the five respondents was that the success of the asset and/or service supplier/user relationship depended on whether both the service provider and service recipient embraced a services culture that involved the interaction of technology, people, and the business of dragline coal mining. In one case, the manufacturer and maintenance service provider actually operated the dragline on behalf of the miner, and thus, carried out 1st-line maintenance (see Fig. 2) activities as part of operational services scope. In other cases, information about dragline condition was provided by the operator to the maintenance service providers as deemed necessary to implement 2nd- and 3rd-line maintenance activities during prescribed shutdowns. In many instances, 2nd- and 3rd-line maintenance activities included removal and replacement of unserviceable parts.

The primary purpose of the maintenance services contracts was to improve the reliability and operational safety of the draglines, whilst mitigating the lack of specialised maintenance skills within the operator organisations. Inevitably, the maintenance agreements involved making replacement decisions, but it was unclear from the interview responses how such replacement decisions were made.

The second empirical evidence arises from a 2018 study designed to ascertain the number of features of servitization and product-to-service transformation that were included in existing commercial agreements. The study entailed reviewing ninety-nine existing commercial agreements against the aforementioned features of product-service

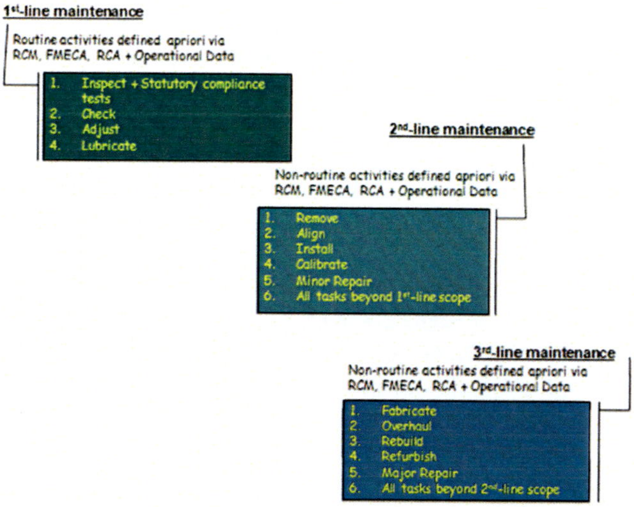

Fig. 2. 1st-, 2nd-, and 3rd-line maintenance activities

systems. The agreements were primarily concerned with the maintenance of turbomachinery in a case study refinery. Each agreement was with a specific vendor, even though a vendor could be involved in up to three specific contracts. Each agreement had been in existence for at least five years. The features of servitization and scope of the corresponding product-service to-service transformation system have been translated as depicted in Figs. 3(i) and (ii).

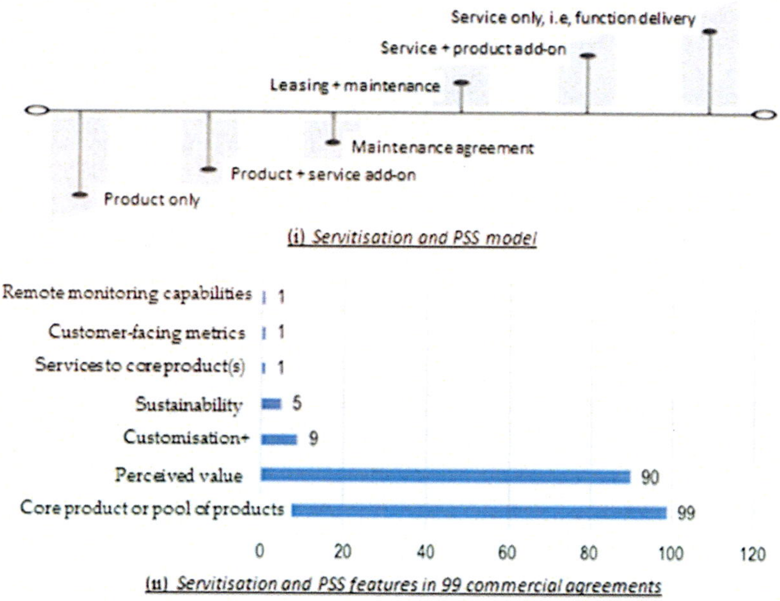

Fig. 3. Product-service transformation through maintenance service offering.

It is interesting to observe in Fig. 3(ii) that the asset (i.e., turbo-machinery), and its perceived value were the predominant features of the product-service system implied in the ninety-nine commercial agreements reviewed. As expected, every agreement offered the turbo-machinery as the core product. Two features, namely *core product* and *perceived value* were offered in ninety-one agreements. The intriguing aspect is that only ten agreements included three or more features of servitization and product-to-service transformation. The observations suggest that the maintenance services agreements reflect the realistic links between the turbo-machinery user and the associated maintenance service providers. The scope of the agreements mostly covered 2nd- and 3rd-line maintenance activities, even though some contracts also included 1st-line activities such as inspection and condition monitoring. The contents of the agreements corroborated the views expressed by the interview respondents, especially regarding roles and responsibilities for replacing unserviceable and obsolete components or parts of an asset.

4 Concluding Remarks

Empirical evidence here provides an indication of the scope of servitization, product-to-service transformation, and product-service systems implicit in existing relationships between users, manufacturers/vendors/suppliers of assets. Taking into consideration that the empirical evidence was gathered in 2014 and 2018, it is remarkable that practitioners appeared to be unconsciously aware of the concepts of servitization and product-to-service transformation. A probable explanation can be adduced from the convention where users first acquire assets from manufacturers, vendors, or suppliers, then secondly consider purchasing bundled-service offerings. The agreements support the convention that asset users tend to outsource associated services not necessarily as an integral part of the product (asset) purchase agreement. The empirical evidence showed that only one of the ninety-nine agreements offered six features of servitization. Eighty-nine agreements were based on the conventional productization business model.

The discourse above conflated a longitudinal review of existing agreements between turbo-machinery stakeholders with interviews of five representatives of mining equipment companies. Based on the empirical evidence examined, the impression is that up to 2018, the prevalent product-service system manifests the traditional and conventional product-oriented business model for the supply side. A possible explanation is that, in the mining and minerals processing environment where the empirical evidence discussed in this chapter was obtained, a legacy of economic sanctions and political isolation provided impetus and supported a tradition where the asset user organisations established extensively resourced in-house maintenance departments with the full range of 1st-, 2nd-, and 3rd–line skills. Hence, the scope for outsourcing maintenance services was narrow and limited to special needs and circumstances. Thus, the empirical evidence more or less highlighted the prevalent culture of deploying 1st-, 2nd-, and 3rd-line maintenance skills in-house within asset user organisations; hence, the apparent reluctance by asset users to embrace the wider scope of servitization offerings. Further empirical evidence beyond 2018 may address the question as to the extent of influence of 4IR technologies and globalization 4.0 business models on servitization, product-to-service transformation and product-service systems. Such evidence can be examined especially with regard to decisions to acquire, utilize (i.e., operate and maintain) and retire engineered assets.

References

1. Rao SK, Prasad R (2018) Impact of 5G technologies on industry 4.0. Int J Wireless Pers Commun 100(1):145–159
2. van der Merwe S, Rada J (1988) Servitization of business: adding value by adding services. Eur Manage J 6(4):314–324
3. Kowalkowski C, Gebauer H, Kamp B, Parry G (2017) Servitization and deservitization: overview, concepts, and definitions. Ind Market Manage 60:4–10
4. Lightfoot H, Baines T, Smart P (2013) The servitization of manufacturing: a systematic literature review of independent trends. Int J Oper Prod Manage 33(11/12):1408–1434
5. Pouria, MH, (2015) Product service system: a systematic review on its definition, design methodologies, value assessment, and guidelines for future research. MSc Thesis, Politecnico di Milano
6. Tucker A (2004) Eight types of PSS; eight ways to sustainability. Bus Strateg Environ 13(4):246–260
7. Wendling, TE (2011) A life contingency approach for physical assets: create volatility to create value. In: Monograph, Society of Actuaries ERM Symposium, Chicago, USA, 14–16 March 2011
8. Ramlal, R, Amadi-Echendu, JE (2018) Product servitization and maintenance contracts. In: Proceedings of IEOM Conference, Pretoria, South Africa, 29 October–1 November 2018

Considerations on Investment and Business Models

Helena Kortelainen(✉), Antti Rantala, and Risto Tiusanen

VTT Technical Research Center of Finland, 33101 Tampere, Finland
helena.kortelainen@vtt.fi

Abstract. Business models are evolving from selling products towards delivering outcome that may be measured as system performance, capacity, and availability. Novel business models may also include elements of sharing value. Machinery manufacturers and suppliers face a rapidly changing business environment and look for major growth in digital solutions, automation and services. The implementation of advanced technologies and business models call also for novel models of sharing risks and benefits. The literature review on advantages and disadvantages of various business models underlines the investment risk. Modelling of the cash flows of different business models provide examples for leveraging initial investments in assets when applying emerging technologies. The business partners also have to consider how their risk landscape changes and what are the preventable, strategy and external risks of the planned business model. The chosen business model poses requirements to the risk management process and highlights the importance of the collaboration and transparency. This paper focuses in the automation options in the transport sector and uses major port terminal as a case example, but the developed methods are applicable also in other capital intensive industries.

Keywords: Investment · Business models · Input-based · Outcome-based · Value sharing · Cash flow · Risk · Port terminal

1 Introduction

Automation and digitalization have changed the work processes, activities, tasks and employee's duties in many industries. In this transformation, transport sector is no exception. In the future logistics systems, machines will be self-steering, and the humans' duties will change from driving the vehicles to monitoring the processes. Frost & Sullivan [1] forecasts that autonomous logistics could extend beyond warehouses to outside logistics and several functions in the transport chain like cross docking and transshipment, could become fully automated and intelligent. However, an automatized fleet requires bigger investment compared to investment in a traditional, manually driven fleet. For instance, Muricy Souza Silveira [2] estimate that the initial investment in an automated port terminal is about 57% higher than initial investment in a manned port terminal with the same capacity Even though the price of automated equipment

A. Crespo Márquez et al. (Eds.): WCEAM 2019, LNME, pp. 10–21, 2021.
https://doi.org/10.1007/978-3-030-64228-0_2

is high, the improvements in safety, reduced unplanned downtime, increased production and improved workforce effectiveness should be taken into consideration when making investment decisions [3]. In a longer term the investments in automation and digitalization may still pay out.

The main criterion in the investment decision making is still often the acquisition price - and not the life cycle cost [4]. The savings incurring over the asset life time - that may extend to several decades - are highly uncertain and difficult to assess in a credible way. Solutions based on digital technologies may have a positive impact on the company´s operations but the monetary benefits across the asset life are often hard to define [5, 6]. Application of a value assessment approach and models improve the ability to communicate the value of digital solutions to the other partners in the business ecosystem. In addition, demonstration and communication of cost savings and benefits can serve as a bridge-builder between technology suppliers, customers and other stakeholders [7].

This paper discusses on the risks and advantages of different business models in capital-intensive industries with the focus in port operations, and introduces an investment model that could help in sharing risks between supplier and customer in a fair way. The paper aims to answer two research questions: (1) *How to leverage the economic barriers that emerge from higher cost of automated solutions?* and (2) *How to share benefits and risks in complex ecosystems?*

2 Study Context and Methodology

The study is a part of the ongoing Finnish national research project 'Operational excellence and novel business concepts for autonomous logistic systems in ports (AUTOPORT)' [8]. The project is a co-innovation project that aims to path the way towards automated operations in ports by developing model-based design, operational excellence, and models for sharing incurring costs and benefits. AUTOPORT project is conducted in close co-operation with the research organizations and the company network that consists of machinery and ICT solution providers and engineering companies.

In our research, we use literature review and content analysis to create knowledge of the risks, disadvantages, benefits and advantages of the different business models to the business partners. We have also developed a simple model and a MS Excel tool that allows us to test and illustrate the impacts of different business models. The modelling aims to highlight the differences in cash flows between discussed business models.

The input data needed for the empirical work is derived from the work of Muricy Souza Silveira [2] that considers the automation options of the Port of Santos in Brazil. In this case, the port assets include 20 Ship-to-Shore Cranes (STS), 5 container spreaders (STS - Spreader), 20 Rubber Tyred Gantry cranes (RTG), 5 RTG spreaders, 10 terminal tractors and 7 trailers, and Terminal Operating System (TOS). In our study, the Port of Santos served as a case example that helps to provide a thorough understanding of the phenomenon in question [9].

3 Business Models for Acquiring New Technologies

Business models describe how organizations create, deliver and capture value [10]. According to Sjödin *et al.* [11] in collaborative business models, more understanding should be developed about how customers and suppliers agree to jointly create value and to share it fairly. Critical point of business models are not only designing the value creation and sharing processes, but ensuring they are adapted and aligned to each other. Appropriate governance mechanisms are needed to ensure that value creation is greater than the cost of realizing that value and that the value surplus is distributed fairly among the parties [12].

In this section, we concentrate on two different business models namely on the input-based model and on the outcome-based model. In a traditional input-based model [13], machine supplier is paid for the product, and perhaps also for the services they deliver to their customer during the product life time. The ownership of the product is transferred to the customer that is also responsible of the operation and on the upkeep of the product [15]. In the outcome-based model [14], customer no longer buys the product but pays for the output that is delivered by the product [15]. The supplier is then in charge for the performance outcome of the product (and service), and is financially responsible for any shortcomings, such as equipment breakdowns. As the paper of Ng *et al.* [16] states, an outcome-based model focuses on the outcome of a system rather than the resources involved in its provision. In such a model, the supplier extends the focus from the delivery and commissioning of a product into the use-phase of the system and takes over the responsibility of the operation and maintenance of the product on their customer's behalf [17]. A number of contract packages could be build (i.e. pay per unit, pay per performance, fixed operations and maintenance fee, etc.) in outcome based business model. A decision as to which business model to use can have a significant impact on the cost and the risk of owning and operating a fleet [18].

For the supplier, both models entail advantages and disadvantages as summarized in Table 1 and 2.

Table 1. Advantages to the supplier.

Business model	Advantages for the supplier	Sources
Input-based model	- Financial risk is shifted to the customer - Responsibilities in operations and maintenance are shifted to the customer	[15]
Outcome-based model	- Barriers to attract new customers are low due to low initial investment by the customer - Possibility to capture larger portion of the value stream and gain more profits - Resilient cash flow and revenue streams - Possibility to develop long-term business relationships that lock out competitors - Possibility to gather data from operations	[15, 17, 19, 20]

Table 2. Disadvantages to the supplier.

Business model	Disadvantages for the supplier	Sources
Input-based model	- Less potential for innovation due to limited access to monitor and gather the data	[15]
Outcome-based model	- Financial risk due to retaining ownership of the system - Financial responsibility of any shortcomings such as equipment breakdowns - Responsibility of the product life cycle - Challenging to estimate suitable price-level for the service	[11, 15, 21]

The business model has also impacts to the customer. An input-based business model is a familiar way of realizing an investment. However, it is bound with major financial risks that may become an obstacle when considering novel technologies. The advantages and disadvantages of an input-based and an outcome-based business model from the customer point of view are summarized in Table 3 and 4.

Table 3. Advantages to the customer.

Business model	Advantages for the customer	Source
Input-based model	- Ownership of the property rights - Simple and familiar business model	[15]
Outcome-based model	- Various costs and activities are shifted to the supplier - Reduction of risks and barriers of acquiring new technologies - Supplier has incentives to improve system performance and reduce overall expense	[15, 19, 20]

Table 4. Disadvantages to the customer.

Business model	Disadvantages	Source
Input-based model	- A major capital investment causes a financial risk - Unknown expenses may occur (maintenance, repair, etc.)	[15, 19]
Outcome-based model	- Limited control of the operations management - Dependency of the suppliers performance - Increased complexity of the business environment	[15, 19, 20, 22]

4 Business Models and Risk Sharing

New business models that are based on complicated inter-organisational systems for innovation, development, common offering and performance metrics entail considerable uncertainties [22]. Understanding prevailing and novel risks, and designing the mechanisms and actions for governing and controlling risks are crucial for any successful business. The success of a business model innovation depends on the company's ability to recognize that the planned activities are more uncertain, complex, and therefore also riskier than current operations, and on the company´s ability to cope with these process characteristics [22].

The ISO/IEC Risk management standard series (e.g. [23] and [24]) define risk as the effect of uncertainty on objectives. Risk is often expressed in terms of a combination of the likelihood and consequences of an event. Enterprise Risk Management (ERM, e.g. [25]) emphasise the need to embed risk management systems within business processes. ERM focuses traditionally in a single company and existing assets, but the framework helps also to identify and manage multiple inter-organisational risks, segmented mostly to four core risk groups: strategic, operational & cultural, financial and hazard risks [26]. For a company investing in assets, the investment assessment usually deals with considerable uncertainties as highlighted also in the Table 2 and 4.

Risks can be categorized as preventable, strategy or external [27]. The management and control of preventable risks require standard operating procedures that help to avoid or eliminate the occurrence of negative events in a cost-effective way. Strategy risks are voluntarily accepted risk in order to generate superior returns from the strategy, and the control models include interactive discussions inside the organisation and stakeholders about the strategic objectives and necessary trade-offs. External risks arise from events outside the organization and are beyond its influence and control. The risk mitigation then concentrates in reducing the impacts should a risk event occur. Building up scenarios help to prepare for external risks.

As companies are increasingly relying on different collaborative arrangements in their business models, they become also more dependent on other companies capabilities and resources. This makes their situation more unpredictable regarding possible changes in the business environment. Inter-organizational networks increase interdependencies and this fact creates further challenges for managing risks [28]. The threat of increased responsibility and loss of control are mentioned also in Table 2 and 4. The networks also lead to second and third-order effects that are absent from a company to company relationship [29]. This requires companies to view their profits and risks not in terms of what they control internally, but in terms of their relational capabilities to the networks in which they are embedded [29] and emphasises the importance of analysing external risk factors affecting the business environment [30].

Figure 1 (next page) illustrates the supplier's increasing risk as a function of the business model. In an input-based business model the supplier carries the risk on the function and compliance of the delivered good at least over the warranty period, but the product liability may be longer. As the supplier takes more responsibility and delivers solutions instead of mere product, it faces increasing risk on the performance of the product in the customer's process and business environment. In an outcome-based model, this responsibility and risk extends beyond the product and its performance towards carrying customer's business risks or at least, towards sharing risks that are external to the supplier company. From the risk management point of view, the importance of thorough understanding the strategy risks and building up scenarios that help to prepare for external risks are of prime importance.

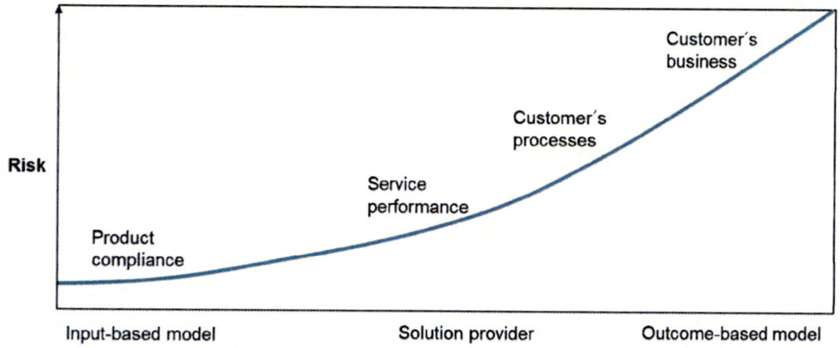

Fig. 1. A schematic presentation on the business model on the supplier's risk.

In inter-organisational relationships, understanding of contractual risks is essential. The risks may be reduced by making visible the advantages from efficient collaboration of the network as well as value destroying effects of actions against the common objectives and principles of collaboration [31].

5 Modelling Cash Flows Generated by Different Business Models

In this chapter we illustrate the cash flows of input-based- and output-based business models, and the cash flows generated by a third option, namely a value sharing contract [32]. Our models represent simplified cash flows of an automated terminal port (Port of Santos), including all assets (spreaders, trailers, gates, etc.) and software, which are estimated to cost $249.330.000 as an investment, and to generate $230.000.000 profit in 10 years [2]. The revenue from a terminal port is generated by the delivery and the reception of containers, and from the associated tasks like inspections and weighing.

We have chosen a 10 year time frame for our study as the life time expectation of the major port equipment is 10 years or longer, and we have also made an assumption that no major renewals or upgrades are necessary during that time period. In addition, our models don't take discounting, inflation or other factors which are usually considered in life cycle costing into account, as the models are generated only to demonstrate the risk and revenue sharing logics of different business models instead of estimating accurate values.

5.1 Input-Based and Outcome-Based Business Models

The cash flows generated by input- and outcome-based business models are presented in Fig. 2 and Fig. 3 The "building cost" includes all costs incurred during the early stages of the product life cycle and those of building up the delivered system [13]. In our model, the building cost is estimated to be 70% of the acquisition price as the supplier is expected to charge a certain profit on the building cost. In addition, the annual revenues are assumed to be same every year in order to keep the models simple.

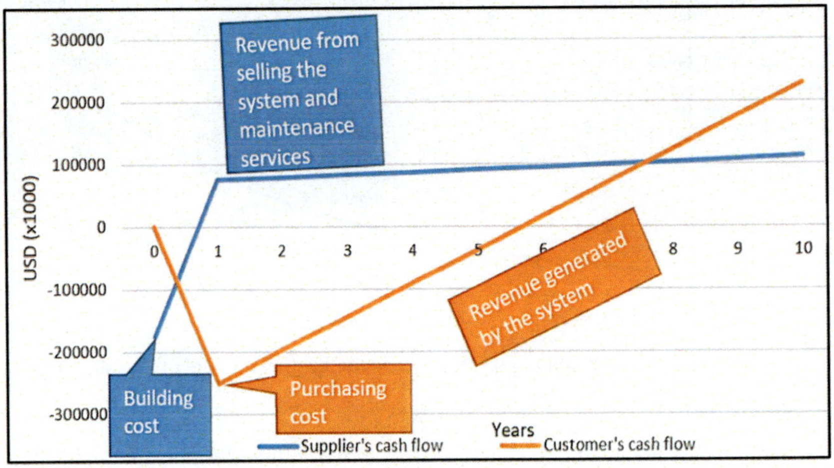

Fig. 2. Cash-flow of input-based model.

In the traditional input-based model, the customer orders an equipment, machine or system, and the supplier delivers it. The customer makes the initial investment and pays the acquisition price of the system in one payment or in several installments. The ownership of a product transfers from the supplier to the customer. In addition, supplier usually offers some kind of maintenance service and gets annual revenue from the customer [7]. In this example, the supplier gets instant profit, whereas customer has to wait about six year payback time before the investment turns profitable.

In outcome-based model, supplier builds the system on its own account and gets the revenue from its performance units instead of selling the product. The supplier who used to make money on the products, maintenance and spare parts, will now have to consider these items as costs, because the revenue depends entirely on the delivered outcome [11]. The customer starts paying to the supplier as soon as the supplier starts to provide the outcome as defined in the contract. In this example, the customer doesn't possess any investment risks, whereas the supplier's breakeven point is after five years. These examples above show why risk sharing business models are needed.

The challenge between the supplier and the customer is the investment risk. The supplier might prefer the input-based model to secure the revenue and to avoid disadvantages mentioned in the Table 1. On the other hand, the customer could prefer the outcome-based model in order to reduce or avoid the investment risk, and to gain other advantages summarized in the Table 2.

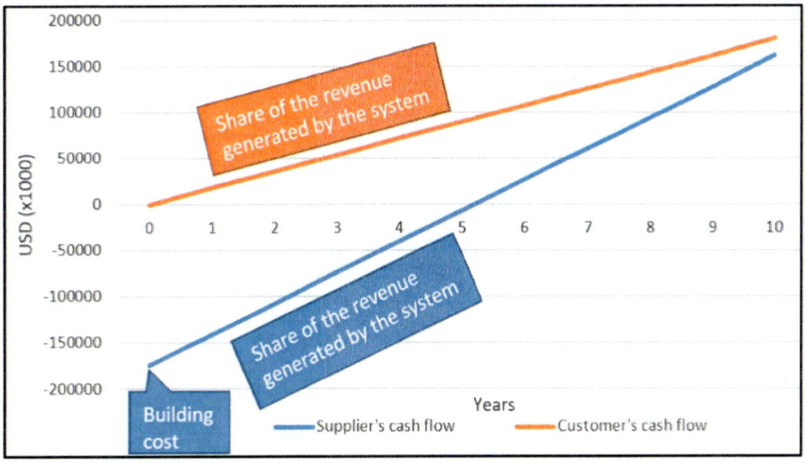

Fig. 3. Cash-flow of output-based model.

5.2 Solution for Risk Sharing

A solution that takes the interests of both parties into consideration is a value-sharing contract [32] that aims to share the investment risk between both parties. Such contract lowers the purchase price of a system in return for a proportion of the future value generated during its operational life. Cash-flow of the value-sharing contract is presented in the Fig. 4.

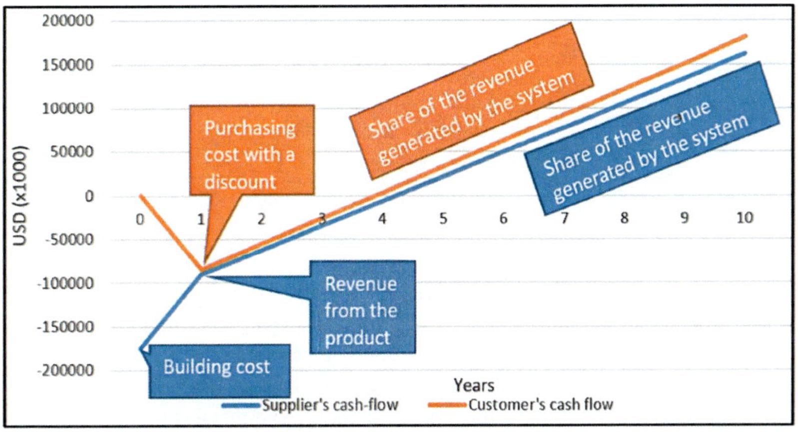

Fig. 4. Cash-flow in value-sharing contract.

In a value-sharing contract, supplier provides the system, and customer purchases it with a discount price but commits into value sharing. The future revenues generated by the system are shared between the supplier and the customer. This would share the investment risk between both parties, and could leverage the economic barriers that emerge from higher cost of new technologies. In this case, both parties would make the investment profitable in about four years.

5.3 Risk Management Framework

For a supplier, the transition from input-based business to more complicated inter-organisational systems gives rise to a variety of new risks. The changing risk landscape must be carefully taken into consideration during the transition process and a risk management process covering the whole range of company internal and preventable risks, strategy risks and external risks is needed. In the steps of the transition process one must also consider the different risks that are connected for example to customer relationship, contracts and responsibility, and business environment of the customer and to second and third-order effects in the network. The risk management framework presented in Table 5 highlights some characteristics of the risk management process in different business models.

Table 5. Risk management framework, adapted from [31].

Business model	Risk management characteristics
Input-based model	Company internal risk management process for identifying, describing, organising and executing risk management and control
Outcome-based model	Inter-company cooperation is needed. Risk environment is more complex and entails external risk factors. Sharing of risk knowledge and cooperation among business partners is needed in order to assess and manage risks
Risk sharing solution	Common guidelines and methods for risk assessment, risk management and knowledge sharing are needed for ensuring an adequate level of risk management and risk sharing between business partners. Risk management practices are improved by collaboration practices

In an input based model, both business partners have their own risk management processes according to the ISO/IEC, ERM or some other common risk management standard that meets the business requirements. The outcome based model necessitates close inter-company collaboration. The companies owning and operating major assets are familiar with investment appraisal methods and risks, but for the supplier these risks may emerge as external risks that are difficult to control. For the customer, the actions of the supplier may also build up external risks, e.g. in a case of bankruptcy and losing control of the assets. A value sharing contract aims at sharing risks but it also calls for a common inter-organizational approach for risk management and control.

6 Conclusions and Discussion

Automatized and digitalized assets and systems are changing the logistics operations towards more cost-effective, reliable and safe direction. However, the high acquisition cost of the solutions applying novel technology may pose a barrier that prevents customer companies to invest in automatized machines. Novel and innovative solutions are needed to share the risk. The paper contributes to the discussion on the novel business models by addressing two issues, namely *How to leverage the economic barriers that emerge from higher cost of automated solutions?* and *How to share benefits and risks in complex ecosystems?*

In this study, we presented two simplified business models (input- and outcome-based), and illustrated the advantages and disadvantages of those models from supplier's and customer's perspective. We have also developed a calculation model and a MS Excel tool, and applied the tool in calculating the impacts of different business models. From the suppliers' point of view, major disadvantages in the outcome based model include the financial risk due to retaining ownership of the system and the responsibility of the product life cycle in uncertain business environment. The outcome based model could attract new customers as for the customer, the initial investment is low. However, this model has major disadvantages and risks to the supplier. The presented cash-flow models for the case company Port of Santos in Brazil clearly indicate that the customer carries the financial risk in the input-based model, whereas supplier carries the financial risk in the outcome-based model.

The third simplified business model that is based on value sharing could offer a potential solution. A value sharing model could help to solve the problem arising from the economic barrier and to leverage the high cost of automatized machines. A value sharing model would decrease the initial investment required from the customer and this could ease the investment decision. Continuous cash flow would make it an attractive solution to the supplier too. A value sharing model would help to share benefits over the time in an ecosystem. However, the suppliers have to be willing to wait for the profits and also in this model the supplier partly shares the risks prevailing in the customer´s business environment.

When several companies are collaborating, they can have value creation opportunities that they couldn't achieve on their own. Thus it is important to make sure that each party captures fair share of the jointly created value. For that reason the value sharing arrangements should be considered at the early stages of the negotiations in order to secure each parties' fair share of the jointly created value. Different kind of value sharing models and can be made to clarify what share each party will get from the jointly created value.

The transition from input-based business to more complicated inter-organisational arrangements gives rise to a variety of risks that are external to the supplier, and the supplier do not have the means for controlling them. On the other hand, the outcome-based model poses also novel risks to the customer from losing the control of the assets to a bankruptcy of the supplier. In addition to the binary relationship, novel business models may involve several organisations. Thus the risk management requires collaboration from all parties.

Even a simplified calculation example show how important it is to build up scenarios and to test the planned business model. For decision makers - let it be a supplier, a customer another partner in the business ecosystem - a calculation model helps to create understanding on the economic consequences on a long run and to test different scenarios. Modelling also contributes to the risk management and helps to discuss and deal with risks by business partners. From the risk management perspective, incorporation of the sensitivity analysis would be a necessary advancement to the model. More research is needed for better understanding of the risks, risk sharing and risk control in inter-company relationships. Further elaboration of the models and tools with industrial stakeholders is required to test the applicability of the ideas in the real business environment.

Acknowledgements. The research group have received funding from Business Finland that is the Finnish government organization for innovation funding and trade, travel and investment promotion. VTT Technical Research Center of Finland and the companies participating in the AUTOPORT project have been co-funding the study. The authors wish to thank all these organisations for their support.

References

1. Frost & Sullivan (2016) Future of logistics - unbundling of the supply chain, Frost & Sullivan
2. Silveira, RMS (2018) Container terminal automation in Port of Santos - a business case analysis. Master's thesis, Center for Maritime Economics and Logistics, Erasmus University Rotterdam. http://hdl.handle.net/2105/43890. Accessed 1 July 2020
3. Sims S (2016) The benefits of calculating ROI to measure a facility's performance objectives. Control Eng 63(10):49–50
4. Uusitalo T, Hanski J, Reunanen M, Kunttu S (2014) Support for life cycle decision-making in sustainable manufacturing – results of an industrial case study. In: Grabot B, Vallespir B, Gomes S, Bouras A, Kiritsis D (eds) APMS 2014, vol 439. IAICT. Springer, Heidelberg, pp 162–169. https://doi.org/10.1007/978-3-662-44736-9_20
5. Kortelainen, H, Saari, L, Valkokari, K, Federley, M, Heilala, J, Huusko, J, Viljamaa, E: Beyond IoT business. White paper. https://cris.vtt.fi/en/publications/beyond-iot-business. Accessed 7 Aug 2020
6. Räikkönen, M, Välisalo, T, Shylina, D, Tilabi, S (2015) Supporting asset management decision-making – new value creation perspective. In: Proceedings of the 10th world congress on engineering asset management (WCEAM). Lecture notes in mechanical engineering, Tampere, no 28–30. Springer, Cham, pp 479–486
7. Räikkönen, M, Keski-Rahkonen, J, Kortelainen, H, Tikkanen, M, Valkokari, P, Vehanen, A, Pirttikangas, S (2020) Towards a framework for assessing the customer value of digital solutions. In: eProceedings of the 30th European safety and reliability conference and the 15th probabilistic safety assessment and management conference. Research Publishing Services
8. AUTOPORT. https://autoport.fi/. Accessed 8 July 2020
9. Yin R (1994) Case study research: design and methods, 2nd edn. Sage Publishing, Beverly Hills
10. Teece DJ (2010) Business models, business strategy and innovation. Long Range Plan 43(2–3):172–194

11. Sjödin D, Parida V, Jovanovic M, Visnjic I (2020) Value creation and value capture alignment in business model innovation: a process view on outcome-based business models. J Prod Innov Manage 37(2):158–183
12. Chesbrough H, Lettl C, Ritter T (2018) Value creation and value capture in open innovation. J Prod Innov Manage 35(6):930–938
13. Ulaga W, Reinartz WJ (2011) Hybrid offerings: how manufacturing firms combine goods and services successfully. J Mark 75(6):5–23
14. Visnjic I, Jovanonic M, Neely A, Engwall M (2017) What brings the value to outcome-based contract providers? Value drivers in outcome business models. Int J Prod Econ 192:169–181
15. Azarenko A, Roy R, Shehab E, Tiwari A (2009) Technical product-service systems: some implications for the machine tool industry. J Manuf Technol Manage 20(5):700–722
16. Ng I, Xin Ding D, Yip N (2013) Outcome-based contracts as new business model: the role of partnership and value-driven relational assets. Ind Mark Manage 42(5):730–743
17. Davies D (2004) Moving base into high-value integrated solutions: a value stream approach. Ind Corp Change 13(5):727–756
18. Vorster, M (2014) Lower equipment acquisition risk, not cost. Construction equipment, April 2014
19. Tukker A (2004) Eight types of product-service system: eight ways to sustainability? Experiences from SusProNet. Bus Strategy Environ 13(4):246–260
20. Baines, T, Lightfoot, H (2013) Made to serve what it takes for a manufacturer to compete through servitization and product-service systems. Wiley, Hoboken, p 53
21. Välisalo, T (2003) From availability performance guarantees to production capacity sales. Promaint, pp 10–12. (In Finnish)
22. Taran Y, Boer H, Lindgren P (2013) Incorporating enterprise risk management in the business model innovation process. J Bus Models 1(1):38–60
23. ISO/IEC 31010 (2019) Risk management - risk assessment techniques
24. ISO 31000 (2018) Risk management – guidelines
25. Committee of sponsoring organizations (COSO) (2014) Enterprise risk management: integrated framework. COSO, New York
26. CAS (Casualty actuarial society) (2003) Overview of enterprise risk management. ERM community, Arlington
27. Kaplan RS, Mikes A (2012) Managing risks: a new framework. Harvard Bus Rev 90:3–13
28. Hallikas, J, Varis, J (2009) Risk management in value networks. In: Supply chain risk. International series in operations research and management science vol 124, no 1, pp 35–52
29. Kleindorfer, PR, Wind, Y (2009) The network imperative: community or contagion? In: Kleindorfer, PR, Wind, Y, Gunther, R (eds) The network challenge: strategy, profit and risk in an interlinked world. Wharton School Publishing, pp 3–24
30. Uusitalo, T, Palomäki, K, Kupi, E (2014) Managing risks in service value networks. In: Proceedings of the sixth world congress on engineering asset management (WCEAM). Lecture notes in mechanical engineering. Springer, Heidelberg, pp 239–321
31. Ahonen T, Reunanen M, Ojanen V, Pajari O, Lanne M (2010) Maintenance communities – a new model for the networked delivery of maintenance services. Int J Bus Innov Res 4(6):560
32. Davies A, Brady T, Hobday M (2006) Charting a path toward integrated solutions MIT Sloan. Manage 47(3):39–48

Establishing the Value System Through Long Term Planning

Ype C. Wijnia(✉)

AssetResolutions B.V., PO box 30113, 8003CC Zwolle, The Netherlands
ype.wijnia@assetresolutions.nl

Abstract. Large organizations, especially those in relatively stable environments like infrastructure managers, are in general managed incrementally with implicit objectives. The rationale for this is quite clear. If the performance of the asset base was adequate in the past, doing more or less the same as last year cannot be that wrong. In volatile times (i.e. the required rate of change exceeds the adaptive capabilities of incremental change) this strategy fails and explicit planning and prioritization (the basic idea of asset management) are needed. A prerequisite for such an exercise is the existence of a shared organizational awareness of what is right (e.g. documented in mission, vision, policy and strategy and the value system as captured in the risk matrix). However, given the incremental past, it is not uncommon to have multiple conflicting versions of the truth. Resolving this directly is almost impossible, as decision makers would (and should) require a reasonable estimate of the consequences for any decision they make. With regard to redefinitions of the value system these consequences are highly uncertain. In this paper, we present an approach which we used to help decision makers specify their value system. We start with modeling the asset base in respect to the dominant change agent (in many cases ageing), so that we can prognosticate cost, performance and risk for diverse concepts of what is right. This helps the decision makers understand what the long term consequences will be of the implicit values they impose on the organization by means of ambitions and objectives, and thus helps them understand if their objectives are right. This approach has been successfully applied in a variety of depths (with associated effort) in several organizations.

Keywords: Asset management · Risk management · Value system

1 Introduction

Large organizations, especially those in relatively stable environments like infrastructure managers, are in general managed incrementally with implicit objectives. There are several explanations for this. First of all, these organizations tend to be old, in the range of decades to centuries. This allowed for the evolution of a complex organization with a finely tuned set of coordination mechanisms [1]. Abrupt changes in the course of the organization may result in a total disruption of the system with its associated risks. Secondly, because of their long existence, their vision may have been realized. Think

A. Crespo Márquez et al. (Eds.): WCEAM 2019, LNME, pp. 22–33, 2021.
https://doi.org/10.1007/978-3-030-64228-0_3

of running water, sewage, electricity: at the start there was competition with existing unsafe alternatives (wells, latrines and oil lamps), but currently virtually everybody in western society has access to them, without the end of product lifecycle in sight. In these circumstances, doing more or less the same as last year cannot be that wrong. In volatile times, however, the required rate of change may exceed the adaptive capability of these organizations. A different management approach is then needed, with more attention for explicit objective setting and planning to achieve these objectives. It is therefore no coincidence that major developments in asset management occurred when the managing organizations were confronted with large external pressures [2].

Explicit objective setting requires a shared organizational understanding of what is right, i.e. the core values of the organization. In its most generalized form this is formulated in the vision and mission, as they state where the organization wants to be and what it does to get there. Other sources for the core values may be the strategy, policy or decision guidelines including the risk matrix. Yet, in the organizations discussed in this paper, vision and mission are not necessarily maintained, (re)developed or deemed relevant, because the fundamental objective already has been achieved and the organization is running in its business as usual mode. It may even be the other way around: in the evolution of the organization several inconsistent or even conflicting value systems may have been developed. Engineering typically is ruled by technical norms (design rules, material standards, functional requirements) whereas finance is ruled by the financial norms like the generally agreed accounting principles (GAAP). These normative systems do not match and may even cause intra organizational distrust: Finance thinks engineering is gold plating the system by constantly upgrading assets that are in good working order, whereas engineering thinks finance is mortgaging the future by not providing the means for maintaining the asset base.

It is virtually impossible to resolve such a conflict of worldviews directly. Decision makers should require an estimate for the consequences of any decision, but in case of changing the rules for decision making it is unclear what the consequences are or who is able to assess them. It would require stepping back to the reasons the existing norms were put in place, but that is not a normal routine for people trained to work within the system.

In this paper we present an approach that helped decision makers in several organizations to understand and specify their value system. In this approach, we start by modeling a decision problem that is large enough (i.e. impacting the entire asset base for a prolonged period[1]) for which the organization is uncertain what the best action is. The problem we typically select is ageing and replacement of the asset base, but other examples can be found in the energy transition [3, 4] or climate adaptation [5]. The model should be able to prognosticate the long term consequences in terms of cost, performance and risk with reasonable accuracy for a variety of strategies. Strategies are in this paper regarded as sets of decision rules and thus are linked to core values. The results of these strategies then will be presented to the decision makers, so that they can judge the acceptability of the outcomes of the tested strategies and thus the appropriateness of the associated value system for the organization. We end the paper with some lessons learnt and an future outlook.

[1] Local and/or short lived problems fall within the realm of resilience.

2 Understanding the Complexity of Asset Replacement

Asset replacement planning may seem a very straightforward exercise to the decision makers: determine the remaining life for all assets and make a plan to replace them just beforehand (or as close as possible) within the organizational constraints like budget and capability. Unfortunately, in reality asset replacement planning tends to be much more challenging. There are several reasons for this unavoidable complexity:

1. Diversity of drivers for asset replacement
2. Diversity of the asset base
3. Uncertainty of technical end of life
4. Subjectivity in functional end of life
5. Instability of input constraining strategies

2.1 Understanding the Drives for Asset Replacement

Structuring a problem field is regarded as a challenge in general [6]. An approach that has proved useful in asset risk management is that of the risk process [7]. In this view, risks do not appear out of nowhere, but are the result of a cause (disturbance) working on an entity (asset), whose reaction (failure mode) results in undesired consequences. Using asset failures as central point (the top event in a BowTie analysis) allows for a thorough identification of all relevant aspects.

The top event driving asset replacement can be framed as asset inadequacy, though it may be useful to distinguish technical and functional variants. Technical inadequacy means that the asset is not able to perform the function it was designed to do, whereas functional inadequacy means that the asset is expected to perform functions it was not designed for. For both varieties the concept of magnitude has meaning, as not every manifestation of inadequacy is equally severe. Total inadequacy would for example be a fatal failure of the asset, or changing regulations making the operation of the asset illegal. Minor inadequacies are more like discolorations or not being the newest generation anymore.

Asset inadequacy can result from diverse sources. For technical inadequacy, main drivers are external damages (lightning strikes, storms, floods, accidents, attacks) and ageing/wear & tear. Functional inadequacy typically results from changes in external requirements (norms, standards and regulations), changes in the use of the assets (peak loads and operating hours) and technological advancement (energy efficiency, functionality). The preventative strategy should also be regarded as a driver for functional inadequacy. Assets that exceed their operating criteria (like age and condition, but also repairability[2]) will be clearly identified for a planned intervention before they are not able to perform their function anymore. Depending on the severity of the inadequacy several interventions may be implemented, ranging from maintenance and repairs to revisions, replacements and upgrades. In some cases it may be necessary to construct new assets. If these interventions are the result of asset failures, in general there are

[2] Minor technical inadequacies are usually repaired, but this is no longer possible if the asset is no longer supported by the supplier and spare parts are no longer available.

unplanned consequences in terms of safety, reliability, reputation and costs. If interventions are scheduled, there are only planned cost. A generalized model of asset inadequacy (=the need for an intervention on the asset) is shown in Fig. 2.

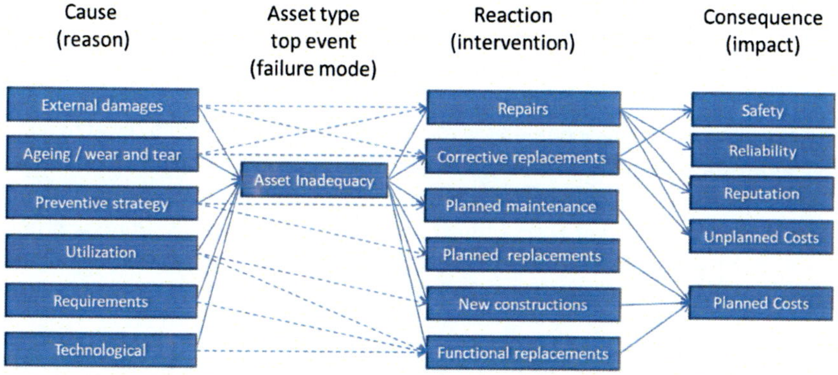

Fig. 1. BowTie diagram of Asset Inadequacy. The dashed lines indicate typical responses.

Which variety of inadequacy is dominant depends on the characteristics of the assets [8]. For passive assets (roads, railways, waterways, pipelines, cables), technical inadequacy is by far the most relevant, as they only have to exist to perform their function. For active assets, it depends. For long lived assets with mature technology, it tends to be technical as well (rotating equipment). But for short lived assets with rapidly developing technology, the asset is outdated before it shows any sign of wear and tear (software, mobile phones)[3].

2.2 Representing the Diversity of the Asset Base

Asset ageing and replacement can be modeled at several levels of aggregation [9]. At the most abstract end of this range, the decision problem would be modeled with averages over the entire asset population and lifespan. The asset base would thus be treated as if only one type of asset was in use with a uniform distribution of asset age over the lifespan. This may seem a very rough approximation, as most asset bases consist of assets varying wildly in their attributes like lifespan, cost and failure consequences. For example, software has a typical technical lifespan of 5 years, electronics 10–15 years, rotating equipment 15–25 years and constructions 25–100 years. But if the asset base can be grouped into asset systems that are replaced as a whole this may be an adequate representation. Furthermore, the ages of the assets may not be uniformly distributed over their lifespan. If the assets were constructed in a short period, this will return in the replacement needs.

To deal with the variability in asset attributes, the asset base could be split up into more types. A first step would be to use the mentioned 4 classes, but the effort could

[3] It is for asset bases vulnerable mostly for functional inadequacy that the simple approach for replacement planning would work.

extend to modeling each individual asset if a very high resolution would be required. Modeling some 10–100 different types would normally be accurate enough[4] [7].

The issue of the uneven age profile requires turning the calculation into a simulation, either continuous or discrete. To achieve a high accuracy it may be tempting to use short timesteps, but one has to bear in mind that strategic decision making is on long term trends and not about short term variations. Typically, a timestep of 1 year is accurate enough even for the short lived assets. Whether or not the simulation needs to be discrete depends on the amount of memory the assets have. If for example the lifespan of an asset is determined by accumulated damages which depend on the level of conservation, changing the maintenance policy will have a different effect on old assets than on young assets. Furthermore, continuous modeling allows the existence of fractional assets. In large bases consisting of 100 s of instances, this is a reasonable approximation, but with a small number of assets (say a shipowner operating 10 vessels) having half an asset fail may raise some eyebrows.

2.3 Uncertainty in the Technical End of Life

Technical end of life can be caused by two drivers: External damages and ageing/wear & tear. External damages typically do not depend on the asset age, and thus do not play a direct role in planning for end of life.

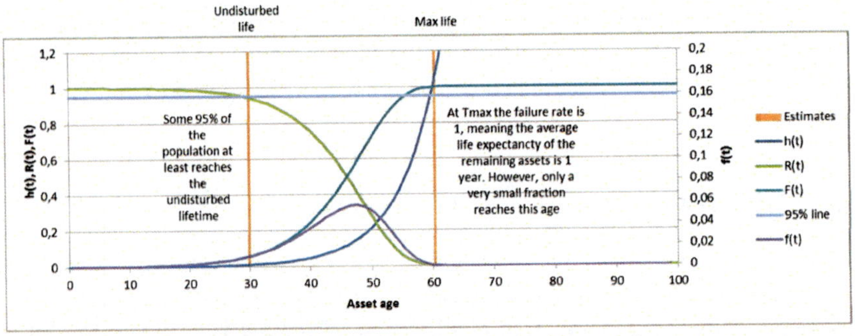

Fig. 2. Distribution of asset life, h(t) = $h(t) = h_0 e^{(c_1 t)}$, $R(t) = e^{-\frac{h_0}{c_1}*(e^{c_1*t}-1)}$, $h(T_{max}) = 1$, $R(T_{und}) = 0{,}95$

With regard to Ageing and Wear & Tear, the technical end of life failure for assets typically follows a Weibull like distribution as shown above (the purple line f(t)) and will not be a fixed, predetermined moment[5]. This means that for any planned replacement moment, there is always the potential for the asset failing prematurely, with the associated extra costs and other consequences. In general, this risk can be neglected within the

[4] Asset bases seem to follow distributions like the 80/20 rule, with a small fraction of types accounting for the majority of costs, risk and performance.

[5] This holds for simple assets. For complex assets in which every part can be replaced there often is no age related end of life

undisturbed lifespan. To reach the technical lifespan of the assets it is in general necessary to maintain assets [10, 11], otherwise ageing might speed up after the initial conservation (paint, lubrication) is eroded away as shown in Fig. 3.

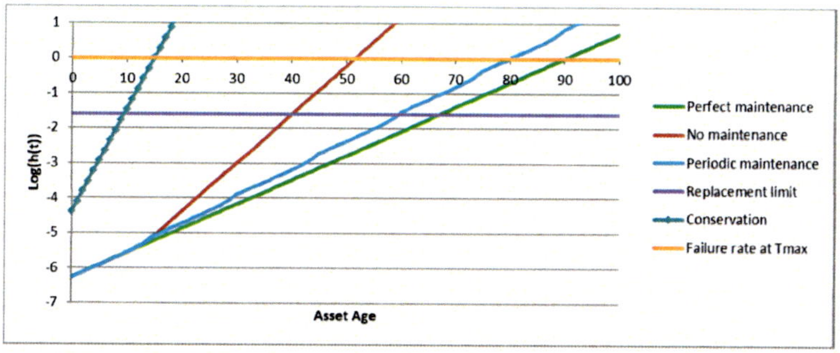

Fig. 3. Failure rate development for an asset for several conservation regimes. In this example, the maximum life (Tmax) of a perfectly maintained asset is 90 years, dropping to a little over 50 if no maintenance is applied (ageing speeds up after the conservation has eroded away).

2.4 Subjectivity in the Functional End of Life

Remaining useful life can be much shorter than remaining technical life. However, where the end of technical life is the result of a fatal failure and thus an objective (albeit uncertain) characteristic of the asset, functional inadequacy is much more subjective and context dependent.

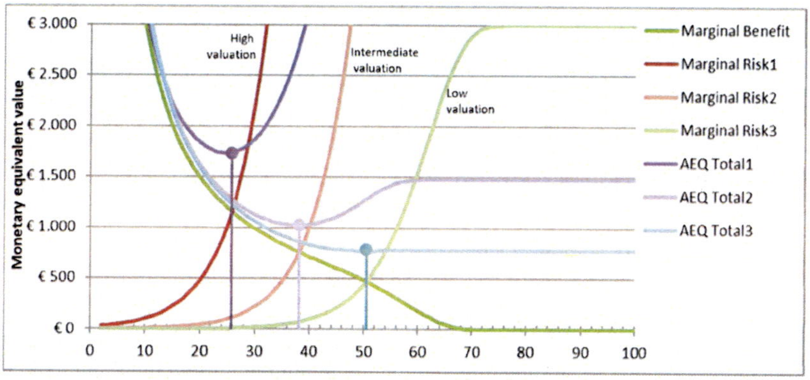

Fig. 4. Lifecycle optimization for different valuations of risk based on the annual equivalent cost (AEQ) over the lifecycle of the asset. At higher valuations (risk 1) the asset is replaced earlier than at low valuations (risk 3). For risk 2 the risk is valued equal to the investment (quite common in practice) and the optimal cycle is some 120% of undisturbed life.

This is because in deciding upon replacement the marginal benefit of postponing the replacement is compared with the marginal cost. The marginal benefit is spreading the investment over 1 more year and thus is declining (first by 1/t, at the end of life corrected for the survival rate R(t)). The marginal cost is the risk of running the asset for one more year, and goes up because of the increasing conditional failure rate. If risk is valued differently the optimal cycle is impacted, as shown in Fig. 4. The valuations are an order of magnitude apart. The valuation of non-financial risk can be captured by means of a well-designed risk matrix and replacing the non-financials with their financial equivalent effects [7, 12].

2.5 Instability of Outcome Constraining Strategies

A relevant problem for asset replacement planning is the potential instability of input constraining strategies like implementing a budget limit. This is because corrective replacements (which are unavoidable in any strategy) cost more than preventive ones and thus can eat away budget for preventative work. As a result, the asset base will age more and the likelihood of failure increases, thus eating away more budget and so on. In the long run, this can mean that the cost of all required expenditure (repairs and corrective replacement) exceeds the available budget [7, 10, 12]. However, this spiral of decline may only happen after a significant time delay. To avoid mortgaging the future, asset replacement planning therefore should be looking at the entire lifecycle of the asset base, which may be much longer than planners are used to.

3 Strategic Options in Asset Replacement

Many decision makers are used to steer their organizations by input constraints (budget, employees) and outcome targets (performance, risk). It is clear that some combinations are infeasible in any asset base (in extremis a no cost, no risk and maximum performance requirement). In general, decision makers appreciate these limitations and are willing to accept risk to reach the desired balance of cost and performance. However, because of the potential instability of input constraining strategies their understanding of what is feasible may be outdated very fast. Once the spiral of decline manifests itself, it is in general very costly to escape, if possible at all. This is because a preventive program would have been implemented on top of corrective work already straining the organization. The financial funding thus may not be the most relevant problem.

To deal with this mismatch between normal management practices and a rapidly changing context, it is vital to understand cost also as an outcome of operating an asset base to realize value. Decision makers thus should focus more on defining their concept of value than on second guessing asset management decisions. If the value system is clear, asset managers will be able to derive a long term strategy by consistently applying these decision principles to risks in the asset base. To help decision makers develop their understanding of their value system, they should at least consider the extreme options listed below.

1. Corrective: Asset are replaced when they fail (hence the name run to failure). The core value is to make maximal use of the asset. This strategy is typically applied if failure consequences are low or delayed.

2. Cost based: This is a formal optimization with regard to the costs of failure versus replacement (for examples see [13]). Assets are replaced when the expected extra cost of operating the asset one year longer equals the benefit of postponing the replacement by one year. The core value behind this strategy is the financial bottom line. Anything that does not end up in the financial report is not relevant.
3. Risk based: This is also a formal optimization, but then with regard to all failure consequences. To allow optimization, non-financial consequences are monetized. Assets are replaced when the total monetized risk for the next year equals the total benefit of postponing. According to cost benefit theory, if every risk decision is made using the same monetization factors, maximal value is delivered to society [14]. The core value therefore is social welfare.
4. Depreciation based: Assets are replaced when they are fully depreciated. As a result, total depreciation of a fixed asset base has always the same value. To prevent large dis-investments (assets failing before fully depreciated) the replacement age is typically close to the undisturbed lifetime. Core value in this strategy is financial predictability.
5. Condition based: assets are replaced when they reach a certain degree of degradation, to be assessed in time or use based inspection or monitoring. If detection is early enough, replacement should take place before actual failure. Core value in this approach is therefore the precautionary principle, as assets are prevented from failing.

In reality, the distinction between strategies may not always be clear. For example, strategies 2–5 will move towards a corrective strategy if exposed to a tight budget limit. The strategies can also approach each other in certain respects. If the non-financial consequences are very low, strategies 2 and 3 will be very similar. And if failure consequences are limited, both 2 and 3 will be like the corrective strategy as the optimum will only be reached at a high conditional failure rate (h(t) in graph 1) when most of the population is already replaced correctively. The condition based strategy in extremis will replace the assets just before failure, combining core values of 5 and 1.

4 Application

The approach was applied for a wastewater transportation and treatment infrastructure, operated by a waterboard situated in the Netherlands. The organization had been working on condition based maintenance and replacement practices, yet it still experienced surprises with regard to very costly asset failures, requiring additional budget. The board wanted to know what maintenance and replacement budget was needed so that no more surprises would occur.

4.1 Description of the Asset Base

The wastewater infrastructure consists of 3 subsystems: collection, transportation and treatment. Collection is organized by the municipalities (and thus out of scope for this study), transportation and treatment are the responsibility of the waterboards [15]. The wastewater system caters for over 800.000 inhabitants and consist roughly of 350 km of pipelines, 80 wastewater pumping stations and 20 treatment plants. The asset base consists of some 10.000 managed objects and has a replacement value of about € 850 M. About half of this value is in the treatment plants, with the pipelines accounting for some 40% and the wastewater pumping stations for 10%. Some 85% of the asset value is in civil engineering assets (=concrete), about 10% in mechanical engineering assets (rotating equipment) and less than 5% in electrical engineering assets (controllers). The mechanical and electrical assets are in general mounted on civil asset and may be replaced several times over the lifespan of the civil assets. Lifespans in this asset base are very long, especially given the dominance of civil assets. The total annual maintenance and replacement budget was about € 13M.

4.2 Resolution Model

The approach was applied at two levels of aggregation. First, a high level continuous simulation (addressing all causes of Fig. 1) was used to develop and validate the failure models of the diverse asset types under the assumption of optimal maintenance, given the current age profile (see also [12]). Because of the desired information on maintenance and replacement, and the desire to know where the largest problems would occur, this model was further developed into a discrete simulation of all individual assets in their current state. The impact of conservation on the life span was modeled in line with Fig. 3. The objects were grouped into some 80 different types at different levels of aggregation (asset, equipment, part), with as a rule of thumb the civil objects forming the assets, mechanical objects the equipment and the electrical objects the parts. These levels had a downward dependency: if an asset would be replaced, all associated equipment and parts would be replaced as well. Lifespan in the asset base ranged from a few years (parts like controllers) to more than 100 years (assets like filtration tanks). The timestep in this simulation was set at 1 year, time horizon at 100 years. Events on the assets in this simulation were based on a random number generator. To allow comparison between strategies, fixed sets of random were used by seeding the random number generator. Some 20 sets were used, enough to test the variance between runs of the simulation. Variance between different sets of random numbers were limited though. The large number of objects meant that any variation would average out. Using one simulation per strategy therefore would be good enough for comparison.

4.3 Output

Results were originally presented in the timestep of the simulation, that is with one year resolution. However, this provoked questions that were beyond the reach of the model, like why does a specific year have so much large incidents. We therefore presented the output as averages per 5 year period. In Fig. 5, two example outputs are shown for comparison.

4.4 Results

The table below contains the key performance indicators for the asset base on which the strategies above were applied in the most fundamental approach. These are the NPV

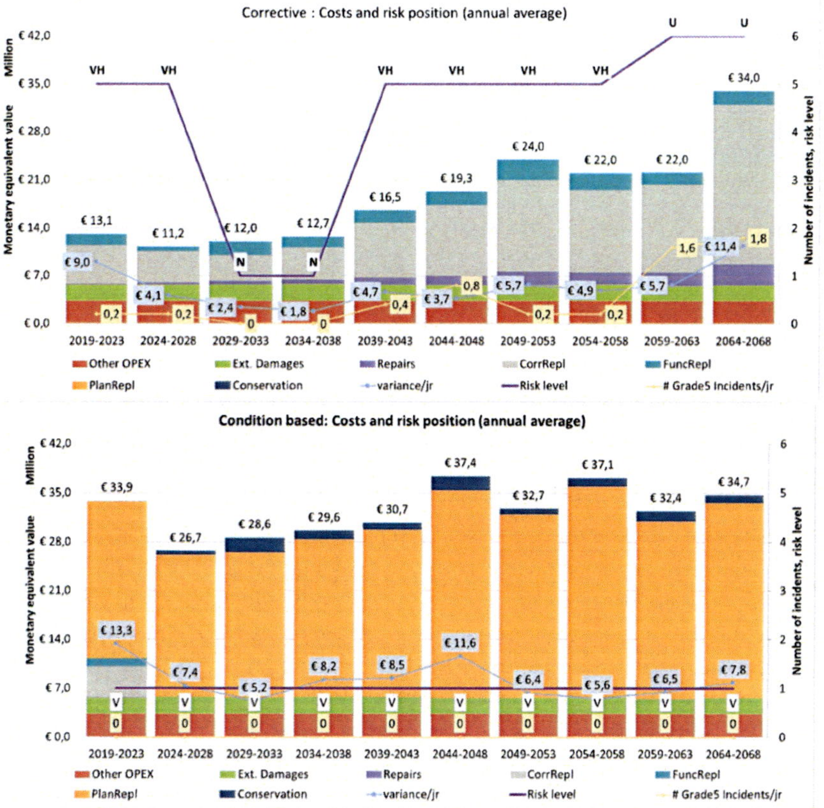

Fig. 5. Model output for costs and risk for two extreme strategies: corrective replacement (high risk, low cost) and condition based replacement (high cost, low risk).

financial (present value of total expenditure), the NPV risk (non-financial impacts), NPV total (sum of previous 2), Condition deficiency (average value loss compared to new assets), the value position (NPV total plus condition) and number of severe incidents. Best (green) and worst (red) strategies are highlighted per indicator. As expected (following the core value) the cost based strategy is cheapest in financial terms, the risk based strategy is cheapest over all values, and the condition based strategy has the lowest risk and the lowest number of incidents. It also has the lowest condition deficiency, but that is no necessity but dependent on the intervention level of the condition. The depreciation based strategy is the most expensive, and the corrective strategy has the highest risk. What is surprising though is that in terms of the value position all strategies are comparable, perhaps with the depreciation based strategy as an exception (but that was more used as a benchmark, not as a serious option). This could be explained by the relatively low additional costs over all values for corrective replacements.

Table 1. Comparing results of the strategic options

Strategy	unit	Corrective	Cost based	Risk based	Depreciation based	Condition based
NPV Financial	M€	603	564	580	*1362*	1034
NPV Risk	M€	*301*	266	245	209	175
NPV Total	M€	904	830	825	*1571*	1209
Condition deficiency	M€	*487*	473	462	207	171
Value position	M€	1391	1303	1287	*1778*	1380
# severe incidents	#	*123*	86	82	6	0
Year to year surprises	M€	*6*	*6*	*6*	2	1

4.5 Findings

As stated earlier, the decision makers were mostly concerned with the large number of surprises in expensive corrective replacements. The observed behavior could only be explained by a corrective strategy. If the organization would be implementing a condition based strategy, these surprises simply would not happen without serious defects in assessing the condition. A closer inspection revealed that the failing assets in general had be identified as vulnerable, but that their replacement was postponed by decision makers challenging the need for replacement as it was not broken yet. The decision makers were apparently acting according to the core value of making maximum use of the asset and thus the corrective strategy. Given the relatively low risk of failure, this is a reasonable strategy (see Table 1), but it requires the willingness to absorb year to year variance. Cost based or risk based optimizations would not bring much improvement in this sense (corrective would be optimal for a number of assets), but a condition based strategy would result in a dramatic reduction of severe incidents and unexpected expenditure, albeit at a significantly higher planned expenditure. Confronted with these results, the decision makers thus became aware that their top level decisions drove the organization precisely in the direction they did not want to go. The optimal strategy would be a mix of condition based, risk based and cost based strategies for different asset types, but more condition based budget would become available.

5 Conclusion and Discussion

By modeling the consequences in terms of cost, risk and performance for a variety of strategies in a strategic decision, decision makers can become aware of the direction in which they lead the organization. If this dialogue is held early enough, corrective interventions can be made. The strategies represent a core value and thus a value system. In that sense, long term planning helps establishing the value system which fits the organization best given its context. This approach has been applied at different levels for different asset systems (all in natural monopolies), always with similar results. Whether the approach also works in a commercial setting or for more complex problems (like the energy transition and climate adaptation) has to be evaluated in future research.

References

1. Mintzberg (1983) Structures in five: designing effective organisations. Prentice Hall, Englewood cliffs
2. Woodhouse J (2014) Asset management is growing up. In: Tutorial at the 9th WCEAM, Pretoria
3. Verzijlbergh R (2013) The power of electric vehicles. Faculty of Technology, Policy and Management. Delft University of Technology, Enschede
4. Veldman E (2013) Power play. In: Electrical engineering. Eindhoven University of Technology
5. Hanski J (2019) Supporting strateguic asset management in complex and uncertain decision contexts. School of Engineering Science. Lappeenranta-Lahti University of Technology, Lappeenranta
6. Morgan MG et al (2000) Categorizing risk for risk ranking. Risk Anal 20(1):49–58
7. Wijnia YC (2016) Processing risk in asset management: exploring the boundaries of risk based optimization under uncertainty for an energy infrastructure asset manager. Technology, Policy and Management, Department of Engineering Systems and Services, Section Energy & Industry. Delft University of Technology, Delft
8. Van der Lei T, Wijnia YC, Herder PM (2010) Towards an integral asset management framework using asset characteristics, asset environment and lifecycle phases as leading principles. In: The 5th world conference on engineering asset management. Springer, Brisbane
9. Endrerud O-EV (2020) S. Universitetet i, and b.o.m. Universitetet i Stavanger Institutt for maskin, Simulation-based decision support for engineering assets: an agent-based approach for operation and maintenance of offshore wind farms. University of Stavanger, Faculty of Science and Technology, Department of Mechanical and Structural Engineering and Materials Science, Stavanger
10. Wijnia Y (2018) Integrating maintenance into long term planning. In: Lyanage JP (ed) 13th world congress on engineering asset management. Stavanger, Norway
11. Jardine AKS, Banjevic D, Makis V (1997) Optimal replacement policy and the structure of software for condition-based maintenance. J Qual Maint Eng 3(2):109–119
12. Wijnia Y, de Croon J (2019) Strategic asset planning: balancing cost, performance and risk in an ageing asset base. In: Asset intelligence through integration and interoperability and contemporary vibration engineering technologies. Springer, Cham
13. IAM (2014) Asset Management - an anatomy Version 2. The institute of asset management
14. Tengs TO et al (1995) Five-hundred life-saving interventions and their cost-effectiveness. Risk Anal 15(3):369–390
15. Wijnia Y, de Croon J (2015) Designing an asset management guideline for the Dutch wastewater industry. In: Lee WB, et al (eds) Proceedings of the 7th world congress on engineering asset management (WCEAM 2012). Springer, pp 637–646

Asset Management in Electrical Utilities in the Context of Business and Operational Complexity

Dragan Komljenovic[1(✉)], Darragi Messaoudi[2], Alain Côté[1], Mohamed Gaha[1], Luc Vouligny[1], Stéphane Alarie[1], Amira Dems[1], and Olivier Blancke[1]

[1] Institut de recherche d'Hydro-Québec (IREQ), 1800, boul. Lionel-Boulet, Varennes, QC J3X 1S1, Canada
{komljenovic.dragan,cote.alain7,gaha.mohamed,vouligny.luc, alarie.stephane,dems.amira,blancke.olivier}@ireq.ca
[2] Hydro-Québec TransÉnergie, Direction Gestion des actifs, Complexe Desjardins, 19th Floor, Montréal, QC H5B 1H7, Canada
messaoudi.darragi@hydro.qc.ca

Abstract. Major electrical utilities are capital-intensive organizations. They face the renewal of large parts of their assets that reach the end of their useful life. These companies also operate in an increasingly complex business and operational environment characterized by deep uncertainties. The impact of extreme and rare events is increasingly important, but not sufficiently studied. New methodologies and ways of thinking are necessary to elaborate models which adequately support decision-making process. This paper presents a global asset management approach developed for Hydro-Québec TransÉnergie which is applicable for major electrical utilities. It is part of an overall resilience management strategy and takes into account the context of operational and business complexity.

Keywords: Asset management of electrical utilities · Complexity · Modeling · Simulation

1 Introduction

Electrical power grids are part of national critical infrastructures. Modern electrical utilities are capital-intensive enterprises. The experience feedback and various studies show that their critical systems, structures and components are aging or reach the end of their useful life. Their operating and business environments are increasingly complex involving significant uncertainties (market evolution, changing regulatory framework, new technologies, malicious human actions, climate change, extreme weather events, etc.). Under such circumstances, electrical utilities need to develop various methods supporting the decision-making process in order to ensure their sustainability and performance. In the context of deep uncertainties the resilience has become a popular approach in managing critical infrastructures and power grids [1, 4, 6, 8, 9, 10, 12]. This concept has

A. Crespo Márquez et al. (Eds.): WCEAM 2019, LNME, pp. 34–45, 2021.
https://doi.org/10.1007/978-3-030-64228-0_4

been developed in recent years and includes economical, technological and organizational resilience [6, 9, 12]. One of key components of the overall resilience management is related to asset management (Fig. 1). It is worth highlighting that R&D and innovation play an essential role in this dynamic context together with other tightly interdependent activities presented in Fig. 1.

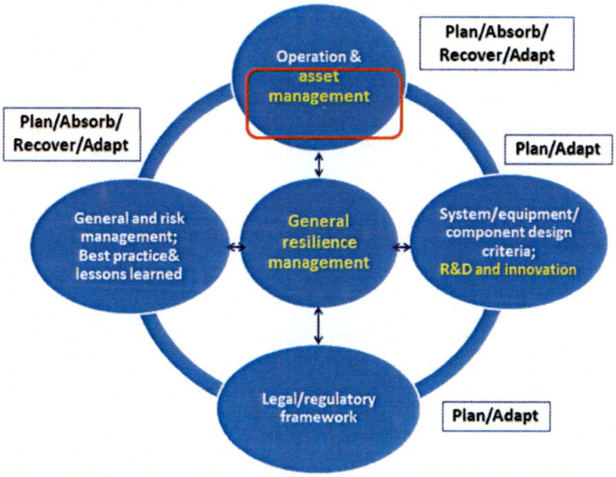

Fig. 1. Overall resilience management and the role of asset management

Asset management is defined by the ISO Standard 55000 as a set of *coordinated activities of an organization to realize the value of assets* [7]. These activities have their own specificities and their particular models that interact and generate a flow from raw data to knowledge and then to appropriate actions. Their interactions and interdependencies are also complex and characterized by significant intrinsic uncertainties [8].

Hydro-Québec TransÉnergie (HQT) as the operator of one of the largest North-American electrical power grids faces similar challenges like other modern electrical utilities. The paper presents a global asset management approach elaborated at HQT which is being improved and implemented through a research project. The overall methodology aims at providing enhanced methods and tools to support a sound decision-making process. This way, it ensures the sustainability and resilience of the organization while remaining economically competitive. It integrates various fields of expertise as well as a data-driven IT system and knowledge base.

2 Global Asset Management Model at HQT

The HQT global model of asset management is presented in Fig. 2. It has been developed at HQT and its improvement is now conducted through a close collaboration between HQT, Hydro-Québec's Research Institute (IREQ) and several universities under a R&D project named "PRIAD"[1].

[1] A French acronym for a robust decision-making support tool in asset management.

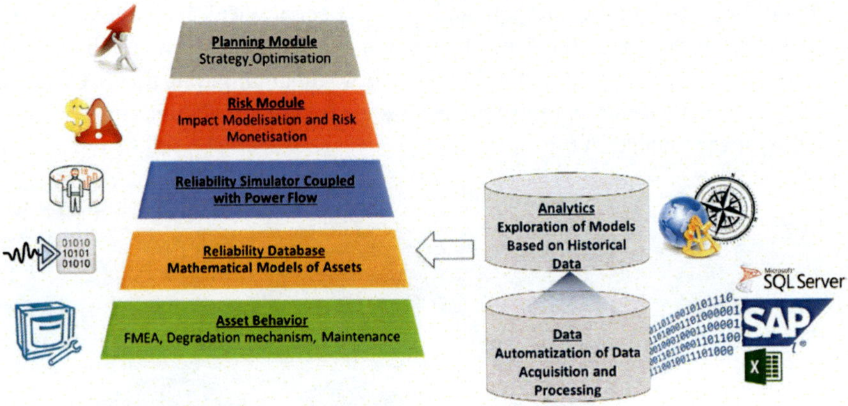

Fig. 2. Global asset management model at Hydro-Québec TransÉnergie (adapted from [18])

The model is composed of five main steps and also an integrated data analytic platform as well as an analytics module which allows acquisition, exploration and analysis of historical data. The main steps, designed to maximize the impact of any knowledge regarding assets' behaviour through a set of reliability focused computing tools include: a) asset behaviour, b) reliability database, c) transmission grid reliability simulator coupled with a power flow software, d) risk module and e) planning and optimization module (Fig. 2). The functional relationship between these modules is shown in Fig. 3. They are shortly described below.

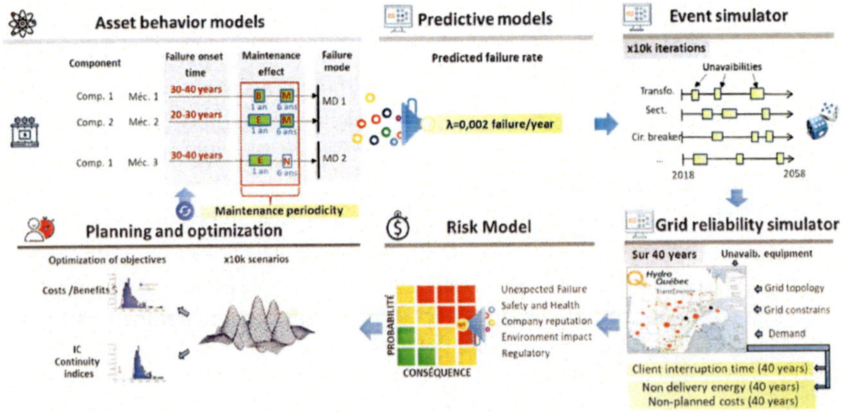

Fig. 3. Functional relationship between the modules of the global model (adapted from [18])

2.1 Asset Behaviour Models and Reliability Database

Asset behavior models aim to predict the effects of time-based preventive maintenance strategies on the reliability performance of different asset classes. A hybrid theoretical approach was developed. The models aim to represent the physical degradation's process of each asset class by combining both expert knowledge and advanced statistical analyses of failure data. Unlike statistical approaches, physic-based models add more robustness when dealing with few or bad data. It is also easier to interpret for experts and decision-makers. The approach was inspired by various recent works in the field [1, 11] and was adapted to the operational context of Hydro-Québec.

To model the effects of time-based maintenance strategies, the first step consists in building the degradation model for each asset class. For this, it is necessary to identify all degradation mechanisms of each component which can lead to different failure modes. It is performed using Failure Mode and Effects Analysis (FMEA). Then, the experts must estimate the inherent propagation time of the degradation mechanisms between the asset commissioning and its failure if no preventive maintenance is carried out. This propagation time of a degradation mechanism DM_1 to its failure mode F_1 is denoted t_{C_1,DM_1,F_1} and is called the failure onset time. C_1 is the component where the degradation mechanism process DM_1 is occurring. While considering biases and uncertainties present in the elicitation process, the failure onset time t_{C_1,DM_1,F_1} has been estimated using a bounded interval. Figure 4 illustrates this degradation model.

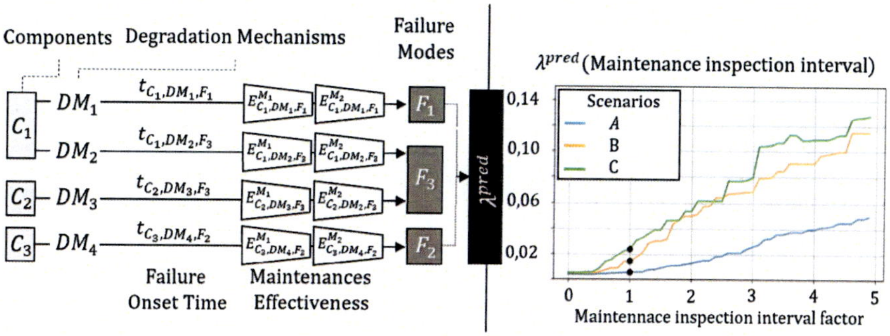

Fig. 4. Asset behavior predictive models fitting process [2]

As a second step, experts are asked to estimate the effects of maintenance actions on the propagation of degradation mechanisms. This consists in estimating the maintenance task effectiveness to detect and stop the degradation mechanism defined as $E^{M_1}_{C_1,DM_1,F_1}$ where M_1 is the maintenance activity. The model is based on a strong hypothesis that a remedial maintenance activity will be carried out within a reasonable time after a preventive maintenance action (inspection) detects an anomaly caused by the degradation mechanism. The degradation models represent the knowledge acquired in physics-of-failure of each asset class. Consequently, they make it possible to predict the failure rate of asset classes for each scenario of various long-term systematic maintenance strategies. It is a work in progress and around 70% of critical assets has been analyzed.

In order to validate of predictive models, it is necessary to fit failure rate prediction with observed failures under past maintenance practices. Thus, a fitting process has been developed using optimization methods and advanced statistical analysis.

The first step consists in estimating the rate of repairable failures for an asset class under the actual maintenance strategy. To improve the fitting process proposed in this paper, the failure rate must be estimated for all components' and failure modes' combinations. Blancke et al. [2], the data inconsistencies of failure events must be considered in the failure rate estimation process. Thus, the confidence interval on the estimated failure rates is inferred from the acquisition processes based on expert knowledge. The values of $\hat{\lambda}^{obs}$ need to be grouped in the fitting process as not all observations are considered in the physical model. The asset reliability models will be stored in a centralized reliability DataBase (DB) (Fig. 2). This centralized DB will ensure consistency and traceability of the data for engineers, analysts and simulation tools. The concept focuses on assets while building key elements of the DB. It will record and classify historical events for each asset including management strategies, attributes, status, risk profile, localization, reliability models or other relevant information. This database will be hosted in a cloud-based infrastructure to allow multiple users to access it throughout the enterprise.

Then, the fitting process consist in an optimization model that aims to match statistical failure rates to the physic-based predictive models Côté et al. [3]. More precisely, it must determine the failure onset time for each degradation process considering intervals given by experts. Blackbox optimization is used to effectively solve complex engineering problems. Figure 5 depicts the predictive model fitting process.

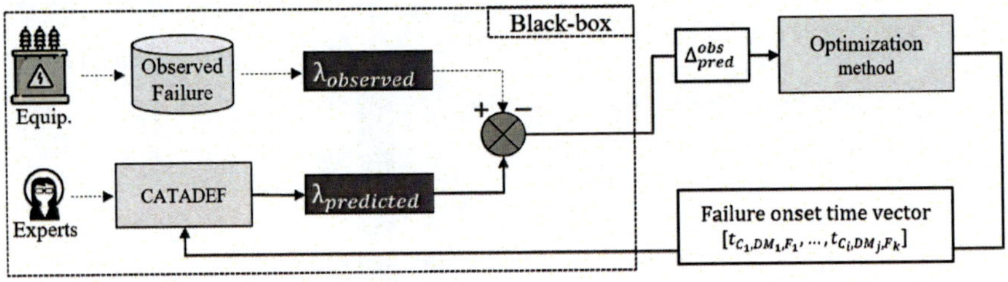

Fig. 5. Asset behavior predictive models illustration [3]

2.2 Grid Reliability Simulator Coupled with Power Flow Software

In order to have a better understanding of the transmission network and its capacity to be resilient to unplanned and planned interruptions, we developed a versatile and scalable simulator. The expected performance of the transmission network is evaluated by using a Monte Carlo approach that runs a high number of simulations to mimic the transmission network behavior.

The main objective of our simulator is to determine how the unavailability of one or several equipment affects the network reliability and how much it affects the power flow. It aims to simulate the expected performance of the HQT transmission network by calculating indicators such as outage durations (i.e. downtime) and the energy balance for a given substation or the entire grid.

Our main simulator is composed by two autonomous simulators that work in a complementary way. The first is an events simulator. It simulates parallel discrete events and its main objective is to evaluate the impact of multiple equipment unavailability on the network. Actually, there are two types of events that can be simulated: the unplanned events (i.e. outages) and planned events (i.e. maintenances). Fundamentally, the difference between maintenance events and outage events is their deterministic nature, hence their ability to be rescheduled under certain conditions or constraints. However, outages are invariant and cannot be postponed. The stochastic nature of frequencies and the durations of the events are described by statistical distributions. In order to generate discrete events according to a given statistical distribution, we use the inverse cumulative function. A random uniformly distributed number U[0, 1] is drawn and the timestamp of the event is calculated according to Eq. 1 in case of a Weibull distribution. For example the inverse cumulative function of a Weibull distribution with the parameters (λ: scale & k: shape) is defined as:

$$U = 1 - e^{-\left(\frac{X}{\lambda}\right)^{k}} \tag{1}$$

Thus, the time to fail in the Monte Carlo simulation is given by:

$$X = \lambda \cdot [-ln(1 - U)]^{\frac{1}{k}} \tag{2}$$

The second simulator is based on the Contingency Analysis (CA). It quantitatively assesses the electrical impact of equipment unavailability on the whole transmission network. The simulator relies on a commercial power-flow engine (software), PSSe, and employs a CA algorithm to identify the events causing power system violations or interrupted energy [5]. In the current version of the CA algorithm the stability and thermal limits are considered and our advanced algorithm can overcome some violations by taking several remedial actions like dispatching energy generation or load shedding. The generation dispatching is prioritized over load shedding for economic reasons. Our CA algorithm can analyse several contingencies for the whole transmission network within a reasonable amount of time.

Moreover, as the complexity of the system to simulate increases, the needed number of MC simulations grows rapidly. In order to overcome long time processing our simulator can record and memorize partial results. In fact, during MC simulations the same events occur several times. To this end, we preprocess and save the electrical impacts of one (N-1) and two contingencies (N-2). This simple approach reduces the processing time by 90%. In fact, according to our experiments 93% of all the generated events are N-1 and N-2 contingency stimulation and only 6% of simulations implies N-3 contingencies.

The processing time for each scenario is summarized in Table 1. In order to keep results simple, the processing time is obtained by using one main CPU. Note that the processing time is not correlated to the simulations number but to the simulated events. Thus, with the increasing number of the simulated events, the processing time gets higher as expected.

2.3 Risk Model

The risk model uses relevant inputs from previous models. It aims at quantifying and translating them into monetary value where possible. Risks are expressed through relevant costs and potential losses. The following risks are considered: operability, reliability,

Table 1. Use cases and processing time

Use case	Equipment	Number of simulations	Simulated events	Simulations in Days
Substation	80	20,000	5.60E+06	0,23
Corridor	400	100,000	1.40E+08	5,83
Main network	2,000	500,000	3.50E+09	145

availability and maintainability (RAM), environmental, financial, regulatory, occupational safety and health (OSH), reputation of the company, and any other risks that may be considered relevant. The interdependence between various types of risk is also analyzed in order to optimize risk management strategies. Figure 6 shows the workflow in the risk calculation that uses the Monte Carlo simulation that is part of the global process depicted in Fig. 3.

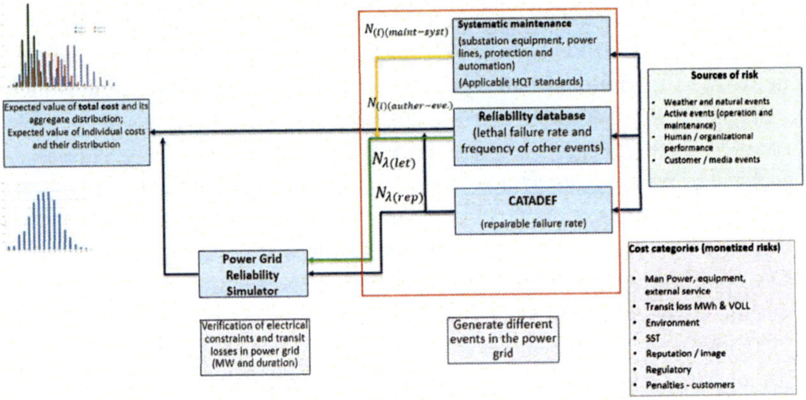

Fig. 6. Global risk module

Costs of preventive and planned systematic maintenance activities are calculated simultaneously with costs generated through reparable and lethal failures. The power grid reliability simulator verifies potential interruptions of energy delivery at the system level which is translated in monetary losses expressed through VOLL (Value of Lost Load). The Monte Carlo simulation gives an output expressed as the expected value and aggregated distribution of individual costs and losses as well as the total cost. Various maintenance strategies have an impact on the failure rates, which are calculated in the asset behaviour models generating various impacts at the system level, which translates into complex cost distributions (Fig. 6). These results are sent to the planning and optimisation model in order to improve and optimize maintenance strategy.

2.4 Planning and Optimization Model

All the above models will be put together to set optimal maintenance strategies for the assets. This will be done by Blackbox approaches since the objective function and the constraints (if any) cannot be reduced to analytical expressions.

The evaluation of the performance of strategies generated by the solver along the optimization process is described below and refers to Fig. 3. Let x be the current strategy under consideration. Formally, x is a set of values that completely defines the underlying strategy. The values of x are injected into the degradation behaviour models to get the corresponding failure rates. These rates are used by the MC simulator to generate scenarios of functionality losses, for which a reliability power flow is next done and all undesirable situations (previously defined by the asset managers) are measured and quantified. This is translated in different metrics such as economic value, financial costs and risk exposures. The computed metrics are then returned to the solver, which generates a new x to evaluate. The process is hence repeated until a local optimum is found.

As shown in Table 1, the amount of calculation to perform for evaluating a given x is large, enough large to make any optimization non-trivial. Fortunately, since MC simulations are independent, the processing time can be accelerated by using multiple CPUs. For example, processing time for the main network can be reduced from 145 days to less than a day by using 150 CPUs. Even then, limiting the total number of evaluations to 2,000, it will take five and half years to proceed an optimization. One can also speed up the optimization by performing evaluations in parallel. To limit it to only one week, 300 evaluations must be carried out simultaneously. Considering that 150 CPUs are required per evaluation, we need to hold 45,000 CPUs to do this. To limit to one month, it is 10,500 CPUs.

The resource requirement here is so great that one must carefully choose the algorithm that will be used. At these scales, it will make the difference between being able to optimize or not. The number of available solvers is however important [13]. Usual metaheuristics such as Genetic Algorithms (and the alike) and Particle Swarm Optimization (PSO) should be kept off since they are largely based on randomness and work with candidate solution populations. Consequently, they spend time and effort to evaluate uninteresting solutions. If one wants to converge on a local optimum with such expensive evaluations, one has to consider algorithms that, by their own conception, try to reduce as much as possible the number of evaluations to perform [14]. Pattern search methods (such as GPS, MADS, HJDS) and trust-region methods (such as DFO, NEWUOA, BOOSTERS) are to be considered [15]. Such methods are currently under investigation.

Note that the evaluations are not only computationally expensive, they are also noisy since they rely on MC simulations, which means that evaluating the same x twice will generate different $f(x)$ values. This must be taken into account otherwise the optimization process will converge on a non-optimal point due to a favorable bias from the Monte Carlo procedure. This arises because (i) whatever the sampling, there is always a non-zero probability that a Monte Carlo simulation provides an estimate far from the expected value, (ii) optimization algorithms take deterministic decisions based on single values, and (iii) the number of evaluations performed during an optimization is large enough for (i) to occur.

If one wants to reduce the noise of f, one can simply increase the number of samples. People often consider that 10,000 samples are enough to get a good empirically acceptable estimation of $f(x)$. However, in the recent work [16], it has been shown that to be able to discriminate solutions when approaching a local optimum, it is rather 1,000,000 samples per evaluation that should be considered, and this was on a tiny maximization problem with only five random variables [17], which took 3403 evaluations and 25 days to solve. Even though, the optimal value of f is overestimated by 1.5% such sampling ensures an accuracy of 1%. Instead of increasing the sampling, one can also decide to fit a smooth function through the noisy evaluated points. In the same work [16], it has been proven useful but still required 100,000 samples per evaluation to do better than above on the tiny problem, which took 21,847 evaluations and 16 days to solve. The approach investigated in [16] is about starting the optimization with a reasonable number of samples and to compare the values $f(x)$ returned by the Monte Carlo simulation with their distribution (or an estimated of it). If it appears that two points x cannot be differentiated due to overlapping distribution, more samples are then performed to refine the distributions. By doing this throughout the optimization, calculations are only limited to interesting points. Preliminary results on the tiny problem show that the proposed approach reached the optimal region in less than 6 h, but then stagnated trying to refine indistinguishable solutions.

The relevance of above strategy optimizations will depend on the ability of FMEA to predict correctly the failure rates for maintenance strategies which have not been experienced before, and hence, the right setting of onset times by the calibration process. We also need to be able to properly quantify the risks that will arise from such maintenance strategies. This is particularly important since optimization problems usually involve saturated constraints at the optimality. We must then expect that the optimal strategies will intentionally cause failures to the limit of what we are ready to tolerate, leaving no room to face it. If the risk associated with a maintenance frequency is underestimated, the optimizer will unduly stretch it, which will lead to an uncontrolled increase in corrective maintenance in a few years. If the risk is overestimated, we will miss opportunities to reduce our costs in preventive maintenance without increasing failures. The challenge here is to define an optimal maintenance strategy that will balance costs between preventive and corrective actions at the system level when it will be deployed in the field.

2.5 Basic Principles Related to PRIAD's IT Architecture

Historically, asset investment planning such as CAPEX and OPEX was carried out mostly manually. Information was drawn from the company's databases, such as SAP, or from Excel files prepared manually. The data was then integrated into an MS-SQL database. Once the inputs were established, ETL tools (mainly Pentaho and T-SQL) are used to carry out various transformations in order to feed a predictive financial simulation software. The software prediction results were then analyzed and formatted to establish the future budgets for asset investments and maintenance required by the "Régie de l'énergie " the Quebec Energy Regulator.

In order to facilitate the work on data acquisition and analysis, all information was gathered in a single database, including ETL jobs, for each year. This way, upon verification and requests by the regulator for a given year, it is easy to retrieve information through the database. On the other hand, when the validation request is more complex, for example if we were asked to use a more recent processing algorithm on old data, this approach becomes inadequate. These requests can result in several methods which must also be archived with their data.

Furthermore, the system must be able to load historical data that was used at the request time, much like a tax computation system. On the other hand, we must be able to adjust the processing and the datasets to establish different prediction without altering system integrity.

The following principles were used in solving these issues while overhauling the system:

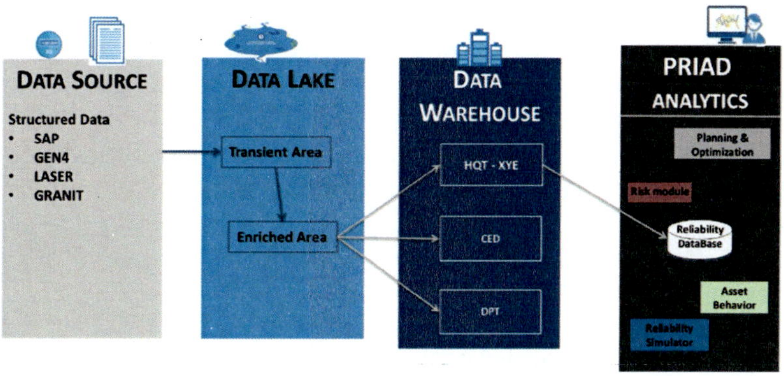

Fig. 7. Data and processing architecture

- Automation of all acquisition processes to avoid manual data acquisition or transformation. The automation process can be quite challenging in order to ensure data quality: cleansing, normalization, coherence, evolution in data systems through time, etc.
- Clear separation of input data, intermediate results related to different computing functionalities and output data. This data organization greatly eases the implementation of new modules for the analytics requirements.
- Using databases with time stamp on each record, which improve data retrieval using time-based meta-data.
- Separation of data transformation methods and analysis, in order to respond more easily to different analysis scenarios. This way, one can easily carry out studies combining the data analysis methods of a given year with the data set of another year.
- Versioning of data analysis methods in order to allow selecting them according to each analysis.
- Modular approach to analytical treatments and the use of an appropriate technology, micro-services, which is very well suited for this kind of system with multiple yet interconnected processing units.
- Use of a messaging integration bus, a proven technology to ensure the decoupling between different modules, which facilitates the coordination of processing and maintenance of such a complex system. In principle, each micro service listens to messages

of interest to itself and can interact with other modules by making requests through the messaging system.

The data and processing architecture are depicted in Fig. 7.

3 Conclusions

The global asset management approach elaborated by HQT aims at maximizing the value from its assets, optimizing resources and ensuring its overall sustainability and resilience in the context of the complex operational and business environment under deep uncertainties. In this regard, an important research project is undertaken to improve existing tools and provide a strong scientific and technical basis for a rigorous decision-making process. The project is ongoing, but results achieved so far are quite encouraging. Its strength comes from i) a highly skilled team involved in the project, ii) a multidisciplinary approach and iii) a long-term vision of the organization. It takes into account and manages the complexity of the operational environment and performs analyses at the component, equipment and system levels. This way, it is reasonable to expect obtaining insights and foresights that enable defining an optimal asset management strategy for the future.

The need to optimize maintenance strategies at the system level will lead to the development of improvements in all the PRIAD modules. These will be largely stimulated by the heaviness of optimizations, but not only. Some improvements will seem unnecessary or unjustifiable when considering each module individually, but will make sense once integrated with the other modules. Despite the difficulties that this causes, the integration of the modules within an optimization frame allows to identify the design weaknesses, but above all, to target the corrections to be made.

References

1. Bertling L, Allan R, Eriksson R (2005) A reliability-centered asset maintenance method for assessing the impact of maintenance in power distribution systems. IEEE Trans Power Syst 20(1):75–82
2. Blancke O, McCarthy G, Payette M, Bibeau L, Boudreau J-F, Coté A, Komljenovic D, Messaoudi D, Souilah B (2020) Performance of reliability models for repairable systems: dealing with information uncertainty, 2020 proceedings annual reliability and maintainability symposium (RAMS). Palm Springs, CA
3. Coté A, Blancke O, Alarie S, Dems A, Komljenovic D, Messaoudi D (2020) Combining historical data and domain expert knowledge using optimization to model electrical equipment reliability. In: 2020 International conference on probabilistic methods applied to power systems (PMAPS), Liege, Belgium
4. Cox LAT (2012) Confronting deep uncertainties in risk analysis. Risk Anal 012(32):1607–1629
5. Delavari A, Prevost J, Gaha M, Dems A (2020) Hydro-Quebec's new approach for asset management. le Power Systems Control and Management, y Data Science and High Performance Computing, PES 2020 General Meeting Conference (IEEE), Montreal

6. Hosseini S, Barker K, Ramirez-Marques, JE (2016) A review of definitions and measures of system resilience. Reliab Eng Syst Saf 145:47–61
7. The Institute of Asset Management (December, 2015) Asset management - an anatomy V3, 84 p
8. ISO 55000 (January, 2014) 2014 Asset Management – Overview, Principles and Terminology, ISO Standard
9. Komljenovic D, Gaha M, Abdul-Nour G, Langheit C, Bourgeois M (2016) Risks of extreme and rare events in asset management. Saf Sci 88:129–145
10. OECD/IEA (2015) Making the energy sector more resilient to climate change, Paris. https://webstore.iea.org/making-the-energy-sector-more-resilient-to-climate-change. Accessed 15 June 2020
11. Electrical Power Research Institute (EPRI) (2017) Preventive maintenance basis database (PMBD), quick reference guide in French, Palo Alto, CA, 000000003002011383
12. Zio E (2016) Challenges in the vulnerability and risk analysis of critical infrastructures. Reliab Eng Syst Saf 152:137–150
13. Rios LM, Sahinidis NV (2013) Derivative-free optimization: a review of algorithms and comparison of software implementations. J Glob Optim 56:1247–1293
14. Audet C, Hare W (2017) Derivative-free and blackbox optimization. Springer series in operations research and financial engineering. Springer
15. Conn AR, Scheinberg K, Vicente LN (2009) Introduction to derivative-free optimization, MPS-SIAM series on optimization, MPS-SIAM
16. Alarie S, Audet C, Bouchet P-Y, Le Digabel S (2019) Optimization of noisy blackboxes with adaptive precision [technical report], Les Cahiers du GERAD, G-2019-84, HEC Montréal
17. Browne T, Iooss B, Le Gratiet L, Lonchampt J, Remy E (2016) Stochastic simulators based optimization by Gaussian process metamodels – application to maintenance investments planning issues. Qual Reliab Eng Int 32(6):2067–2080
18. Côté A, Messaoudi D, Komljenovic D, Alarie S, Blancke O, Boudreau J-F, Gaha M, Truchon E, Pelletier S (2019) Development of an Asset Management Decision Support Tool for Hydro-Québec TransÉnergie. In: Proceedings of 2019 CIGRE CANADA Conference. Montréal, Canada

Combining Quality of Service and Quality of Experience to Visualize and Analyze City Services

Gabriela Gongora-Svartzman[1](✉) and Jose E. Ramirez-Marquez[2]

[1] Carnegie Mellon University, Pittsburgh, PA 15213, USA
ggongora@cmu.com
[2] Stevens Institute of Technology, Hoboken, NJ 07030, USA
jmarquez@stevens.edu

Abstract. Research is rich in analyzing performance and quality of city services (e.g., the frequency and timeliness of the), while the quality of experience is usually overlooked. This research combines the quality of service and the quality of experience into a single multi-faceted framework. The proposed framework incorporates visualizations and implements urban analytics to compare QoS and QoE. Discrete event simulations and Markov Chain Models are created to model city services' behavior and simulate events. At the same time, resilience metrics are used to evaluate the quality of the service. Data mining of social media, natural language processing, and emotion analysis techniques are employed to measure the QoE. The framework is implemented to analyze how services react to different types of service disruptions (e.g., weather, maintenance, attacks) through time and their effect on customers. This research's main findings are related to the differences in resilience of the services during disruptions, while QoE reveals additional issues associated with their use. The proposed framework is illustrated through transportation services, providing insights on their performability and customer's perception of the service. Future work involves collecting data on other services to test the framework further, as well as improving the layers of superimposed data for explorations and visualizations, thereby creating a decision-making tool for stakeholders.

Keywords: Smart cities · Urban analytics · Visualization · Simulation

1 Introduction

Over the past decade, there has been increased interest in smart cities. Washington DC, in the USA, has taken several steps forward to become a smart or more responsive city [1], through its program Grade.DC.gov, which allows citizens to evaluate the quality of a city service (e.g., the Department of Motor Vehicles). New York City, Boulder, Pittsburgh, San Francisco, and Austin are additional examples of smart cities using existing data to facilitate citizens' lives and collect their insights on city services through different initiatives.

A. Crespo Márquez et al. (Eds.): WCEAM 2019, LNME, pp. 46–54, 2021.
https://doi.org/10.1007/978-3-030-64228-0_5

City services need to be recorded, analyzed, and evaluated to increase citizens' quality of life. Transportation services are part of the core infrastructures of any city and, therefore, should be considered. Their evaluation can involve measuring their reliability [2–4] or their traffic management [5, 6] or even by analyzing how resilient their services are to disruptions [7–11]. These performance and quantitative aspects are referred to as the quality of the service (QoS). Customers could still be dissatisfied with the service [12, 13] even when the QoS reports numbers that are adequate for transportation agencies. A decade ago, customer satisfaction and perceived service opinions could only be achieved through customer surveys [14]. Still, in recent years social media has proven to be a reliable indicator of the quality of experience (QoE) [15–18, 24]. In literature, both the quality of experience and quality of the service have been explored separately.

This research proposes to analyze in a single framework the QoE and the QoS, with the following specific objectives: 1) Understand the quality of city services (QoS), 2) Analyze the Quality of Experience (QoE) towards city services, and 3) Create a framework that allows the comparison of Quality of Service and Quality of Experience of city services through urban analytics and visualizations.

2 Previous Work Done

The quality of transportation services has been a concern for the past decades, but social media's incorporation as a solution to this concern has been a recent adaptation. Sasaki, Osaki, and Matsuo [22] explore the train system in Tokyo, Japan, by mining Twitter in search of accidents and reports in the train lines. Their initial work can become an early delay sensor. Seven years later, Ni et al. [23] use seasonal auto-regressive algorithms and moving averages in time series in turnstile data to predict passenger flows in combination with social media. The authors mentioned provide proof that social media should be used to research traffic incidents and transportation services.

Sentiment analysis and topic modeling have become leading techniques to evaluate customer behavior and opinion mining. Gu, Quian, and Chen [15] use social media to detect issues, i.e., incidents on highways, and provide a comprehensive evaluation of the events. This analysis had additional details and experiences not recorded in the official 911 feed for the cities analyzed. Ahmed et al. [25] implement sentiment analysis for smart city services, highlighting the uses of gathering data within a city to detect and fix city issues. Zhang et al. [27] use Twitter to measure concentrations during surges in North Virginia. They use clustering and correlation techniques to detect traffic patterns and disruptions through Twitter concentrations.

More recent studies focus on using social media and sentiment analysis to gather customers' opinions on services and policies related to the services. Mogaji and Erkin analyze customer experience in UK train services by performing a sentiment and polarity analysis on twitter related feeds [28]. The polarity of the customer's opinions is used to measure the quality the transportation service is proving.

Chakraborty and Sharma also use sentiment analysis on customer tweets, but target Delhi's transportation system and focus on evaluating the transportation policies and their effects [29]. Mounica and Lavanya recognize the use of social media for assessing transportation services [30]. Their study focuses on the emerging trends and possible methodologies to incorporate different social media types into analyzing both the quality of the transportation services and the quality of experience using them.

The literature shows effective social media uses to detect incidents or disruptions in services, mainly transportation services. Nevertheless, these analysis are done separately from the quality of the service, therefore creating an opportunity for further research.

3 Proposed Framework

The overall framework is described in Fig. 1. The author proposed the framework to explore, analyze, and assess the quality of service (QoS) separately or jointly with the quality of experience (QoE) of a service. This framework has four sections, where section one refers to the city services and collecting data from them. Data sources can be the agency providing the service, up to mining social media, to gather customers' perception of a service. Section two relates to the simulations and data manipulations required to assess how the service performs, i.e., the Quality of Service (QoS). In this same section, measures such as reliability, resilience, waiting times are developed. Section three analyzes how customers perceive the service offered, i.e., the Quality of Experience (QoE). This section implements emotion analysis, sentiment analysis, topic modeling, and event detection to address customer's concerns when using a service. Section four combines sections two and three into a visual analysis to evaluate the results over time or a particular event. This last section is crucial to understand and analyze the bridge between QoS and QoE.

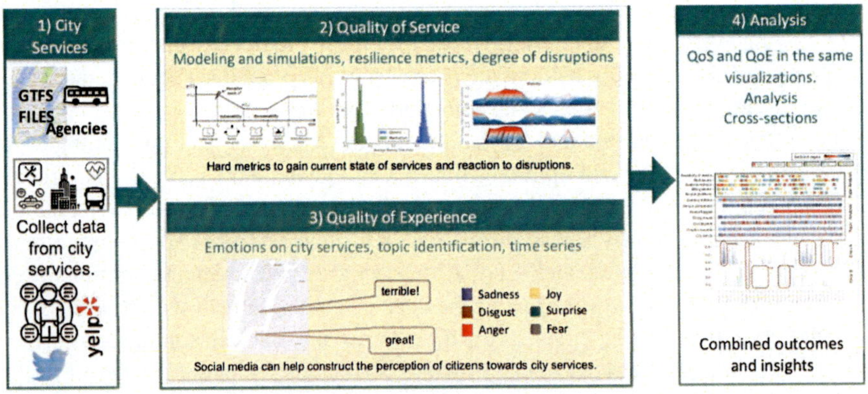

Fig. 1. Overview of proposed framework [21]

4 Case Studies

4.1 Bicycle Sharing Systems

A bicycle-sharing system was chosen for the prosed framework, which was described in the previous section. The specific bicycle-sharing program analyzed was Citibike in New

York City, NY, US. The reliability of Citibike is measured to evaluate the QoS aspect of this service. Measuring the reliability in bicycle sharing systems can be considered through two possible errors: *A* and *B*. Error *A* occurs when a user tries to return a bicycle, and there are no available docks. Subsequently, error *B* occurs when there are no available bicycles. The user must walk to the nearest available station, also leading to user dissatisfaction and defeating the program's primary purpose. Markov Chains and a Discrete Event Simulation (DES) were implemented to measure errors *A* and *B*. The Markov Chains took in data from five years of trips (2013–2018) from Citibike. Using origin-destination pairs and calculating the state stochastic probability, it was possible to estimate – through the power method - the number of bicycles at each station and the frequency of trips between the origin-destination pairs. A DES was created with these calculations, using a trip duration generator, a process generator for the destination, and a customer arrival process generator [21].

According to the framework in Fig. 1, sections three and four still need to be implemented to measure the QoE of Citibike data from Yelp and Twitter. The data is from 2013 up to 2018, covering five years of data. Natural language processing was applied to the Twitter data to clean it, process it, and apply a sentiment analysis by topics. Text collected from Yelp was richer and longer; therefore, an emotion analysis was possible, using Colneric and Demsar's model [19] and implementing topic analysis. Thus, creating an emotion analysis per topic. For topic modeling, k-mean++ was implemented, discovering seven Twitter data issues and five for the Yelp data. With the QoS and QoE done, the last stage in Fig. 1 is joining them, and for this, a new visualization was created. This new visualization followed the idea of a lexical dispersion plot. These plots are used to track words and their frequency throughout a chapter or a whole book. An adaptation of this plot was created to follow sentiments (Twitter) and emotions (Yelp) for each topic and throughout the five years of data. The outcome is shown in Fig. 2.

Figure 2 shows the result of applying the framework from Fig. 1. This visualization provides additional details that could not be considered if only the reliability of the system was measured. These details are exemplified in the variety of topics from Yelp and Twitter, showing that customers not only care about the availability of bicycles and docking stations (errors A and B) but are also concerned about billing, customer services, the appearance of bicycles, among other concerns. Decision-makers can use this framework to analyze what customers express about their service and how the service's performance affects customers, therefore targeting service improvements that will consider both the QoE and QoS.

There are several insights to gather from this visualization (Fig. 2); for example, point of interest one shows that during the first four months of operation of Citibike, the QoS for error *A* was unreliable more than 50% of the time, i.e., there were no docks available. On the QoE aspect, the Twitter analysis shows a negative inclination towards city events and disruptions affecting Citibike. Nonetheless, customers still mention the program as successful and compare it positively to similar services. The Yelp reviews reveal fear towards billing and appearance issues, as well as anger towards customer service. Another example of how Fig. 2 provides insights is by looking at the point of interest number four, which shows the winter of 2015 to 2016, affecting bicycles' reliability. A new topic arises on Twitter, which is road struggles. Yelp offers an adverse

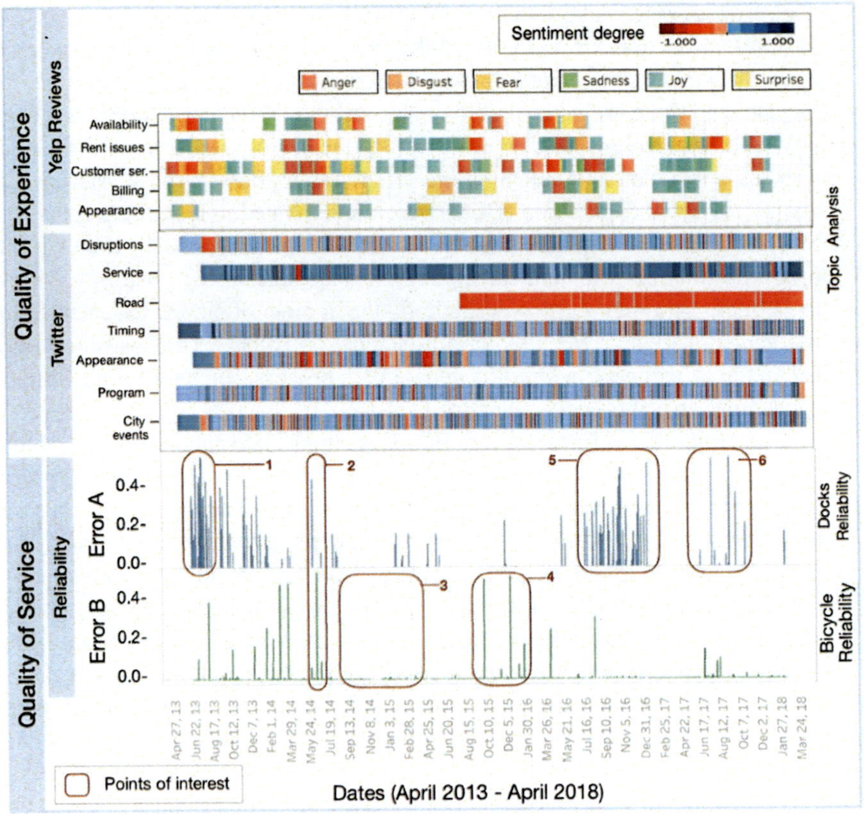

Fig. 2. Comparing QoS and QoE for Citibike [21]

change in the service's availability, renting issues, customer service, and billing issues. As an additional note, the framework proposed in Fig. 1 has also been tested in railroad and subway systems to detect events and measure resilience through QoE and QoS.

This first application of the framework described in Sect. 3 reveals a powerful visual dashboard to compare QoE and QoS in transportation service. This framework can help decision-makers determine areas of improvement for Citibike as a service and areas of improvement of city factors that affect Citibike's services. For example, Fig. 2 reveals negative comments regarding the road conditions and the struggles of riding bicycles around New York City. This should hint government stakeholders to work on policies to improve road conditions and bicycle culture around the city, thereby improving customers' experience riding Citbike.

4.2 Railroad Systems

Railroad systems differ from bicycle-sharing systems as they do not focus on origin-destination pairs but routes and transportation many people simultaneously. Railroad systems are a form of massive public transportation, similar to subways and busses, except that they accomplish longer and further commutes. The framework described in

Fig. 1 can also analyze railroad systems, providing different insights than with bicycle-sharing systems.

Data from New York's Metro-North Railroad (MNR) was collected to analyze a railroad system. This type of transportation system allows the QoS analysis to go further, looking at reliability and all stages of resilience [26]. The data for the MNR came in the form of GTFS files, listings arrival and departures times of the trains. The MNR also keeps a record of all the disruptions occurring in the system, when they started, the reason behind them, and the time of their resolution. This allows for a resilience analysis as part of the QoS. Social media, in particular, Twitter, was used to measure the QoE aspect of this transportation system. This data collection aspect concludes part 1 of Fig. 1.

In this case study, the disruptions and GTFS files were taken to measure the system's visual resilience as part of the QoS (part 2 of Fig. 1). The tweets were examined with natural language processing techniques (NLP) to measure the QoE (part 3 of Fig. 1). The same disruption was measured through the tweets, identifying words that would signal the beginning of a customer complaint (a disruption) and track when the complaints ended (the system stabilized). Unlike the previous case study with Citibike, this case study measures the same aspect for QoE and QoS to compare and improve the system's accuracy.

To exemplify the MNR analysis, Fig. 3 provides a zoom into the New Haven Line, throughout November 15th, 2018. This particular MNR line suffered severe delays due to an unexpected snowstorm on this date. The grey shaded line represents the QoS, measured by the GTFS files. These files are reported by the transportation agency that runs the MNR. The faint pink line shows the difference in delays reported by the customer through the QoE. This difference is significant, indicating that the customers perceive different delays in the system, different waiting times, and the MNR stakeholders should evaluate the measurements they are reporting in their QoS.

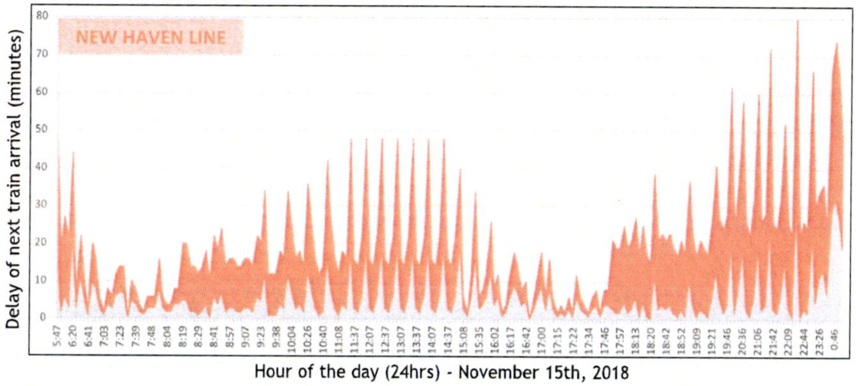

Fig. 3. Comparing QoE and QoS resilience of the MNR through a snowstorm (November 15th, 2018, East Coast, USA) [20]

The importance of this application goes beyond this specific application. This framework allows decision-makers to zoom into any data point within the thirteen months of data to collect and analyze a great variety of disruptions and their comparative QoE and QoS resilience processes. They are thereby allowing to categorize responses to disruptive events and develop a more accurate assessment of resilience in transportation services.

5 Conclusions and Discussions

The work done demonstrates a framework that bridges both QoS and QoE with public transportation services in a city. This framework has additional capabilities such as the combination of metrics to measure the performance of the service, the usage of customer opinion, the understanding of complex measures, such as resilience during a disruptive event, in a visual environment, and the feasibility of multi-service analysis, where different data sources are incorporated. The work done demonstrates the ability to analyze city services from two perspectives. The objective is to include other services, such as busses and railroad trains, to measure resilience during certain events through time in separate transportation services and jointly. Data visualizations and urban computing has facilitated the analysis and exploration of both QoE and QoS into a single framework. Therefore, points of interest and insights can be acquired from the service being analyzed.

The second application of the framework, through a railroad system, provides a different facet of the framework by measuring disruptions both through QoS and QoE and comparing them. This analysis can be two-fold as to 1) verify that the physical infrastructures in transportation systems are reporting and working accurately and 2) obtain insights from the customers using such transportation services, validating the systems and including their perception towards delays in the system.

This framework has the potential to be used as a future forecasting tool. Knowledge from past disruptive events and social perspectives towards city services can help predict how those services will behave in a future disruptive event. For example, suppose the bicycle sharing system is analyzed. In that case, the five years of data account for different changes to the services and the emotion customers expressed towards that change through time. This can be extremely useful in predicting what types of changes to the services will yield higher customer satisfaction and what specific services characteristics need to be improved, i.e., which ones currently have a negative emotional response.

Thereby, this framework's predictive scopes are in terms of recovery time to disruptive events, perception of commute delays towards services, emotions, and sentiments of commuters towards changes in city services, and multi-service response prediction to different types of disruptions. This can be of great use to private and government decision-makers - who are trying to improve current city services - by facilitating their jobs, determining which service characteristics need improvement, the customer's responsibility to those characteristics, and how will that services respond to future unforeseen events.

The framework can work as a sandbox to explore new legislation and budgets to maintain and upgrade city services. Decision-makers can benefit from analyzing past disruptive events and the resilience of transportation services, thus providing them with insights for future events. If enough data were available, this framework could transform into a real-time system rather than a predictive and analytical post-disruptive event system. An example of these framework capabilities would be if different disruptive events were evaluated simultaneously in various transportations services. Suppose the framework was to show that city busses' resilience process is better (i.e., they recover faster) than subway systems. In that case, it might make sense for decision-makers to adapt their policies and deploy more busses when disruptions occur than trying to invest city budget in obtaining more subway cars.

Additionally, decision-makers can assess the severity of disruption against multiple transportation services, therefore acquiring commuters aggregated delays and service options. This is one of many examples where the assessment measures in this research could be of great use to improving city services. The framework itself proves its potential in a world where service improvement is of great importance.

One of the limitations of this work is the availability of data provided by the agencies that control city services. The more data provided, the better the comparison becomes, although simulations and modeling have become essential to fill in these gaps and enable this work to move forward. Customers can serve the purpose of reporting concerns with the service, and those concerns can be interpreted as an event or disruptions in the system. Obtaining the perceived disruptions and their duration would enable a solution for the problem and an avenue to explore QoE against QoS further.

Future work involves using other transportation systems and even different cities to keep testing and expanding the proposed framework. This can yield other areas of improvement for services in smart cities and evaluation platforms for decision-makers.

References

1. Goldsmith S, Crawford S (2014) The responsive city: engaging communities through data-smart governance. Wiley, Hoboken
2. Papazek P, Raidl GR, Rainer-Harbach M, Hu B (2013) A pilot grasp hybrid for the static balancing of public bicycle sharing systems. In: Computer aided systems theory-EUROCAST 2013. Springer, pp 372–379
3. Haider Z, Nikolaev A, Kang JE, Kwon C (2014) Inventory rebalancing through pricing in public bike-sharing systems
4. Nachlas JA (2017) Reliability engineering: probabilistic models and maintenance methods. CRC Press, Boca Raton
5. Haas K (2006) Benefits of traffic incident management. National traffic incident management coalition
6. Li R, Guo M, Lu H (2017) Analysis of the different duration stages of accidents with hazard-based model. Int J Intell Transp Syst Res 15(1):7–16
7. Sheffi Y et al (2005) The resilient enterprise: overcoming vulnerability for competitive advantage. MIT Press Books, vol 1
8. Henry D, Ramirez-Marquez JE (2012) Generic metrics and quantitative approaches for system resilience as a function of time. Reliab Eng Syst Saf 99:114–122
9. Erol O, Sauser BJ, Mansouri M (2010) A framework for investigation into extended enterprise resilience. Enterp Inf Syst 4(2):111–136
10. Murray-Tuite PM (2006) A comparison of transportation network resilience under simulated system optimum and user equilibrium conditions. In: Proceedings of the 2006 winter simulation conference. IEEE, pp 1398–1405
11. Barker K, Ramirez-Marquez JE, Rocco CM (2013) Resilience-based network component importance measures. Reliab Eng Syst Saf 117:89–97

12. Swindell D, Kelly J (2005) Performance measurement versus city service satisfaction: intra-city variations in quality? Soc Sci Q 86(3):704–723
13. Collins C, Hasan S, Ukkusuri SV (2013) A novel transit rider satisfaction metric: rider sentiments measured from online social media data. J Public Transp 16(2):2
14. Musto C, Semeraro G, de Gemmis M, Lops P (2015) Developing smart cities services through semantic analysis of social streams. In: Proceedings of the 24th international conference on world wide web. ACM, pp 1401–1406
15. Gu Y, Qian ZS, Chen F (2016) From Twitter to detector: real-time traffic incident detection using social media data. Transp Res Part C Emerg Technol 67:321–342
16. Sasaki K, Nagano S, Ueno K, Cho K (2012) Feasibility study on detection of transportation information exploiting Twitter as a sensor. In: Sixth international AAAI conference on weblogs and social media
17. Krstajic M, Rohrdantz C, Hund M, Weiler A (2012) Getting there first: real-time detection of real-world incidents on Twitter
18. Atefeh F, Khreich W (2015) A survey of techniques for event detection in Twitter. Comput Intell 31(1):132–164
19. Colneric N, Demsar J (2018) Emotion recognition on Twitter: comparative study and training a unison model. IEEE Trans Affect Comput
20. Gongora Svartzman G, Ramirez-Marquez JE (2019) Listen to the people! Comparing perceived and documented disruptions in public transportation, through quantitative quality of experience, the case study of NYC. In: IEEE international conference on systems, man and cybernetics. IEEE
21. Gongora Svartzman G Ramirez-Marquez JE, Barker K (2020) Social media analytics to connect system performability and quality of experience, with an application to Citibike. Comput Ind Eng, 106146
22. Sakaki T, Okazaki M, Matsuo, Y (2010) Earthquake shakes Twitter users: real-time event detection by social sensors. Proceedings of the 19th international conference on world wide web
23. Ni M, He Q, Gao J (2016) Forecasting the subway passenger flow under event occurrences with social media. IEEE Trans Intell Transp Syst 18(6):1623–1632
24. Sheffi Y (2005) The resilient enterprise: overcoming vulnerability for competitive advantage. MIT Press Books, vol 1
25. Ahmed KB, Radenski A, Bouhorma M, Ahmed MB (2016) Sentiment analysis for smart cities: state of the art and opportunities. In: Proceedings on the international conference on internet computing (ICOMP), p 55. The steering committee of the world congress in computer science, computer engineering and applied computing (WorldComp)
26. Giachanou A, Crestani F (2016) Like it or not: a survey of twitter sentiment analysis methods. ACM Comput Surv (CSUR) 49(2):1–41
27. Zhang Z, Ni M, He Q, Gao J, Gou J, Li X (2016) Exploratory study on correlation between twitter concentration and traffic surges. Transp Res Rec 2553(1):90–98
28. Mogaji E, Erkan I (2019) Insight into consumer experience on UK train transportation services. Travel Behav Soc 14:21–33
29. Chakraborty P, Sharma A (2019) Public opinion analysis of the transportation policy using social media data: a case study on the delhi odd-even policy. Transp Dev Econ 5(1):5
30. Mounica B, Lavanya K (February, 2020) Social Media data analysis for intelligent transportation systems. In: 2020 International conference on emerging trends in information technology and engineering (ic-ETITE). IEEE, pp 1–8

Encouraging Fuel-Efficient Commuting in Young Drivers in Duluth, MN

Turuna S. Seecharan(✉)

Department of Mechanical and Industrial Engineering, University of Minnesota Duluth, Duluth, MN, USA
tseechar@d.umn.edu

Abstract. This work investigates the commuting habits of young (18–24 year old) drivers, in Duluth, Minnesota (MN), with the goal of encouraging sustainable commuting. Sustainable commuting modes are cleaner, cheaper and healthier. Encouraging younger drivers to increase their use of public transportation would contribute to a more sustainable use of transportation infrastructures. Transportation infrastructure asset management is the allocation of available funds across infrastructure classes (e.g. pavements, bridges, signs) or programs (e.g., maintenance, construction). Sustainable commuting attempts to encourage the more efficient and sustainable use of transportation infrastructures. By focussing on younger drivers, it is hypothesized that early adoption of sustainable habits will lead to long-term sustainable habits. The research questions becomes how to encourage young drivers to reduce their dependence on personal vehicles? This work presents the results of a survey of 370 18–24 year old drivers from the University of Minnesota Duluth (UMD) campus to understand their commuting habits and problems with the current bus system. Approximately 46% of UMD students choose their personal vehicle as their primary commuting option. Among these students, 24% cite "inconvenience of the bus schedule" as the reason that prevents them from riding the bus more often even when using the bus can save them more than $2,000 per semester in vehicle maintenance costs. It seems as though poor marketing of bus routes and lack of education toward using the bus system has led to efficient routes being cancelled and lack of ridership.

1 Introduction

Carbon Dioxide (CO_2) levels are their highest in 650,000 years at 411 parts per million (Nasa 2019). The most significant source of CO_2 is the burning of fossil fuels with the industry and transportation sectors as the two largest contributors (Nasa 2019; Ritchie et al. 2017). In many countries around the world, transportation account for more than a quarter of anthropogenic emission and it is growing quickly. Greenhouse gas (GHG) emissions from transport sector has increased, from 1990 to 2015, from 15% to 26% in Europe and from 23% to 27% in the United States (Claflin and Steinwand 2019; EEA 2018; EPA 2019). In 2018, total gross U.S. GHG emissions were 6,677 million metric tons of equivalent (MMT CO_2 Eq.) (EPA). As shown in Fig. 1 below, transportation activities are the largest source of emissions at, approximately, 29% of total GHG emissions in 2018.

A. Crespo Márquez et al. (Eds.): WCEAM 2019, LNME, pp. 55–67, 2021.
https://doi.org/10.1007/978-3-030-64228-0_6

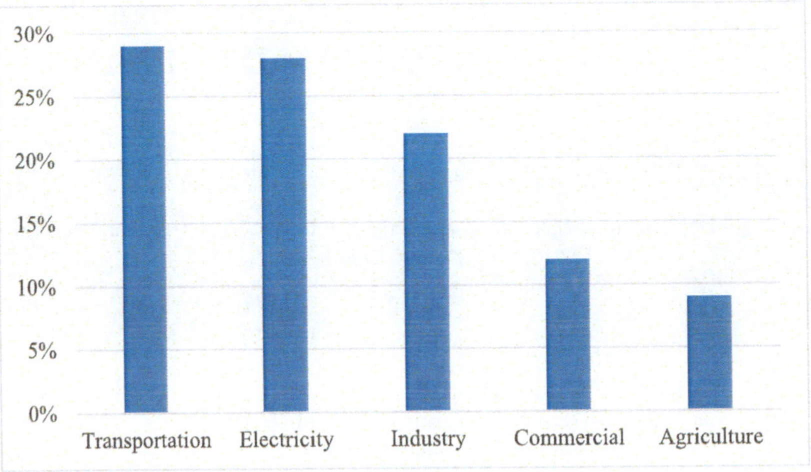

Fig. 1. Source of greenhouse gas emissions.

In the state of Minnesota (MN), more than 70% of emissions from the transportation sector come from light-duty trucks, passenger vehicles and medium to heavy-duty trucks (Claflin and Steinwand 2019). Although the greenhouse gas emissions from transportation have decreased by 8% since 2005, it still accounts for about one-quarter of the GHG emissions in Minnesota. This paper will investigate a behavioural approach toward reducing fuel consumption and, ultimately CO_2 emissions, in the transport sector.

Sustainability implies balancing current and future economic, social and environmental qualities thus meeting the needs of the present without comprising the ability of future generations to meet their own needs (Grant 2020). In addition to engineering or technological advancements, such as hybrid-electric vehicles, behavioural approaches such as choosing eco-friendly vehicles, proper vehicle maintenance or alternative transportation options help reduce personal consumption (US DOE 2019; Wahlberg 2007). Behavioural changes toward sustainable travelling or commuting focus on older fleet drivers and rarely focus on younger drivers (Wahlberg 2007; Seecharan et al. 2016; Zarkadoula et al. 2007). This work hypothesizes that in order to encourage long-term behavioural changes in driving or commuting habits, young drivers should be the demographic of interest. Furthermore, this work attempts to understand the environmental, economic and social factors that prevent young drivers from using the public transit system more frequently.

1.1 Young Drivers - Crash Statistics and Income

In Minnesota, 18–24 year old drivers are the most represented in fatal crash statistics and the also form the demographic with the highest poverty rate (DATA USA 2020). The Minnesota Department of Public Safety (DPS) states that young drivers are most likely to have fatal crashes and are overrepresented in traffic crashes because of their inexperience, immaturity, distractions from their surroundings, night-time driving, speeding, and not using their seat belt (Fig. 2).

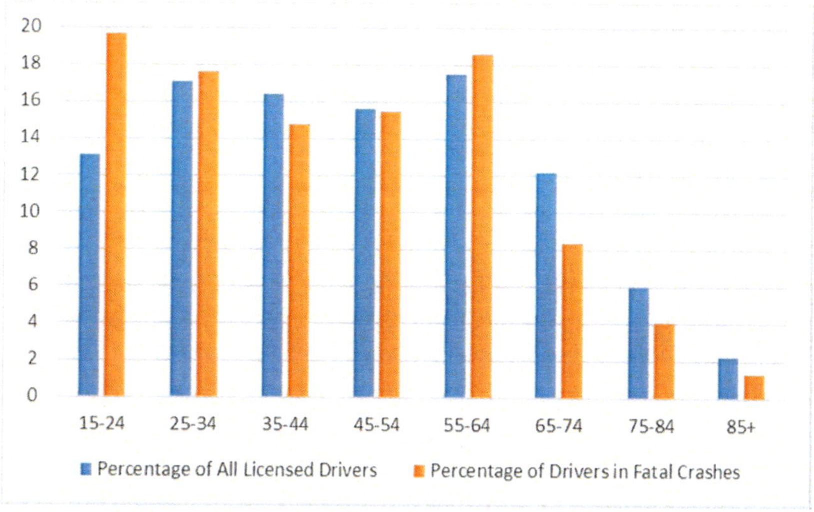

Fig. 2. Licensed versus fatal crash-involved drivers by age-group

Ownership and maintenance of a personal vehicle, although it instils a sense of responsibility, can be expensive to young people between the ages of 18 to 24. This has a big impact on their personal expenses considering that they form the demographic with the highest poverty rate. Males and females aged 18 to 24 residing in Duluth, Minnesota, form the demographic with the highest poverty rate as shown in Fig. 3. They account for just over 20% of the total Duluth population living in poverty.

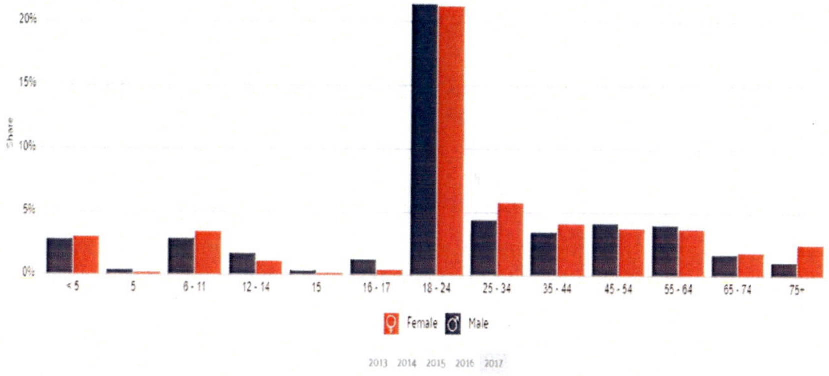

Fig. 3. Share of population that live below the poverty line (DATA USA 2020).

In addition to the reduction in personal expenses, reducing the number of vehicles on roads will save money by heading off the need to spend money on highway expansion, which currently costs anywhere between $2 million to $6 million per mile depending on location, construction codes and size (Elswick 2016). Doing so will also ease congestion, reduce emissions of pollutants that harm public health and alter the climate, and save lives through avoided vehicle crashes along with increased health benefits (Steg and Gifford 2005; Circella et al. 2016; Ewing et al. 2003). From 2014–2018, the economic impact of inattention related serious injuries in Minnesota totalled $91,408,400. From 2013–2017, the economic impact of speed-related injuries totalled $122,193,200 (MN DPS 2019).

1.2 Commuting in Duluth

Between 2001 and 2009, the average number of miles driven by 16 to 34 year-olds dropped by 23%, as a result of young people taking fewer trips, shorter trips and a larger share of trips by modes other than driving (Circella et al. 2016). According to a study of transit riders in the US (Clark 2017), 23% of all transit users were between the ages of 25 to 34. In addition, household incomes less than $15,000 and $100,000 or more are the highest transit users at 21%. The reasons riders give for using public transit are broken into three general categories; reasons of simple preference not involving money, reasons of preference for the economy of using transit, and reasons of need. The largest single category is preference for intangible aspects of travel by public transit (44%) such as convenience, saving time, saving the environment, and avoiding traffic. Neither need nor economy is involved – they simply prefer to use public transit. In addition to those who simply prefer to use public transit, many riders (16%) offer reasons for preferring public transit that have to do with economy, including saving money on parking and gas, and, for a few, taking advantage of an employer transit subsidy. Finally, a substantial number of responses (40%) involved a lack of alternatives.

However, this is not the case in Duluth even when personal costs are high. According to the U.S. Department of Transportation Federal Highway Administration, transportation asset management links user expectations for system condition, performance and availability with system management and investment strategies (Gaj 2007). Understanding the travel behaviour of young drivers in Duluth and attitudes towards the use of transportation should be an important goal for Duluth Transit Authority (DTA) in order to increase students' use of public transportation. Furthermore, facilitating non-auto infrastructure will support evolving travel preferences as observed in Boulder, Colorado (Henao et al. 2015) and lead to a healthier lifestyle Ewing et al. (2003).

In many areas of the United States, cities like Duluth try to provide affordable and sustainable transportation options for residents. The Duluth Transit Authority (DTA) is the main public transit option in Duluth. The DTA provides 23 fixed bus routes that travel in the City of Duluth, Proctor, Hermantown, and Superior. In the Midwest, there are 5 major transit systems; the DTA, St. Cloud Metropolitan Transit Commission, City of Rochester Public Transportation, Sioux City Transit System and City of Fargo Metropolitan Area Transit. From Fig. 4, the DTA has the highest ridership among the five cities but is not the city with the highest population; St Cloud = 68000, Rochester = 117,000, Sioux City = 82,000, Fargo = 124,000, Duluth = 86,000. Therefore, the DTA is well accepted in Duluth. However, from 2015 to 2016, there is a decline in ridership but an increase in Rochester (Steinport 2020).

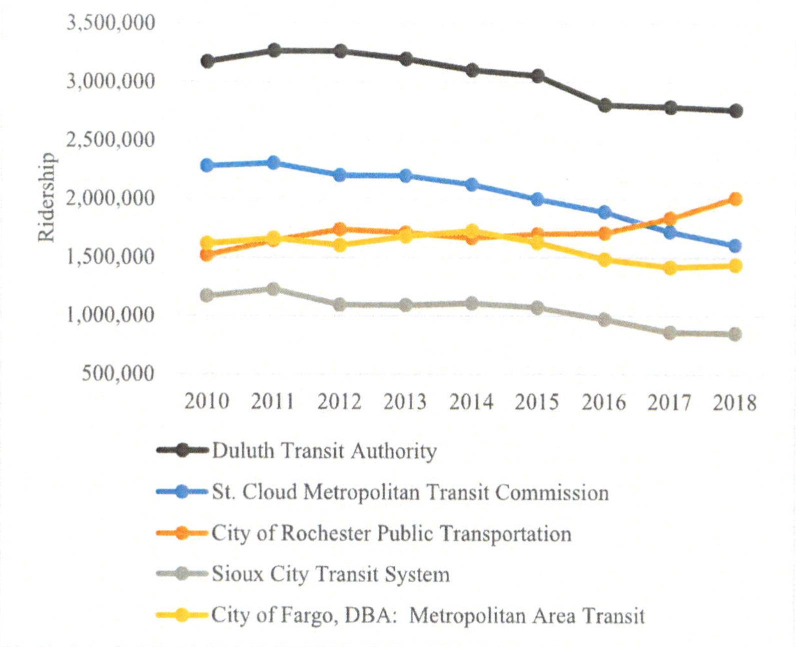

Fig. 4. Ridership comparison among the five major transit systems in the Midwest [21].

A transportation asset management (TAM) system is a strategic resource allocation framework that allows transportation organizations to manage the condition and performance of transportation infrastructure cost effective (Meyer et al. 2010). For transportation systems to grow effectively, asset management provides new insights and tools to help transportation professionals make wise investments that result in improved service and greater cost-effectiveness. Based on an interview conducted with the DTA, the DTA strives to increase the UMD student ridership during the main school session and eventually throughout the calendar year. For the DTA, in 2017, fares revenue totalled $2,500,871 but their operating and maintenance (O&M) costs were $16,605,026. Although the government of Minnesota subsidizes the DTA, increasing their ridership will assist toward their general operating expenses. In order to increase their ridership, it is important for the DTA to understand the needs of their customers.

For a public transportation network, the design of routes is important to transport customers, efficiently, from their home to places of interest. If effective asset management involves the ability to monitor asset performance accurately, then monitoring the performance of the public transportation is important for the longevity of the DTA.

The DTA has a U-Pass program with UMD that allows students enrolled in a program unlimited free rides on the buses (UMD 2016). UMD's published newspaper, The Statesman/The Bark, states that the University pays $400,000 a year to the DTA for the U-Pass program (Kunkel 2012). All UMD students enrolled for a semester with six or more credits are required to pay a $16 transportation sustainability fee every semester that is automatically charged to the students' account. The DTA records the number of rides on their buses as well as if the rider is from UMD based on their card that is swiped upon boarding. Table 1 shows UMD ridership compared with overall ridership. The DTA ridership card swipe information shows that only 35.91% of the university's enrolment in the 2018 spring semester rode the bus at least once, which is just about 4000 UMD students compared to the 11,040 students that was enrolled for that semester.

Table 1. UMD ridership and total systems ridership for the DTA January to April 2018

Month	Total system ridership	UMD ridership	Percent of riders from UMD
January 2018	233,585	41,863	18%
February 2018	229,498	48,843	21%
March 2018	249,550	38,515	15%
April 2018	236,293	44,217	19%

This research aims to understand the travel behaviour of young drivers in Duluth, attitudes toward the use of alternative forms of transportation and the policies or incentives needed to encourage a reduction in the use of a personal vehicle. Young drivers are the focus of this study because it is assumed that they are more likely to make a change in their everyday habits. If they are able to make a change to using an alternate transportation mode like riding the bus, they are likely to continue this lifestyle change throughout their adult life.

The goal would be to leverage opportunities at UMD to further sustainable transportation activities on campus. If a relationship between DTA and UMD is created, DTA could possibly be supported more by the university through funding resources and higher student ridership. This would possibly help the students reduce personal expenses.

2 Methodology

This project was an exploratory study of the commuting preferences of young drivers in Duluth Minnesota. The participants of the study were mostly students from the University of Minnesota Duluth (UMD). The participants consisted of UMD students aged 18 to 24, this age range consists mostly of undergraduate students and a portion of graduate students.

Flyers with information about the study were posted around the university campus. The flyers had a QR code that students could scan which would give them access to the survey and they would be able to participate on their mobile phone. The participants completed an online survey that consisted of 27 multiple-choice questions. The survey gained information on key commute behaviours and attitudes toward the public transportation system in Duluth. The survey took an estimate between 3 to 5 min to complete. As an incentive to participating in the survey, 12 randomly chosen participants received a $10 gift certificate to the on-campus store.

The survey sampled 370 students at UMD and gathered information on preferred mode of transportation, popular destinations within Duluth, how often they used public transportation, popular bus routes, bus wait time, preferred wait time, and reasons that prevent them from using the bus more often. There are other transportation behaviour measures that reflect the use of the bus system in Duluth area. In other to analyse it we asked the respondent what prevents them from using the bus more often and if they would consider taking the bus more often if the issues were resolved. Finally, the respondents were asked to rate on scale of 1 to 5, their experience with the current bus schedule, the length of the transit times and the consistency of the current transit times.

Respondents were able to leave additional thoughts and comments in a short answer style question at the end of the survey. The survey received sample population of 319 usable responses, which ensured a 90% confidence level with a 5% error. All the survey responses were automatically recorded in to google sheets. This survey response sheet was then exported into Excel where all analyses were performed.

3 Results and Discussion

Among the 319 respondents, 46% said their preferred commuting method was their personal vehicle followed by 25% who said they used the bus, 17% chose to walk/bike and 10% carpooled. As expected, the most popular location travelled to by commuters within the age group under study was school. The various locations are shown in Fig. 5 where SuperOne is a local grocery and the Duluth Entertainment Convention Center (DECC) is local arena that hosts sporting events, concerts, expos etc.

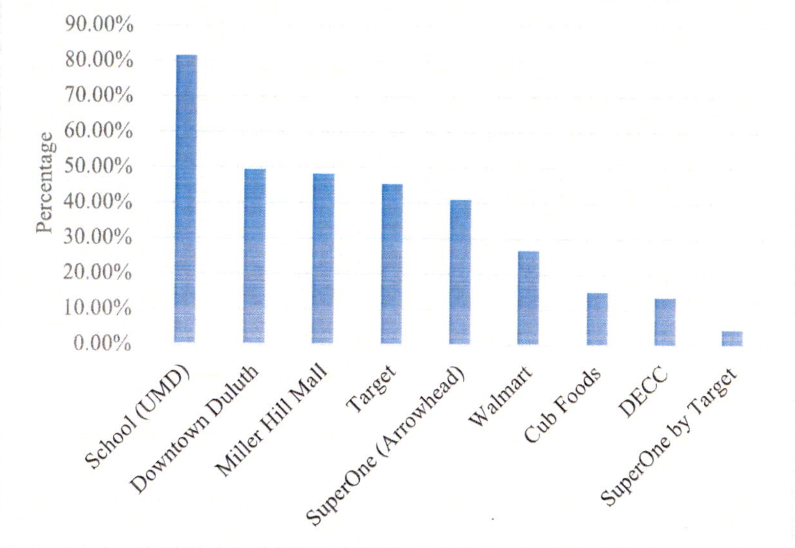

Fig. 5. Popular destinations according to study survey

In order to increase sustainability in the Duluth area and UMD, students should try to live more sustainable lives. A way this can be done is to try to reduce the number of students driving. Students who live far away from campus choose to drive their personal vehicle over alternative forms of transit but that using the bus is still a significant form of transit. The analyses show that transit method varies widely with where students are living with respect to campus. There is a weak positive relationship between higher wages and driving a personal vehicle. The correction coefficient of the students' average commuting method and wage was found to be 0.175. There is a medium positive relationship between owning a personal vehicle and driving it for every trip. As the correlation coefficient done on students' commuting method and personal vehicle ownership was found to be 0.565.

Owning and operating a personal vehicle, for a driver within the 18 to 24 year old age group is costly. Figure 6 shows the comparison among semester income, operating and maintenance (O&M) costs of a personal vehicle along with cost of taking the bus. Among all vehicle types, the O&M costs of a small sedan is the cheapest averaging $6,354, annually, assuming 15,000 miles driven per year (Edmonds 2016). Self-reported semester income from survey respondents varies with age starting at $10.89/h at age 18 to $14.65/h at age 24. Further, insurance costs under the age of 25 are significantly higher. Equations 1, 2 and 3 show how the cost of taking the bus is calculated.

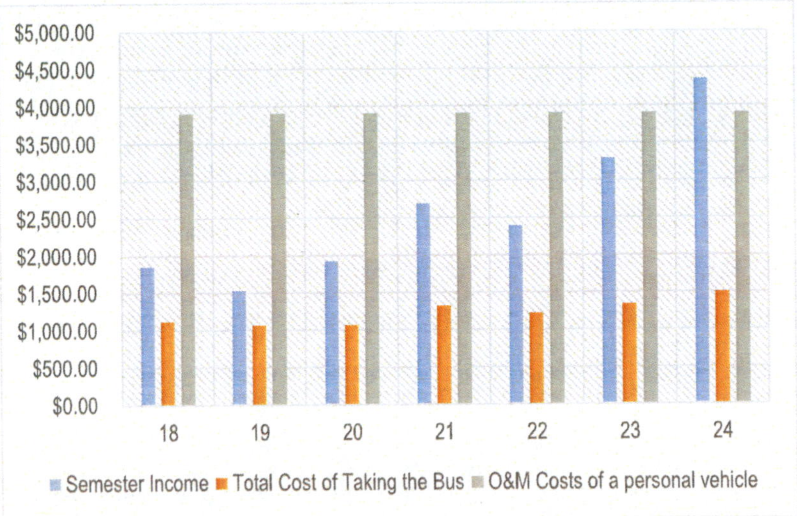

Fig. 6. Comparison among O&M Costs of a personal vehicle, semester income and cost of taking the bus.

$$\mathit{Cost\ of\ taking\ bus} = \mathit{Walk\ and\ Wait\ Cost} + \mathit{Travel\ Cost} \tag{1}$$

$$\mathit{Walk\ and\ Wait\ cost} = \mathit{Wage}(\$/hr) \times \mathit{Walk\ and\ wait\ time}(hr) \times \mathit{Trips\ per\ semester} \tag{2}$$

$$\mathit{Travel\ Cost} = \mathit{Wage}(\$/hr) \times \mathit{Extra\ travel\ time} \times \mathit{Trips\ per\ semester} \tag{3}$$

Even though commuting by bus has potential savings of approximately $2,600, 46% of respondents still choose their personal vehicle. This work attempts to understand why.

The questionnaire also asked for possible reasons why students choose not to ride the bus more often. Figure 7 shows the response distribution for reasons why students do not take the bus more often.

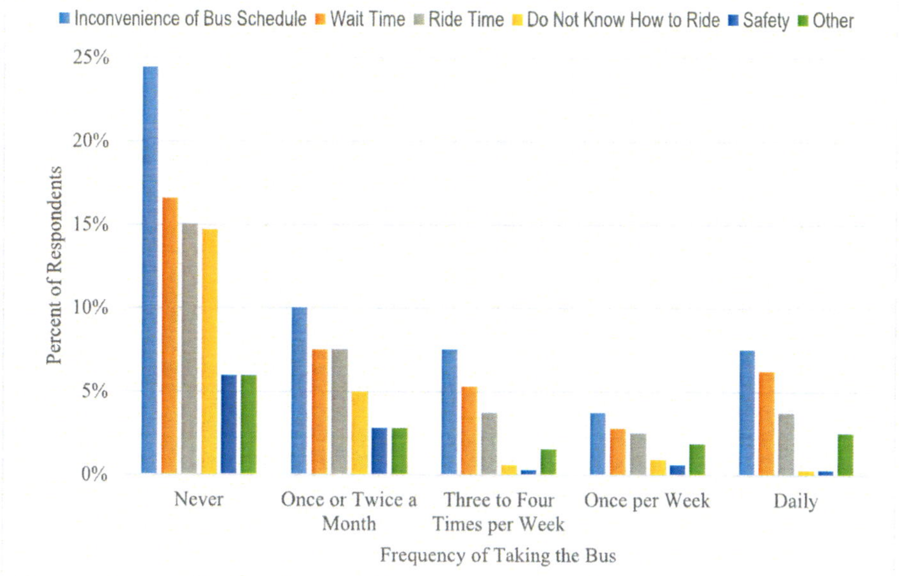

Fig. 7. Reasons why students do not ride the bus more often

Among the respondents who say they never rode the bus, approximately 24% cited "inconvenience of bus schedule" as the major reason why they chose not to ride the bus. Even for respondents who stated they used the bus daily, "Inconvenience of Bus Schedule" was also the most popular fault with the DTA. The second most common reason was "Wait Time" followed by "Ride Time".

The wait time of the bus was the second most answered reason on why the students are not riding the bus. From questionnaire responses, the average time that students were willing to wait for the bus was found to be 7.66 min, compared to the average 6.59 min that student actually wait for the bus. This shows that there is a misconception to how long the students actually wait for the bus. The survey asked the respondents what they would rate the current bus schedule on a 1–5 scale (where 1 is Bad and 5 is Excellent). The average rating of the bus schedule for people who have taken the bus at least once was found to be a 3.612 out of 5. This shows that people who take the bus find the scheduling to be adequate.

For respondents who do take the bus, the questionnaire asked to state the bus route they used most often. Figure 8 shows these results. Routes 13, 11, and 23 all have over 20% of the university student population riding those buses. Route 13 services UMD to the transportation centre in downtown Duluth. The route passes through some of the popular residential areas between UMD and downtown Duluth where students live during the semester. Route 11 also services UMD to the transportation centre but also travels to the SuperOne on Arrowhead road and the other local college, St. Scholastica. Route 23 is a UMD circulator that starts and ends on campus with the primary goal of transporting students to and from campus from their home. Its route travels through student residential areas during morning and afternoon rush hours.

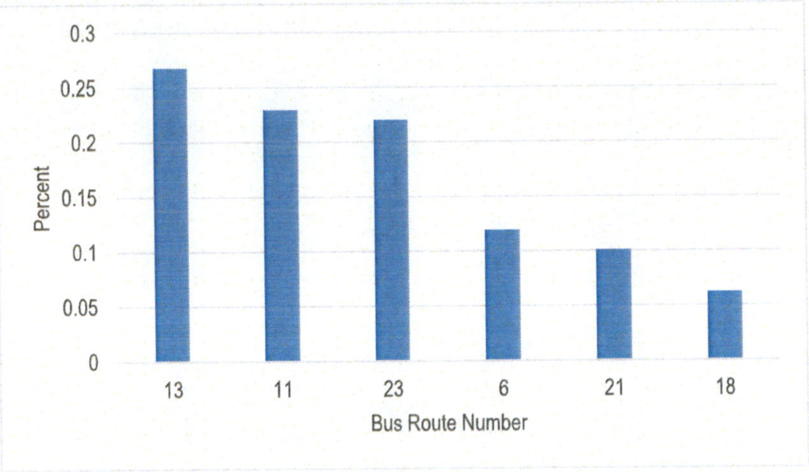

Fig. 8. Most popular bus routes based on questionnaire responses

Table 2. Travel time comparison between car and fastest bus route.

	Travel time from UMD bus hub to popular destinations (minutes)						
Mode of transportation	Downtown Duluth	Miller Hill Mall	Target	SuperOne (arrowhead)	Walmart	Cub foods	DECC arena
Car	10	15	15	6	16	14	16
Bus	24	46	47	12	68	35	24
Difference	14	31	32	6	52	21	8
Route 21	X	26	33	X	X	25	X

According to an interview with the DTA, routes 23, 6 and the 13 are their best performing routes compared to all of their routes in the Duluth-Superior area based on the number of rides per trip. The ridership these routes are 58, 23, and 22 passengers per hour, respectively. The DTA states that these routes run every half-hour and they suspect that this is the reason why their ridership is so high. Although routes 13, 11 and 6 facilitate travel to the other points of interest, a large portion of respondents still do not use the bus.

If we investigate the commute time to points of interest, as shown in Table 2, commuting by bus is, on average, 23 min longer or more than twice the commuting time when using a car. Some respondents have stated that this becomes a problem if they have tight deadlines such as going from school to work. Interestingly, the DTA developed the

Route 21 that was an express route from UMD bus hub to Miller Hall Mall. From Table 2, the times to some popular destinations is comparable to travelling by car. However, the DTA cancelled this route, leading to the increasing suspicion that students do not know about this route. If this route is marketed to the student population and ran on weekends, it would likely see a large increase in ridership.

4 Conclusion

Encouraging sustainable commuting in young drivers in Duluth involves several stakeholders, in particular the customer (young driver) and the DTA. A transportation asset management system is a strategic resource allocation framework that allows transportation organizations to manage the condition and performance of transportation infrastructure cost effectively. The DTA fares revenue is approximately $2,500,000 but O&M costs are approximately $16,600,000 meaning that revenue from fares is not enough to cover their O&M costs. In order to increase the performance of their network, they must better understand the needs of their customers. UMD students, along, account for 25% of all ridership. The UMD circulator is the most lucrative of all bus routes. This route is designed specifically to transport UMD students to campus from home.

This work obtained information on the commuting habits of a sample of drivers from UMD and their use of the DTA. Based on self-reported income, O&M costs of owning and operating a vehicle and the cost of taking a bus (including the cost of waiting and commuting), a UMD student can potentially save $2,600 per semester if they choose to ride the bus instead of using their personal vehicle. However, 46% of students choose not to ride the bus. Among these respondents, 24% cite "inconvenience of the bus schedule" as the reason why. Although this initial study does not dig deeper into this issue, just looking at commute time shows that taking the bus to popular destinations takes, on average, 2.7 times longer than using a personal vehicle.

A bus route that does seem to be lucrative to the DTA and beneficial to students is route 23 since it transports students from their home to campus avoiding the need to pay for on-campus parking that is not reliable. Considering other places of interest, the DTA did have a UMD to Miller Hill Mall express route but this was discontinued due to low ridership costing them approximately $83.98 per rider. Transportation asset management provides for a fact-based dialogue among system users and other stakeholders. The next steps of this research would be to dig deeper into the factors needed for a young driver to reduce their dependence on a personal vehicle and the infrastructure needed from external stakeholders such as the DTA, UMD or the City of Duluth to develop more efficient and sustainable use of transportation infrastructures.

Acknowledgements. The author would like to acknowledge UMD undergraduate students Dallas Gapinski, Jessica Smith and Ryan Sperle for their help in data collection and analysis along with Christopher Belden from the DTA.

References

Circella G, Fulton L, Alemi F, Berliner RM, Tiedeman K, Mokhtarian PL, Handy S (2016) What Affects millennials' mobility? Part 1: investigating concerns, lifestyles, mobility-related attitudes and adoption of technology of young adults in California. UC Davis Institute of Transportation Studies, Davis

Claflin A, Steinwand F (2019) Greenhouse gas emissions in Minnesota: 1990–2016. Minnesota Pollution Control Agency, Saint Paul

Clark HM (2017) Who rides public transportation. The American Public Transportation Association. https://www.apta.com/wp-content/uploads/Resources/resources/reportsandpublications/Documents/APTA-Who-Rides-Public-Transportation-2017.pdf

DATA USA (20 July 2020). DATA USA: Duluth, MN. DATA USA website. https://datausa.io/profile/geo/duluth-mn#economy

Dps.mn.gov. (2019) Teen Driving - FAQ. https://dps.mn.gov/divisions/ots/teen-driving/Pages/faq.aspx. Accessed 7 Feb 2019

Edmonds E (23 August 2016) AAA reveals true cost of vehicle ownership. AAA Newsroom. https://newsroom.aaa.com/tag/cost-to-own-a-vehicle/

Elswick F (5 January 2016) How much does it cost to build a mile of road. http://blog.midwestind.com/cost-of-building-road/

European Environment Agency (11 October 2018) Air quality in Europe - 2017 report. https://www.eea.europa.eu/publications/air-quality-in-europe-2017. ISSN 1977-8449

Environmental Protection Agency (10 April 2019) 2017 National Emissions Inventory (NEI) Plan. https://www.epa.gov/air-emissions-inventories/2017-national-emissions-inventory-nei-plan

Ewing R, Schmid T, Killingsworth R, Zlot A, Raudenbush S (2003) Relationship between urban sprawl and physical activity, obesity, and morbidity. Am J Health Promot, 47–57

Gaj S (2017) Asset management. U.S. Department of Transportation Federal Highway Administration. Washington

Gharaibeh NG, Chiu Y, Gurian PL (2006) Decision methodology for allocating funds across transportation infrastructure assets. J Infrastruct Syst 12(1):1–9

Grant M (2020) Sustainability. Investopedia. https://www.investopedia.com/terms/s/sustainability.asp

Henao A, Piatkowski D, Luckey KS, Nordback K, Marshall WE, Krizek KJ (2015) Sustainable transportation infrastructure investments and mode share changes: A 20-year background of Boulder, Colorado. Transport Policy, 64–71

Kunkel A (08 October 2012). UMD Parking partially funds "free" bus transit program. The Statesman Archives. http://www.theumdstatesman.com/blog/2012/10/08/umd-parking-partially-funds-free-bus-transit-program-2?rq=Parking

Meyer MD, Amekudzi A, O'Har JP (2010) Transportation asset management systems and climate change: adaptive systems management approach. Transp Res Rec 2160(1):12–20

NASA Jet Propulsion Laboratory (28 May 2019). Global climate change: vital signs of the planet. NASA Global Climate Change Web site. https://climate.nasa.gov/

Ritchie H, Roser M (11 May 2017) CO_2 and Greenhouse Gas Emissions. https://ourworldindata.org/co2-and-other-greenhouse-gas-emissions

Seecharan TS, Donmez B, Chen HW, Jardine AKS (March 2016) Simulator-based eco-drive training for fleet drivers. In: Koskinen K et al (ed) Proceedings of the 10th world congress on engineering asset management, pp 545–552

Steinport J (06 June 2020). National Transit Database. https://www.nationaltransitdatabase.org/

Steg L, Gifford R (2005) Sustainable transportation and quality of life. J Transp Geogr, 59–69

UMD (24 August 2016) Bus. Sustainability. d.umn.edu/sustainability/campus-initiatives/transportation/bus

United States Environmental Protection Agency. (n.d.) Data highlights: inventory of U.S. greenhouse gas emissions and sinks: 1990–2018. EPA Web site. https://www.epa.gov/sites/production/files/2020-04/documents/us-ghg-inventory-1990-2018-data-highlights.pdf

United States Department of Energy (US DOE) (23 May 2019). Driving more efficiently. https://www.fueleconomy.gov/feg/driveHabits.jsp

af Wahlberg AE (2007) Long-term effects of training in economical driving. Int J Ind Ergon 37:4, 333–343

Zarkadoula M, Zoidis G, Tritopoulou E (August 2007) Training urban bus drivers to promote smart driving: a note on a Greek eco-driving pilot program. Transp Res Part D Transp Environ 12(6):449–451

anon (2014) Parenthetical referencing. http://en.wikipedia.org/wiki/Parenthetical_referencing

Gaj S (2007) Asset management. Federal Highway Administration, Washington

Is EAM a Corporate Level Strategy?

David E. Mills(✉) and Steve Pudney

Southern Cross University, Coolangatta, Australia
david.mills@scu.edu.au

Abstract. We speculate that COVID-19 will drive engineering asset owners to require governance entities such as boards of management (BoM), to elevate engineering asset management (EAM) to being a corporate level strategy and core competence. We sought to understand whether there was a fit-for-purpose conceptualization of EAM as a corporate level strategy.

Our research considered foundational Asset Management (AM) prescriptions finding that the available AM models have distinct gaps as to the roles of governance entities and of CEO's in asset management where these are actively involved in the policy and practice of asset management. Using the Victorian Government Asset Management Accountability Framework as an example of active top-level involvement we identified a 'bottom-up' approach, incomplete integration with the whole organization and positioning of AM planning subsequent to strategic level planning as barriers to effective corporate level asset management strategy implementation. Responding to those findings we offer a program of future research aimed at developing fit-for-purpose strategic EAM tools to assist those BoM's who seek to elevate EAM to a corporate level strategy.

Keywords: Asset management · Corporate strategy · Engineering asset management · Governance · Integrating strategies

1 Introduction

As the boards of governance of engineering asset-based organizations unpack the implications of COVID-19 for the survival of their organization it is highly likely that owners will have heightened expectations of their organization's asset management competence, seeking performance above cost-effective projects, compliance and sustainability deliverables (Koronios et al. 2007). These heightened expectations upon the Boards of Management (BoM) may elevate engineering asset management (EAM) to being one of the two or three core corporate strategies, and to being a core competence (Prahalad and Hamel 1990). We adopt Beard and Dess's (1981) conceptualization of corporate strategy as determining the scope and resource deployment of an organization's resources to achieve corporate level objectives. The issues then for a BoM are whether there is a fit-for-purpose conceptualization of this top-down EAM, whether there are examples of this approach in practice, what might be the factors which support or hinder EAM being a corporate level strategy integrated with other corporate level strategies and, whether better performance is achieved.

A. Crespo Márquez et al. (Eds.): WCEAM 2019, LNME, pp. 68–79, 2021.
https://doi.org/10.1007/978-3-030-64228-0_7

Our approach to this research was to first interrogate the asset management (AM) and EAM academic literature and then explore the industry and government publications for evidence regarding a strategic and whole-of-organization perspective on optimizing the outcome from engineering assets. Two observations are necessary. Firstly, that EAM is a relatively new discipline, being formally acknowledged only 10 years ago by way of the work of Amadi-Echendu et al. (2010). Secondly, this recent emergence has been 'bottom up' with strategic governance factors being confronted only relatively recently in the work of Mahmood et al. (2015). This recency of both conceptual development and application of theory in practice signals a pressing need for substantial empirical research and consequent development of theory.

Because of this paucity of literature as to theory and practice we sought out an example where expectations as to the AM performance of the governance entity had been specified and where performance had been independently assessed. The Victorian Government's *Asset Management Accountability Framework* (AMAF 2016) is prominent for the government's dissatisfaction with actual asset management performance and consequent escalation of requirements as to accountability, effectiveness and efficiency, placed on the governance entity and the accountable officer. The Auditor-General of Victoria (VAGO 2019) has reported publicly as to the approach taken by governing entities in applying, and assuring compliance with, the requirements. This reporting, and expressed views as to the likely success of some of the approaches taken by the entities made the AMAF case highly suitable for our research.

Our paper has been assembled in the following four sections. The first reports the literature relevant to strategic involvement in EAM. The second reports the evidence as to practice gleaned from the Victorian Government AMAF accountability approach. The third discusses the implications of this evidence for a heightened involvement of BoM in EAM. Finally, the Conclusion section forms and justifies an agenda for future research aimed at more strategic and more effective EAM.

Our research sought to see EAM from the perspective of a Board of Management (BoM), which might reasonably ask the following questions:

1. To what extent do models conceptualize EAM as a corporate level strategy in which the BoM is actively involved?
2. What is the evidence of EAM being applied as a corporate level strategy in practice?
3. What are the factors that support or hinder EAM being an effective corporate level strategy integrated with other corporate level strategies?

2 Literature: Strategic (BoM) Involvement in EAM/AM

We first sought to understand the extent to which AM and EAM prescriptions encompass the governing entity (typically a BoM) and top management and the artefacts of their corporate level governance. We first examined the foundational ISO 55000:2014 Asset Management – Overview, principles and terminology, for such evidence and then explored the article of Mahmood et al. (2015) which lays out the component processes of six AM Capability Maturity Models and provides specific critique of the coverage (or not) of high-level management and strategic governance requirements for effective asset management.

2.1 The Standards

In this paper, ISO 55000 means the entire family of ISO 5500x standards but not including the new standard ISO 55010:2020 Asset management - Guidance on the alignment of financial and non-financial functions in asset management.

ISO 55000:2014 defines asset management as coordinated activity of an organization to realize value from assets and identifies leadership and commitment from all managerial levels as essential to successful asset management. We explored whether, within the Standards there were specifications as to: the roles of the BoM or equivalent governance entity; requirements made upon 'top management' i.e. the chief executive officer and second level executives; and whether EAM/AM is portrayed from the whole-of-organization, 'top down' perspective of a BoM.

Involvement of Governance Entity

There is no mention of the governance entity. There is no separate, specific mention of a chief executive officer or similar role that in terms of corporations law, or governance of public entities, has very specific accountabilities for an immense range of matters. The Standards are no more precise than to use the term 'organization' which, in 3.1.13 it defines as '... a group of people...'.

Governance is addressed but from the perspective of providing assurance that capability of the asset is ensured.

Top Management Involvement

Within ISO 55000:2014 S 2.4.1 specifies that an organization's '...top management, employees and stakeholders should implement ... (AM) ... to exploit opportunities and to reduce risks to an acceptable level.'

S 2.4.2 (c) specifies as a fundamental of AM 'Leadership and commitment from all managerial levels is essential for successfully establishing, operating and improving asset management within the organization'.

S 2.5.3.3 squarely allocates to top management responsibilities for developing AM policy, AM directives together with vision and values that guide policy and practice.

'Top management' is defined in S 3.1.23 of the Standard as the "person or group who directs and controls an organisation at the highest level". It is not clear whether 'Top Management' is intended to include the BoM. The Asset Management – Management systems – Requirements ISO 55001:2014 prescribes in S 5.1 Leadership and commitment, prescribes actions that the 'Top Management' shall take which are quite operational and are the province of management and not the governing entity. It appears that the intent is for 'Top Management to refer to the CEO and second level roles. These second level roles may be the 'Business Leader' role identified by The Institute of Asset Management (2008) as being the top of a hierarchy of AM roles, immediately above the 'Head of Asset Management'.

Whole-of-Organization Approach

We applied a whole-of-organization lens in our examination of the literature, for two reasons. First the governance entity has responsibility for, and command of the resources of, the whole organization. Second, Prahalad and Hamel (1990) conceptualize core competencies as collective learning across the whole corporation and are insistent that a core competence cannot be the result of the efforts of a specialist unit but rather are formed by efforts of the entire organization, orchestrated from the corporate level.

ISO 55000:2014 (S 2.5.2 (b)) envisages benefits to top management from the cross-functional integration and planning process inherent in an asset management system that is integrated with other organizational systems. Further, ISO 55000 observes that not all asset management activities reside within the asset management system. Rather leadership, culture, motivation and behaviour may be managed by the organization using arrangements outside the asset management system (ISO 55000:2014 S2.4.3).

However, whilst ISO 55000 makes clear the need for integration of asset management throughout the organization and its systems the Standards are either silent about the importance of the strategic asset management plan (SAMP) and planning activities relative to the key corporate artefacts, or indicate that AM planning is consequent upon those corporate level activities. Specifically, Annex B 'Relationship between key elements of an asset management system' employs a figure to depict the 'feed down' effect of 'Stakeholder and organization context', then 'Organizational plans and organizational objectives' down to 'Asset management policy' and SAMP with its 'Asset management objectives'. The integration of the asset management system with the key corporate artefacts, namely the strategic plan, financial plan (budget), is depicted as asset management activities not being strategic and not the province of the governance entity.

Consideration of the overall Standards reveals a strong prescription of integration of AM planning and systems with those of the wider organization. The whole-of-organization impact of this integration is, however, diminished by AM planning and SAMP being subsequent to, and driven by earlier corporate level decisions and artefacts.

2.2 AM Capability Maturity Models

Organizations seeking to operationalize asset management policy as proposed by the Standards have increasingly used capability maturity models. These Asset Management Capability Maturity Models (AMCMM) have been used to both specify and measure the organization's performance in each process within the domain of AM. Applying the five dimensions of EAM (Amadi-Echendu et al. 2010) Mahmood et al. (2015) interrogated 6 AMCMM's to determine whether there were gaps in the key process areas specified in each AMCMM.

Mahmood et al. (2015) found that all models had strong asset performance measurement process areas and four started only at the asset management policy and strategy level. The remaining two models started from the perspective of organizational strategic governance (corporate governance, corporate policy and corporate strategy) and then moved on to asset management strategy. Only one model covered leadership, change management, competence management and asset management culture.

Mahmood et al. (2015: 344) concluded that a well-designed asset management maturity model must address the corporate levels of organizational management by integrating corporate planning processes with asset planning processes. The authors proposed an AMCMM in which asset planning processes are separate (in the Temporal dimension) from corporate policy, strategy and governance (in the Organizational dimension), yet reinforced the need for an AMCMM to start from the perspective of organizational strategic governance. Such a conceptualization of AM promises benefit by way of tighter focus of organizational activities and resources, if applied in practice. However, there is no evidence as to whether this is the fact in practice, either in AM or in EAM.

3 Practice: Strategic (BoM) Involvement in EAM/AM

Having interrogated all identifiable academic literature that contemplated the matter of EAM or AM strategic governance we turned to the industry and government publications to understand whether practice reflected the tenets of the Standards and maturity models, or offered a different approach to governance of engineering asset-based organizations. The Victorian Government's Asset Management Accountability Framework (AMAF) (AMAF 2016) is prominent for the government's rejection of then existing standards of actual AM performance and consequent escalation of requirements as to accountability, effectiveness and efficiency, placed on the actual governance entity and the accountable officer. The Auditor-General (VAGO 2019) has reported publicly as to the approach taken by the governance entity in applying, and assuring compliance with, the requirements. The availability of the reporting by the Auditor-General was attractive because of the independence of the performance reviewer, quality assurance and challenge mechanisms in respect of facts and conclusions.

The requirements of the AMAF and the findings and conclusions of the Auditor-General are brought together in the following subsections: Strategic level AM governance requirements and the evidence as to Strategic level AM governance practice.

3.1 Strategic Level AM Governance Requirements

The AMAF was formed in 2016 by the Victorian Government which had concluded in 2013 that '...agencies were not taking accountability for their performance in managing their assets.' (VAGO 2019: 18). The AMAF applies to government corporations, departments and other agencies, thus encompassing some entities that have major engineering assets. The Government sought to achieve both effective management of assets and accountability on the part of entities through a strengthened emphasis upon top management leadership and an attestation by the accountable entity as to compliance with the AMAF. This focus on the top of the organization suggested a re-conceptualization of asset management as a top-down strategic governance activity.

Entity and Accountable Officer Inclusion in AM

AMAF (2016: 10) states that 'Without leadership and accountability... particularly from management and Accountable Officers, an organization's asset management strategy and service delivery objectives may be ineffective'.

The AMAF Guidance (2017: 4) specifies that it is the role of Accountable Officers (Secretaries or CEO's) to '...demonstrate asset management leadership' but does not include leadership in the list of roles specified for Responsible Bodies (departmental Secretaries or Boards). The reasoning behind the Responsible Body not being required to provide asset management leadership is not explained.

Role of the Whole Organization
Management is required to '… drive a culture of continuous improvement in asset management.' (AMAF 2016: 30) with the AMAF (2016: 9) prescribing that effective asset management '… is supported by organizational leaders who promote the principles and policies of asset management…' making clear that such promotion of asset management is to the broader organization and is not limited to asset management professionals. The broader staff are to be informed of both the role of asset management in the organization and their contribution, role and responsibilities for asset management (AMAF 2016: 30).

AMAF Guidance (2017: 15) introduces the concept of 'asset management thinking' and its integration into reporting lines and operating frameworks with the objectives of reducing 'asset management silos' and fostering a collaborative approach across the whole of organization that balances strategic, technical and budgetary considerations. This whole-of-organization approach to asset management is conceptualized in Fig. 1: Asset management levels, provided in the AMAF Guidance (2017: 15).

Fig. 1. Asset management levels (AMAF Guidance 2017: 15)

Importantly, this layered conceptualization allocates to the highest level, the 'Management of the Organization', the role of determining how AM contributes to, and achieves, organizational objectives.

3.2 Strategic Level AM Governance Practice

Entity and Accountable Officer Performance in AM
The Auditor-General (VAGO 2019: 9) found two Agencies had demonstrated better practice by:

- Active involvement of their senior leaders in AM
- Considering the criticality, risk and complexity of their assets

- Having a second level executive oversee the AMAF implementation across the whole-of-organization; and
- 'most importantly', being motivated to improve asset management as they understood its value to the success of service delivery.

The agencies examined by VAGO (2019: 50) were all departments and as such the accountable body is the 'Secretary' and good practice was for Deputy Secretaries to have specific responsibilities for each asset class.

In contrast, VAGO (2019: 36) observed that the evidence was that in three of the five agencies where the assets were complex and higher risk the corporate finance and compliance unit was responsible for driving and coordinating AMAF implementation. This task was found to be difficult for these units if the unit did not have strong support from those with asset management responsibilities or expertise.

Whole-of-Organization Approach
The Auditor-General commended an exemplar department which developed a whole-of-organization asset management plan that:

- Communicated a shared understanding about the purpose, direction and expectations for asset management across different asset classes
- Drove the department to improve capability and change practices in response to identified needs
- Highlighted the role of senior leaders in AM (VAGO 2019: 11).

The 'whole-of-organization' approach is established in the AMAF for the purposes of offering the option to entities (typically the larger multi-function departments) of being able to have an asset management strategy for each major asset class. AMAF (2016: 32) further specifies that the asset management strategy should be at an appropriate level to the organization's size and functions.

However, VAGO (2019: 31–32) observed that departments '...without a whole-of-department asset management plan ... may find it challenging to consistently achieve their asset management and service delivery objectives...'. To illustrate, VAGO (2019: 32–33) gave the example of a diverse human services organization with non-current assets of $30bn which had an asset plan for each of its asset classes. This case was described as a 'bottom-up' approach which missed the opportunities for understanding asset management strengths and weaknesses, improving asset management, and prioritizing and directing efforts across the department. 'Bottom-up' governance is where the executives reporting to the CEO are 'independently-minded' and have a capacity to impose strong discipline upon their CEO, providing advantage, and disadvantage to the organization (Landier and Thesmar 2005).

4 Discussion

We embarked on this research with the objective of understanding whether the EAM body of knowledge was fit for the purpose of guiding the BoM of an engineering asset-based organization that has decided to elevate EAM to the status of core corporate level strategy and core competency. The evidence as to the BoM involvement contemplated by ISO 55000:2014 and Asset Management Capability Maturity Models and the AMAF which applies to entities owned by the Government of Victoria was laid out in the preceding section. We now discuss the key implications revealed by that evidence.

4.1 Current Conceptualization of EAM as a Corporate Level Strategy

EAM developed on the back of ICT inventions and innovations of the 1970's and 80's that were not widely implemented until the 1990's (Hodkiewicz and Pascual 2006). Hodkiewicz and Pascual (2006) report that a number of international bodies representing their national EAM communities emerged from 2004 onwards and that these bodies have since been active in defining required bodies of knowledge, moving on from technical, to decision-making and communication skills and related soft skills. This recency of EAM led us to widen our exploration to include AM. In 2015 Mahmood et al. (2015) examined AM capability maturity models to identify gaps in the process areas specified, identifying the significant absence of consideration of corporate level management and organizational governance components of well-designed AM processes. This research concludes that AM and EAM are moving briskly from being only a function facilitated by ITC to one characterised by active involvement of those governing the organization.

An increased role of the top of the organization is explained in the theory of Mahmood et al. (2015) and demonstrated by the example of the Victorian Government AMAF (AMAF 2016). Yet, whilst the AMAF is fully consistent with ISO 55000:2014 it goes beyond the conceptualization of those Standards which are silent as to the role of the governance entity or accountable officer in AM.

In summary, the Standard, ISO 55000, does not present a conceptualization of AM (or EAM) as a corporate level strategy. Specifically, no involvement of the governance entity or BoM is described. Understanding the need for Standards to specify generalized principles, and acknowledging that the number of engineering asset-based organizations that determine a need for a strategic, corporate level approach to EAM may be a small proportion of the overall population of EAM organizations, it may be appropriate to develop a conceptualization specific to an owner (as was the Victorian Government AMAF) or to an industry e.g. energy generation, water services.

4.2 EAM as a Corporate Level Strategy in Practice

This research, with its approach of only interrogating the academic literature and industry and government publications, had found that the available evidence as to practice was limited. The Victorian Government AMAF (AMAF 2016) was chosen as a case study to inform this research because of the availability of the publicly published VAGO (2019) report as to performance that itself was the subject of quality review processes and based on a deep experience of review of asset management performance. The AMAF (2016) case study provided some assistance to this research but also has limitations as to its applicability. The case study exemplified the unequivocal requirement of the owner that there be effective AM and accountability of the governance entity and accountable officer for such AM. Further, the combination of the AMAF (2016) and VAGO (2019) report made clear that a 'top down', whole-of-organization conceptualization of AM is an inherent part of the strategic, corporate level approach to AM.

However, the AMAF (2016) has limited generalizability as the governance entity had no choice as to the adoption (or not) of the AMAF and there is no clarity as to where the AM policy and strategy were positioned and integrated vis a vis other corporate, strategic policy and strategies. As the government entities and accountable officers were compelled to be actively involved in AM it can be assumed that there was no decision of the governance entity, say a BoM, to adopt AM as one of two or three core corporate strategies and as a core competence (Prahalad and Hamel 1990) directed to achieving organization objectives. Further, despite the attention given by the governance entity, CEO and second level executives, there was no indication that the AM policy and strategy had been elevated to being key corporate artefacts.

4.3 Barriers to EAM Being a Key Corporate Level Strategy

The approach taken in this research was not intended to develop an exhaustive list of barriers to the adoption of a corporate level (Viljoen and Dann 2000) AM strategy, but through the examination of practice inform the design of future research. The themes as to findings regarding barriers to EAM being a key corporate level strategy were, 'bottom up' approach to AM, AM activities not being integrated across the systems and activities of the whole organization, and positioning of AM policy and integration with corporate, strategic policies.

'Bottom up' Approach

A 'bottom up' approach to the structuring of the governance of asset management was criticised by the Victorian Auditor-General (VAGO 2019) who made clear their view that an organization without a whole-of-organization AM plan would find difficulty in achieving AM and service delivery objectives because of missed opportunities for understanding AM strengths and weaknesses, improvement of AM and prioritization of AM efforts across the whole organization.

This example, plus the Auditor-General's according exemplar status to the entity which developed a whole-of-organization asset management plan which communicated purpose, direction and expectations across different asset classes, suggests that the Auditor-general saw benefit in a 'top-down', whole-of-organization approach to AM.

ISO 55000 does not convey a 'top-down' approach to AM. The evidence is that the Auditor-General believes that where an organization seeks to introduce a strategic, corporate level approach to AM then the practical implementation should be enacted 'top-down' commencing with the development of a whole-of-organization strategic corporate level AM plan.

The literature as to the 'bottom-up' and 'top-down' approach to business planning has been summarised by Babafemi (2015) who found the bottom-up approach is said to offer strategies that are consistent with customer needs and expectations but have the disadvantage of corporate business directions being substantially influenced by people who are unaware of the internal and external business environments. In contrast, Babafemi (2015) found that the literature characterised the 'top-down' approach as producing plans that are truly corporate in scope but which may not meet the reality of internal capability, stakeholder credibility and cultural fit. It is thus necessary to establish whether there is evidence of a 'top-down' approach to AM that has increased efficiency and effectiveness, and to identify the extent to which the two approaches are applied.

Incomplete Integration of AM with Whole Organization

The ISO 55000 conceptualization of integration looks from the asset management system outwards to the other systems of the organization. The AMAF (2016: 30) goes further towards a whole-of-organization perspective requiring the organizational leaders to promote the principles and policies of asset management to the broader organization and specifying to each member of the broader staff their contribution, role and responsibilities for asset management. This approach is explained as 'asset management thinking' directed at reducing 'asset management silos' and fostering a collaborative approach across the whole of organization (AMAF Guidance 2017: 15).

Increased integration throughout the whole organization is held, in the Standards, the AMAF (2016) and the observations of the Auditor-General regarding the problems associated with a 'bottom-up' approach, to contribute to efficiency and effectiveness of AM. It would seem that: integration of systems and activities is in itself a key area of focus for organizations seeking improved AM performance; integration can be advanced by a top down approach to AM strategy, planning and implementation; and integration of systems and activities across the whole organization is a key feature of a strategic, corporate level approach to AM.

Integration of Corporate Level AM Strategy With Other Corporate Strategies

The Standards present a strong prescription of integration f AM planning with that of the corporate level, but AM planning and the SAMP are depicted as being driven from above by earlier corporate level decisions and artefacts. The matter of the integration of a corporate level SAMP with the other corporate level strategies and plans is not explained in the Victorian Government example, nor in the literature. Similarly, the benefit of integration on a whole-of-organization basis assumed in the Victorian model was not quantified or otherwise established.

As such integration has not been explored by the AM and EAM literatures we turned to the literature surrounding the progressive integration of organizational sustainability with strategic management theory, seeking an indication of the scope of the task. Engert et al. (2016) identified that over a period of 25 years 118 peer reviewed articles had collectively identified factors and issues that support or hindered integration of a corporate sustainability strategy with the organization's strategic management. Engert et al. (2016) ordered these findings into eighteen categories which they offered as a basis for managerial decisions that could be considered in order to ensure or promote success in the integration process. Engert et al. (2016) did not empirically test these areas or categories of issues.

For our EAM research some categories of issues such as risk management, economic performance and cost reduction resonate with typical EAM/AM corporate objectives suggesting that the larger body of literature regarding the integration of corporate sustainability literature can inform the design of future research into the matter of integration of EAM at the strategic, corporate level.

5 Conclusion

If EAM were to be a core corporate level strategy then a governance entity such as a BoM will need guidelines and models as to fit-for-purpose Strategic EAM policy, strategy and practice. Our research identified that the family of ISO 55000 Standards do not adequately address that need and that there are a number of barriers that must be addressed in the design and implementation of policy and practice. Some barriers were signalled rather than resolved by the exploration of the Victorian Government AMAF (AMAF 2016) confirming the need for substantial empirical research and consequent development of theory.

We identified that such empirical research might address the following:

- Where the artefacts of EAM policy, strategy, planning and activity are positioned within the hierarchy of artefacts.
- The roles of the BoM and CEO
- The positioning of EAM leadership
- The positioning of EAM related measures e.g. availability, cost.
- Whether there was a 'top down' approach to EAM policy, strategy, planning and activity.
- Whether there had been a chronological progression of EAM in the organization.

In addition, consideration of the report by Engert et al. (2016) as to the factors that the literature asserts supports or hinders the integration of a strategy of corporate sustainability into the strategic management of the organization has highlighted that our research must be aware of myriad factors specific to the case study organizations and industry.

We suggest that the research task will for each case study organization be substantial and that within the immense scope of EAM organizations there are diverse needs. Further, the availability of informants for interviews or surveys and the availability of internal archival records and publicly available planning artefacts suggest that public organizations might be a suitable starting point for research. Public organizations that have substantial engineering assets which dominate their core service delivery purpose, such as water systems, appear suitable.

Our research approach would be to first develop the fields of evidence we seek and investigate the viability of that information being available from water services in a specific country or countries. The research methods would be public document analysis, informant interviews, archival study, and thematic analysis utilizing NVIVO. Our objective is to develop a model and guidance as to the configuration of EAM as corporate level strategy and practice.

In doing so we will have both advanced EAM theory in respect of the role of EAM as an organizational strategy and provided to BoM's and CEO's tools to assist them in achieving increased performance from EAM.

References

Amadi-Echendu JE, Willett R, Brown K, Hope T, Lee J, Mathew J, Vyas N, Yang BS (2010) What is engineering asset management? In: Definitions, concepts and scope of engineering asset management. Springer, London, pp 3–16

AMAF (February, 2016) Asset management accountability framework, state of Victoria (Department of Treasury and Finance)

AMAF Implementation Guidance (March, 2017) Asset management accountability framework – implementation guidance, state of Victoria (Department of Treasury and Finance)

Babafemi ID (2015) Corporate strategy, planning and performance evaluation: a survey of literature. J Manag Policies Pract 3(1):43–49

Beard DW, Dess GG (1981) Corporate-level strategy, business-level strategy, and firm performance. Acad Manag J 24(4):663–688

Engert S, Rauter R, Baumgartner R (2016) Exploring the integration of corporate sustainability into strategic management: a literature review. J Clean Prod 112:2833–2850

Hodkiewicz MR, Pascual R (2006) Education in engineering asset management – current trends and challenges. Paper presented at the international physical asset management conference, Tehran, August 28–31

ISO 55000:2014 Asset management – overview, principles and terminology

ISO 55001:2014 Asset management – management systems requirements

ISO 55002:2019 Asset management - management systems - guidelines for the application of ISO 55001

Koronios A, Nastasie D, Chanana V, Haider A (2007) Integration through standards – an overview of international standards for engineering asset management. In: Second world congress on engineering asset management and the fourth international conference on condition monitoring, Harrogate, June 11–14

Landier A, Thesmar D (2005) Bottom-up corporate governance. Working paper, New York University. https://archive.nyu.edu/jspui/bitstream/2451/26412/2/FIN-05-011.pdf

Mahmood MN, Dhakal S, Wiewiora A, Brown K, Keast R (2015) A comparative study on asset management capability maturity models. Int J Strat Eng Asset Manag 2(4):328–347

The Institute of Asset Management (2008) Asset Management competency requirements framework. Part 1. The Institute of Asset Management, London

VAGO (2019) Compliance with the asset management accountability framework, Victorian Auditor-General's Office, May 2019, PP No 27, Session 2018-19

Viljoen J, Dann S (2000) Strategic Management, 3rd ed. Frenchs Forest

Risk Management Issues

Planning Major Overhaul and Equipment Renovation Based on Asset Criticality and Health Index

Antonio de la Fuente[1(✉)], Adolfo Crespo[1], Antonio Sola[2], Antonio Guillén[1], Juan Gómez[1], and Joe E. Amadi-Echendu[3]

[1] Department of Industrial Management, University of Seville, Seville, Spain
antoniodela84@gmail.com, adolfo@etsi.us.es, ajguillen@us.es, juan.gomez@iies.es

[2] Ingeman, Association for the Development of Maintenance Engineering, Seville, Spain
asrasrasr@telefonica.net

[3] Department of Engineering and Technology Management, Graduate School of Technology Management, University of Pretoria, Pretoria, South Africa
joe.amadi-echendu@up.ac.za

Abstract. This chapter explains the process to follow in order to schedule major overhauls and renovation of capital intensive, physical assets. The methodology followed considers the importance of the assets for the business (asset criticality) and the asset condition projections over time and according to existing operational and maintenance records.

The current models for the calculation of Life Cycle Cost analysis (LCC) ignore as the aging of assets, in particular conditions of geographical location, operation and maintenance. That's why there is a need to identify critical failure modes of the important assets of any organization, and how the condition of the systems and its evolution until the end of its useful life can determine the cost of the life cycle. The contribution of this document consists of the use of criticality analysis and asset health index tools, to determine the life cycle cost analysis, based on the impact of each failure mode and the aging of the assets. The methodology provides an objective point of view in order to justify major overhaul, asset renovation, or the extension of useful life.

The above methodology is applied to assets of an Oil & Gas company. As a result, an economic profile is obtained up to the 2035 horizon of two different strategies, the first one is the result of the LCC standard analysis without taking into account the aging of the equipment, the second one takes into account the methodology proposed in the paper, which it allows to know when to advance or delay major maintenance to minimize the impact caused by the increase in the frequency of failures of assets, linked to aging. The application of the previous model to a series of critical complex assets of an oil and gas company is included.

Keywords: Assets management · Capital investment · Operation & maintenance decision-making and assets health

A. Crespo Márquez et al. (Eds.): WCEAM 2019, LNME, pp. 83–90, 2021.
https://doi.org/10.1007/978-3-030-64228-0_8

1 Introduction

Nowadays, the use of tools for decision making about long-term renewal and replacement of equipment for organizations is sufficiently extended. Using simulation models, it is possible to estimate the economic cost of an asset over its useful life and to estimate the time for maintenance interventions, major overhauls and/or refurbishments, as well as replacement where necessary [1]. Often, the challenge is how to take into account the large number of variables that must be included when estimating the real impact (in terms of total cost and total value) of an asset over its useful life.

This paper focuses on the application of asset health index as a tool which considers different sources of data and information, for example, operation and maintenance histories, control system, asset condition reports, location and environmental conditions, etc. The first part deals with the introduction of concepts, the second part corresponds to the development of the methodology and the third part is the analysis of the results of application to a case study in an oil and gas organization.

2 Risk Indicator and Asset Health Index

A common definition of risk (associated with failure) is the probability that a failure will occur with undesirable consequences thereof. It is not intended to deal with risk as an expected value, but as a probability of undesirable consequences [2]. According to [3], risk may be expressed as follows (i referred to event i):

$$R = \sum_{i=1} Pfi \times Cfi \quad (1)$$

Where:

- R is the risk,
- Pfi is the probability of failure
- Cfi is the consequences of the undesired event.

A risk indicator (RI) is a practical tool for identification of the optimal time for replacement or repair of assets in order to minimize the operational cost of maintenance. To achieve success in asset management, organizations with a large volume of assets, have to know the risk inherent in the operation of maintenance and how it will change throughout the life cycle [4].

Asset Health Index (AHI) is an asset score, which is designed, in some way, to reflect or characterize the asset's condition and represent a practical method to quantify the general health of a complex asset. The tool provides an objective point of view in order to justify, the extension of an asset useful life, or in order to identify which assets from a fleet are candidates for an early replacement as a consequence of a premature aging [5]. The application procedure for calculating the health index is based on 5 consecutive steps, which are listed below [6].

Step 1. Asset's selection, definition of category and sub-category. Information capture to estimate time to overhaul, major repair or replacement, and obtaining the estimated normal life of the asset. The estimated normal life of the asset is a data that, in general, comes from the technical direction of the company, considering the experience accumulated so far and the information provided by the different manufacturers.

Step 2. Impact's evaluation of load and location factors by type of asset, technical location and estimated life expectancy. The load factor (FC), as well as the location factor (FE), are inherent in the technical location. The FE depends on the distance to the coast, altitude above sea level, annual average of outside temperature, exposure to corrosive atmosphere agents, exposure to dust in suspension, etc. The FC measures the load request that is made on the equipment in that location, in front of the maximum admissible load. The Estimated Life of the asset is the quotient that results from dividing the estimated normal life obtained in step 1, between the product of the location and load factors (Eq. 2).

$$Estimated\ life = \frac{Estimated\ normal\ life}{F_E \times F_C} \tag{2}$$

Step 3. Calculation of the aging rate. The aging rate is the parameter of the model that allows us to express mathematically this mode of behavior, and consider the different phenomena that the asset can suffer throughout its useful life, such as corrosive phenomena, wear, oxidation of oils, breakage of insulation, etc. The equation for its calculation is the following (Eq. 3).

$$\beta = \frac{\ln \frac{H_{new}}{H_{estimated\ life}}}{Estimated\ life} \tag{3}$$

Where:

- β: Asset aging rate.
- Estimated life: Time calculated in step 2.
- HI new = 0,5; Health value corresponding to a new asset.
- HI estimated life = 5,5; Value of health corresponding to their estimated lifetime.

Step 4. Obtaining the Initial Current Health Index. For the calculation of the initial current health index (HIi) of an asset is used the following (Eq. 4), where the age of the asset is the current age (in units of time) and the aging rate β is calculated in step 3.

$$HI_i(age) = H_{new} \times e^{\beta xage} \tag{4}$$

Step 5. Evaluation of the impact of health modifiers, reliability and calculation of the Asset Health Index (AHI). For any asset, the Asset health index will be determined by its status, operating conditions and reliability conditions at the time of evaluation. For the determination of the current health index, the following (Eq. 5) is used.

$$AHI(age) = HI_i(age) \times MS(age) \times MF(age) \tag{5}$$

Where:

- HIi: initial current health index calculated in step 4.

- MS: Modifier of the health of the asset (condition and operation).
- MF: Reliability modifier of the asset.

The first range is between the values AHI = 0.5 and AHI = 4. The behavior of the equipment in this range is similar to that of a new equipment. The second range, which is comprised between the values AHI = 4 and AHI = 6, corresponds to the period in which the first symptoms of wear on the equipment begin to appear. In this range, the value corresponding to AHI = 5.5, is the value of the health index equivalent to the normal life expected for the type of equipment under study, as well as each subsystem thereof. The methodology assumes that exceeding the value of AHI = 8, the equipment is at the end of its useful life.

3 Life Cycle Cost Approximation Based on Criticality Analysis and Asset Health Index Risk Indicator and Asset Health Index

The methodology used is based on the techniques of operational reliability and with an approach based on the integral management of the assets [7], the methodology tries to maximize the profitability of the organization and reduce the operational risks (reliability and availability) and financial related to the cycle of life of these assets. The following figure (see Fig. 1) shows the methodology used in this paper for the identification of optimal planning for asset´s replacement overhaul, taking into account the criticality analysis and the asset health index during the life cycle cost.

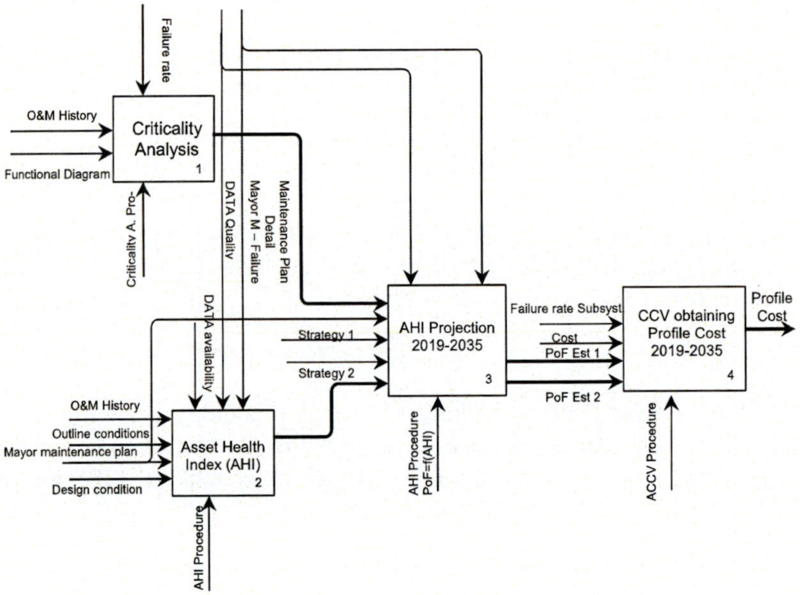

Fig. 1. Calculation procedure for optimizing replacement and overhaul in the life cycle cost based on criticality and asset health index.

First, the criticality analysis of the assets is done [3, 8, 9], based on the importance of the function performed by them so that the organization meets and reaches its objectives. Together with the criticality analysis of the assets, the consequences of the different failure modes and failure rate are also identified, making use of the operation and maintenance histories.

The next step is to calculate the AHI according to the methodology previously proposed [6]. This step is the most important for the success of the application of the process, it requires carrying out the following activities:

- Identify the degradation factors according to their physical location in the plant, exposure to environmental agents, proximity to the coast, to other industrial plants, altitude above sea level, etc.;
- Identify possible health modifiers of the assets;
- Identify possible reliability modifiers based on the investigation of historical failure modes.

Once the AHI and behavior during their life cycle has been calculated, the next step is the AHI projection to 3035, in two different scenarios. The methodology proposed by the authors of the standard [10] is the projection of the AHI and linking frequencies of each failure mode to the value of the index through the use of a multiplier factor that is proportional to the aging of the asset. The maintenance cost will be affected by these new frequencies. On the other hand, new strategies for overhaul could be proposed, in order to avoid the increase of failure frequencies, by "reset to zero" aging. The results in this step are strategy 1, which considers a maintenance plan based on the hours of operation and recommendations of the manufacturers, in which it is possible to increase the probability of the different failure modes resulting from the premature aging of the active. This is associated with an increase in corrective maintenance costs during the life cycle of the asset. On the other hand, in strategy 2 it has been proposed to perform overhaul and replacement when the AHI projection reaches the value of 5.5, thus avoiding the multiplier factor of the probability of failure, keeping constant the corrective maintenance costs during the life cycle of the asset.

Nowadays, and regarding the a.m. consolidated Asset Management Framework [11], the use of tools like LCC for decision making regarding long-term renewal and replacement of assets is quite extended. In this paper, the LCC of an asset [12] is determined by computing the different costs accrued to the holder, by the asset considered, throughout its useful life and after it, until its total elimination or including a residual value that would have for the possessor, in its accounting once its useful life has ended. This cost includes the purchase value and that of all the studies, works, developments, etc., that would be necessary, until it is put into operation and the start of the operation and maintenance phase and all subsequent to this moment. On the other hand, the effects of the inflation and capitalization discount rates to be used are considered in these costs, in order to update them to date.

4 Results Discussion and Case Study Application

The application of the previous model, has been made in a complex asset of an oil & gas company, specifically, for a 4 cranks compressor API 618, with a CAPEX recommended that consist in an overhaul every 10 years ($150.000), an operation and efficiency study every 5 years ($2.500) and minor maintenance every 2 years ($30.000). For OPEX calculation, the sum of the operating costs and corrective maintenance derived from the functional failures are considered, which increase as the asset ages. Likewise, when the overhaul is carried out, the "reset to zero" of the asset is considered, so the probability of the failure decreases as well as the value of the AHI. Table 1 shows how the LCC study of strategies S1 and S2 includes reinvestments in capital (Capex) and operating and maintenance costs (Opex) expected for an analysis horizon that ends in 2035, when considering inflation. The table shows the detail of the expense profiles per year, for the analyzed asset. The values in the table are obtained from the investment costs, the expected operation and maintenance costs, and the costs associated with the risk of failure and subsequent repair. The difference between strategy 1 and strategy 2 is to change the planned times for overhaul, to minimize the risk associated with a higher AHI, minimizing the cost of repair due to failure.

Table 1. Results of strategies 1 and 2 of the LCC of a complex asset taking into account the index of criticality and AHI.

	2019	2020	2021	2022	2023	2024	2025	2026	2027	2028	2029	2030	2031	2032	2033	2034	2035	
S1	5	33	0	35	0	43	0	39	202	0	7	44	0	47	0	58	0	
	252	259	267	275	283	292	317	350	398	328	338	348	359	369	380	392	404	
S2	5	33	0	35	0	43	0	196	0	42	7	44	0	47	0	58	0	
	252	259	267	275	283	292	300	309	319	328	338	348	359	369	380	392	404	NPV
S1	257	292	267	310	283	335	317	389	600	328	345	392	359	416	380	450	404	$2.704
S2	257	292	267	310	283	335	300	505	319	370	345	392	359	416	380	450	404	$2.647

In Strategy 1, future overhaul and major revisions are scheduled to be performed based on the hours of operation probably in 2027. On the other hand, due to the increase in AHI as a consequence of the physical location of the asset, load factors and health modifiers, studied from its installation to the present, the probability of failure modes will increase exponentially before performing the overhaul, with the increase of the corresponding OPEX.

For strategy 2, it is calculated assuming that the planned preventive inspections and maintenances are being carried out and adapting the major maintenance plan, to the times that allow maintaining a controlled level of reliability, preventing the asset's failure rate from increasing beyond the standard values considered (AHI ≤ 5.5). This strategy allows keeping the OPEX values constant, advancing the planned CAPEX investments over time.

5 Results Discussion and Case Study Application

The development of this study allows defining a long-term Maintenance Plan (for the period 2019–2035) and an expected expense profile associated with it, with the purpose of maximizing profitability and reducing the technical and financial risks related to the life cycle of assets.

The study has been supported in:

- Available data related to the behavior of the assets since their commissioning.
- Information provided by equipment and system manufacturers.
- Industry standard reference information for such equipment and systems, and the use of criticality analysis methodologies and asset health indexes.

At that moment, it is where, for instance, a proper definition of the Assets Health and its measure by an index or a risk indicator comes into play as a support tool [5]. This example is also a simple application that justifies the proposed arguments regarding risk and how industry 4.0 may help enhancing the quality of any asset function along its life cycle.

The AHI plays an important role in the digitization of assets, it is a very useful tool for asset management, both for overhaul planning and for the extension of the useful life. For a correct implementation, it is important to take into account the unification of all sources of information related to operation and maintenance, facilitating the integrity of the index.

The main contribution of this paper is obtaining the Asset Health Index to date and projecting its value in the future, adapting the maintenance plan to the times that allow maintaining a controlled level of reliability, preventing increase the equipment failure rate beyond the standard values considered (AHI ≤ 5.5). This will be fundamental for the management of the critical assets of any organization.

References

1. Durairaj SK, Ong SK, Nee AYC, Tan RBH (2002) Evaluation of life cycle cost analysis methodologies. Corp Environ Strategy 9(1):30–39
2. Nilsen T, Aven T (2003) Models and model uncertainty in the context of risk analysis. Reliab Eng Syst Saf 79(3):309–317
3. ISO 31000 (2010) Risk management. Principles and guidelines. International Standards Organizations
4. Adams J, Srinivasan R, Parlikad AK, González-Prida V, Crespo AM (2016) Towards dynamic criticality-based maintenance strategy for industrial assets. IFAC-PapersOnLine 49(28):103–107. https://doi.org/10.1016/j.ifacol.2016.11.018
5. De La Fuente A, González-Prida V, Guillén A, Crespo A, Sola A, Gómez J, Moreu P (2018) Strategic view of an Assets Health Index for making long-term decisions in different industries. In: Haugen et al. (eds) Safety and reliability – safe societies in a changing world. Taylor & Francis Group, London, pp 1151–1156. ISBN 978-0-8153-8682-7
6. De la Fuente A, Candón E, Martínez-Galán P, Crespo A, Sola A, Moreu P (2018) Asset health index method for a process pumps fleet. Industria química 57:38–43 ISSN 2340-2113. (Special ACHEMA 2018)

7. Institute of Assets Management (IAM). Subjects 6 and 7: Capital Investment Operation and Maintenance Decision-Making. Versión 1.0, June 2016. https://theiam.org/knowledge/projects/ssgs/subjects-6-and-7-capital-investment-operation-and-maintenance-decision-making/
8. Aven T, Andersen HB, Cox T, López E, Greenberg M, Guikema S, Kröger W, Renn O, Zio E (2017) Risk analysis foundations. society for risk analysis (SRA)
9. Crespo Márquez A, Moreu de Leon P, Sola Rosique A, Gómez Fernández JF (2016) Criticality analysis for maintenance purposes: a study for complex in-service engineering assets. Qual Reliab Eng Int 32:519–533
10. UK DNO Common Network Asset Indices Methodology. Health and Criticality. Version 1.1, 30 January 2017
11. Crespo Márquez A (2007) The maintenance management framework: models and methods for complex systems maintenance. Springer, London
12. ISO 55002 (2013) Asset management - Management systems – Guidelines for the application of ISO 55001. International Standards Organizations

Graph Theory and Its Role in Vulnerability Evaluation of Infrastructure Networks

Seyed Ashkan Zarghami[1] (✉) and Indra Gunawan[2]

[1] Torrens University Australia, Adelaide, Australia
ashkan.zarghami@laureate.edu.au
[2] The University of Adelaide, Adelaide, Australia

Abstract. Infrastructure Networks (INs) have reached a level of complexity where conventional vulnerability analysis methods cannot fulfil the challenges for the management of this increasing complexity. There is, therefore, a need for a complementary approach through which the structural complexity and the level of interaction among the components are studied. In this context, researchers have increasingly looked into graph theory for help in understanding the structure, efficiency, and vulnerability of INs. The desire to employ the graph theory has resulted in the proliferation of a wide range of topological metrics. To gain a better appreciation for how various graph theory quantities have been used for vulnerability evaluation of INs, this chapter documents a variety of concepts from graph theory in one place and presents an overview of the application of graph theory in the field of INs. It also reviews the conceptions of the four widely used vulnerability-averse indicators of INs, which are reliability, resilience, robustness, and redundancy. The conventional graph theory methods are criticized on several grounds and the future evolution of graph theory tools is forecast.

Keywords: Graph theory · Infrastructure networks · Vulnerability

1 Introduction

Infrastructure Networks (INs) have dictated the living conditions of the human being as well as the economy of human life. However, the historical records highlight a high prevalence of disruption and dysfunction of the performance of these networks due to operational threats and disruptive events. INs consist of multiple interconnected interacting components, in which failure of any of these components may lead to system failure. Viewed from this perspective, an IN serves as a good example of complex systems. The complexity of INs along with the extensive societal dependence on these networks emphasizes the importance of evaluating the vulnerabilities (Johansson et al. 2013). The vulnerability of these networks is major concern for researchers in ensuring public health, safety, and societal welfare.

Much effort has been devoted to the development of a vast array of methods for vulnerability analysis of INs. These methods typically follow two rather different approaches: (1) a functional-based approach; (2) a topological-based approach.

A. Crespo Márquez et al. (Eds.): WCEAM 2019, LNME, pp. 91–101, 2021.
https://doi.org/10.1007/978-3-030-64228-0_9

The functional-based approach is concerned with various functional attributes of INs. Examples of the functional attributes are pressure head in water distribution networks, the time to travel a given length of a road in transportation networks, and the voltage level in power networks. The topological-based approach evaluates the functionality of a network based on its layout. In this approach, the importance of a component in a network, as well as the consequences of its failure, are investigated (Zarghami and Gunawan 2020b). Accordingly, in this approach, vulnerability is assessed by adopting various tools from graph theory.

The literature on topological vulnerability analysis of INs covers a broad spectrum of graph-theoretic metrics, each of which captures a particular structural attribute of the network. Moreover, researchers have adopted various vulnerability-averse indicators of INs as indirect measures of vulnerability. The aim of this chapter is threefold. First, presented in Sect. 2, this chapter reviews the conceptions and applications of the four widely used vulnerability-averse indicators of INs, which are reliability, resilience, robustness, and redundancy. Second, in Sect. 3 and 4, it documents a variety of concepts from graph theory in one place in order to provide an overview of the applications of graph theory in the field of INs. Third, in Sect. 6 and 7, this chapter seeks to address the shortcomings of graph-theoretic methods, and subsequently, to identify an agenda for future research.

2 Vulnerability Analysis of INs

In the context of INs, the term vulnerability has been widely used to refer to the inability of a network to cope with disruptive events. Vulnerability is an *ex ante* notion (Naudé et al. 2009), which implies that it should be estimated based on possible predictions. In light of this, research attention lies in quantifying topological vulnerability by predicting how a network responds to disruptive events based on its existing structure. As such, researchers have attempted to employ a number of vulnerability-averse indicators that account for the network response to disruptions. These indicators have been widely used in the INs literature as the countermeasures of vulnerability. In what follows, the four most widely used vulnerability-averse attributes are discussed.

2.1 Reliability

The concept of reliability refers to the ability of INs to provide continuity of operations (Zarghami et al. 2020). Hashimoto et al. (1982) provided two definitions of reliability as the probability that a system functions in a satisfactory state, as well as the probability of no failure within a given operating period. A completely satisfactory IN should create a continuity of operations in order to provide important services throughout its design period as well as in the face of all hazards. The process of determining how well an IN can satisfy this goal is the theme of reliability analysis.

2.2 Resilience

The literature presents several definitions of resilience. Most studies define resilience by looking into the performance of a system before and after the occurrence of a disruption

(Liu and Song 2020). Herrera et al. (2016) defined resilience as the ability if a system to maintain its required level of operational performance when a disruption strikes. In a work by Chang and Shinozuka (2004), a resilient system is defined as a system that possesses three features: reduced failure probability, reduced failure consequences, and reduced time to recovery. In a similar vein, Cimellaro et al. (2016) highlighted three characteristics of a resilient system as 1) the ability to avoid a shock, 2) the ability to absorb a shock if it occurs, and 3) the ability of a fast recovery after the occurrence of a shock.

2.3 Robustness

Robustness of an IN is concerned with the ability of the network to tackle with perturbations during its operation (Agathokleous 2017). Carlson and Doyle (2002) described the robustness as the ability of a system to maintain its desirable characteristics when some of its components experience fluctuations in their behavior. Li et al. (2008) defined the robustness as the ability of a system to continue operating despite the failures of a fraction of its components.

2.4 Redundancy

The literature defines the structural redundancy of INs as the existence of alternative paths in the network that can be used when the main path is failed (Gunawan et al. 2017). In fact, the degree of redundancy increases when multiple paths reach the same nodes on the network (Di Nardo et al. 2017). In the research literature, redundancy has been studied in both local and global sense. The local redundancy indicates the reserving capacity of the network elements, whereas the global redundancy refers to the residual network capacity (Ziha 2000).

2.5 Quantifying the Vulnerability-Averse Attributes of INs

In order to quantify the vulnerability-averse attributes of INs, researchers have sought new methods that capture the topological properties of networks. One such method that accounts for the topological characteristics of INs is the graph model that places a special emphasis on the topology of the networks. Graph models seek to answer the following types of questions.

(1) A reliability-related question: What is the probability that the elements of the network remain connected when a disruptive event takes place?
(2) A resilience-based question: What is the association between the network topology and the post-disruption recovery?
(3) A robustness-based question: To what extent the network is capable of withstanding perturbations in structure without significant changes in its operation.
(4) A redundancy-related question: How does the existence of alternative paths reduce the vulnerability of the network?

The next section provides a synopsis of graph theory and outlines a number of graph-theoretic methods and their characteristics to represent the topological properties of INs.

3 A Synopsis of Graph Theory

Graph theory is a branch of mathematics that studies graphs. The use of graphs seems to have been common practice since the 18th century. In 1735, Leonhard Euler, a Swiss mathematician, presented his solution of the *Königsberg bridges problem.* Euler attempted to find a route through which each of the seven bridges of Königsberg could be crossed once and only once (Wilson 2013). The Euler's treatment of *Königsberg bridges problem* led to the first mathematical theorem in graph theory. This theorem facilitated the growth and development of a vast array of graph-theoretic methods. In graph-theoretic methods, a network is mapped into a graph of nodes and links. The nodes and links can represent all types of real-world physical networks. This desirable feature has triggered interest in using graph theory in the field of INs.

A graph representation of an IN provides an abstract modelling context in which nodes and links represent a collection of identifiable assets as well as the relationships among them. Table 1 shows the examples of nodes and links for various INs.

Table 1. Examples of nodes and links in graphical representations of INs

INs	Nodes	Links
Water distribution networks	Reservoirs, Consumers	Water pipelines
Gas distribution systems	Gas wells, gas generators, consumers	Gas pipelines
Power transmission networks	Power plants, consumers	Transmission lines
Supply chain networks	Suppliers, manufacturers, distributors, retailers, customers	Material flow, information flow
Transportation networks	Airports, seaports, cities, rail stations	Air routes, highways, railways

The study of graphs has become a fundamental concept in INs research. With the development of several graph-theoretic metrics, the application of graph theory in vulnerability analysis of INs has continued to grow over the past two decades. The desire to employ the graph theory methods for INs has resulted in the proliferation of a wide range of topological metrics. In order to better follow a broad spectrum of the graph- theoretic methods, the following section is devoted to organize these methods into a classification scheme based on their similar features.

4 Graph Theory Methods

The graph-theoretic methods aim to provide numerical indicators to quantify vulnerability and possible operational consequences of an asset failure (Yazdani and Jeffrey 2012). A key focus of graph-theoretic methods is on quantifying different vulnerability-averse attributes of INs by capturing various topological properties of these networks. By applying graph theory, researchers examine how the topology of an IN is associated with its reliability, robustness, resilience, and redundancy to ensure that the network can deal with disruptions.

The works on the application of graph theory in vulnerability analysis of INs fall into six broad categories. (1) Articles that perform path analysis of networks. (2) The literature that deals with spectral measures in which the algebraic representation of a network, as an adjacency matrix, is used. (3) Methods that are concerned with geometrical distance among nodes. (4) Papers that attempt to evaluate the redundancy of networks. (5) Articles that study the relationship between the clustering of a set of components and graph theory. (6) Research that integrates graph theory methods with other scientific approaches. Table 2 shows some of the most commonly used graph invariants that set a basis for the development of a great number of graph theory methods.

The graph invariants, presented in Table 2, can be classified into two groups: 1) local metrics that quantify the vulnerability of a network at specific locations; and 2) global metrics that measure the vulnerability of a network as a whole. To gain some insight, the values of global and local graph invariants are calculated for an illustrative IN, taken from Zarghami et al. (2018a). Figure 1 depicts the example network with seven nodes and eight links.

Table 3 reports the overall vulnerability of the example network using various global metrics including *edge density* (E_d), *algebraic connectivity* (λ_2), *meshedness coefficient* (r_m), and *clustering coefficient* (CC_T). Table 4 shows the vulnerability values of the links in the example IN using local metrics including *degree centrality* (C_{deg}), *betweenness centrality* (C_B), *closeness centrality* (C_C), and *eigenvector centrality* (C_e).

As can be seen from Tables 3 and 4, each graph invariant captures different topological properties of a network. Thus, in order to provide a richer insight to the vulnerability analysis of INs, the juxtaposition of graph invariants should be considered.

4.1 Path-Based Measures

Path-based measures identify the importance of a node based on the number of paths or shortest paths that pass through it. The underlying concept of all path-based measures is that if many paths would have to pass through node *i* in order to reach other nodes of a network, then i is an important junction node of the network. The examples of these metrics in this category are stress, betweenness, K-path centrality. Applications of these measures span much of the spectrum of INs with examples of wireless sensor networks (Jain 2016), water distribution networks (Agathokleous et al. 2017), and urban complex transportation networks (Liu et al. 2019).

Table 2. The most commonly used graph theory metrics

Metric	Mathematical expression	Parameters	Category
Edge density	$E_d = \frac{2m}{n(n-1)}$	m: Number of edges n: Number of nodes	Density-based
Clustering coefficient	$CC_T = \frac{1}{n}\sum \frac{2t_i}{k_i(k_i-1)}$	n: Number of nodes k_i: Degree of node i t_i: Number of triangles attached to node i	Clustering
Meshedness coefficient	$r_m = \frac{m-n+1}{2n-5}$	m: Number of edges n: Number of nodes	Density-based
Degree centrality	$C_{deg}(i) = \frac{k_i}{n-1}$	k_i: Degree of node i n: Number of nodes	Distance-based
Closeness centrality	$C_C(i) = \frac{n-1}{\sum_j d_{ij}}$	n: Number of nodes d_{ij}: Shortest path length between node i and j	Distance-based
Betweenness centrality	$C_B(i) = \frac{1}{(n-1)(n-2)} \sum_{s \neq r \neq r} \frac{n_{s,r}(i)}{n_{s,r}}$	$n_{s,r}(i)$: Number of shortest paths between s and r passing through i $n_{s,r}$: Number of shortest paths between s and r n: Number of nodes	Path-based
Eigenvector centrality	$C_e(x) = \frac{1}{\lambda} \sum_{y \to x} C_e(y)$ $\lambda e = Ae$	$C_e(x)$: Eigenvector centrality of node x A: Adjacency matrix e: Eigenvector centrality of all nodes λ: Largest eigenvalue of A	Spectral
Algebraic connectivity	λ_2	λ_2: Second smallest Laplacian eigenvalue of a graph	Spectral

4.2 Spectral Measures

Spectral measures have been used as one of the key evaluation indicators of complex INs. These measures look at the properties of networks based on the left dominant eigenvector of the network adjacency matrix or similar entities. In the global sense, spectral measures determine how well a network is connected by finding the smallest non-zero eigenvalues of its Laplacian matrix. In the local sense, using these measures, a node is an important junction node of a network if it connects to many nodes with high eigenvector centrality. Algebraic connectivity, PageRank, and eigenvector centrality are the most well-known examples of spectral measures. Several varieties of spectral measures have been adopted

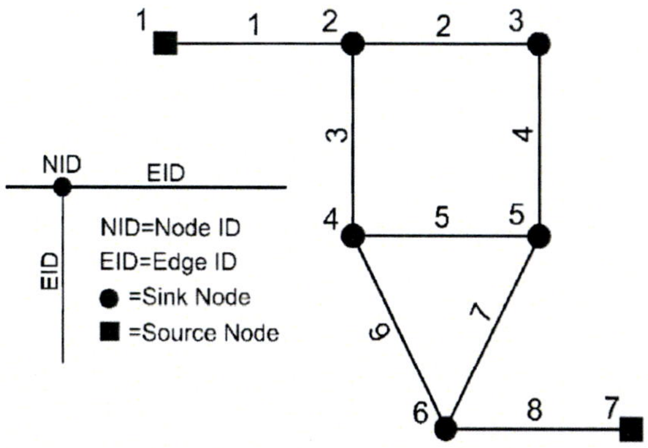

Fig. 1. An illustrative IN (adapted from Zarghami et al. 2018a)

Table 3. The values of global metrics for the example network

Network	E_d	λ_2	r_m	CC_T
IN	0.4667	−0.408	0.2857	0.2778

Table 4. The values of local metrics for the nodes in the example network

Node	C_{deg}	C_B	C_C	C_e
1	0.17	0	0.40	0.29
2	0.50	0.37	0.60	0.75
3	0.33	0.07	0.55	0.66
4	0.50	0.33	0.67	1.00
5	0.50	0.17	0.60	0.98
6	0.50	0.33	0.60	0.89
7	0.17	0	0.40	0.34

for identifying vulnerabilities in airport transportation networks (Wei and Sun 2011), water distribution networks (Zarghami et al. 2019a).

4.3 Distance-Based Measures

In the distance-based measures, the criticality of a component is a function of distances. Indegree, closeness, and Lin's index are examples of this category. These measures quantify the number of nodes that are located at every distance as a measure of importance. The distance-based measures have been applied in a wide variety of real-world networks such as maritime logistic networks (Honglu et al. 2018), water distribution networks (Giustolisi et al. 2019).

4.4 Density-Based Measures

Density-based measures quantify the extent to which the nodes in a network are connected. The density-based measures have been widely adopted to evaluate the overall redundancy of real-world networks such as power transmission networks (Li et al. 2016). The examples of graph invariants that fall into this group are edge density and Meshedness coefficient.

4.5 Clustering Methods

Clustering methods attempt to group components of a network based on their similarity in such a way that those components of each cluster exhibit relatively great similarity, whereas relatively great differences are observed between the components of different groups. Clustering analysis has enjoyed its successful application in various networked infrastructures such as water distribution networks (Diao et al. 2014) and road networks (Han et al. 2017). It attempts to determine how integrated or fractured the overall network is.

4.6 Integrating Graph Theory Methods and Various Scientific Approaches

In recent times, researches are recognizing the effectiveness of combining the conventional graph theory methods with a broad spectrum of various scientific approaches such as fuzzy logic (Neumann 2016), and entropy theory (Zarghami et al. 2019b). It is now an accepted fact that combining graph theory methods with other approaches will be necessary for capturing the complexity inherent in INs.

5 Practical Implications of Graph-Theoretic Methods

The following practical implications emerge from applying the graph-theoretic methods to IN.

First, INs are the critical assets of a country. Disruption to the services provided by these strategic assets would have debilitating effects on public health, safety, security, and economy. As such, these facilities must be maintained to ensure continuity of their operations. Maintenance plays a significant role in ensuring the availability and reliability of these valuable assets and carries substantial costs. A significant percentage of maintenance costs can be avoided by implementing an effective predictive maintenance strategy. Applying graph-theoric methods to the vulnerability analysis of INs provides an opportunity to identify and prioritize the vulnerabilities. This, in turn, enables the network operators to establish and implement a sound predictive maintenance strategy. A cost-effective maintenance strategy based on the topological vulnerability analysis of INs increases the current and future operational reliability while optimizing the cost of maintenance activities (Zarghami and Gunawan 2020a).

Second, graph invariants can be used as decision support tools in designing INs. The graph-theoretic metrics provide baseline information for INs design. From early on, a given graph metric can be set as a baseline threshold in pre-structuring phase. Therefore, any network layout within the baseline threshold can be granted further considerations, otherwise, it can be excluded from further investigation.

6 A Critique of the Conventional Graph-Theoretic Methods

The graph theory methods presently used for vulnerability evaluation of INs suffer from various drawbacks. In the sequel, we point out some of the basic limitations of the current graph theory tools.

First, a graph is an abstract representation of a network, through which the network is simplified as a set of linked nodes. This overly simplistic representation of networks does not entirely capture the topology of real-world networks. For example, valves in water distribution networks or links in maritime and air networks cannot be clearly defined by a graph. This, in turn, hinders a realistic characterization of INs.

Second, the task of identifying the most important component in a network, using graph invariants, presents confusion. This is because each graph theory tool identifies a different component as the most important component.

Third, each IN exhibits properties not shared by other networks. This raises concerns about the effectiveness of applying the conventional graph theory methods since these methods lack an effective adherence to the features of a particular IN.

Fourth, in the case of weighted networks where an attribute is used to weight different nodes and links, the traditional graph theory metrics are not effective enough.

Fifth, as reported in Tables 3 and 4, each graph invariant is intended to quantify a vulnerability-averse attribute of INs by capturing a particular topological property. Therefore, none of these metrics supplants the others. It is thus apparent that the sole utilization of a single graph invariant can be deceptive.

Sixth, since the focus of graph theory tools is on interconnections between network components rather than on the components themselves, the stand-alone use of these tools yield insufficient information with respect to the likelihoods of components becoming inoperable (Bloomfield et al. 2017).

Seventh, graph theory tools are static, meaning that these tools are only effective for demonstrating a snapshot of a network at different discrete times. In actual reality, the topological characteristics of INs may vary over time in response to various internal and external factors (Zarghami et al. 2018b).

7 Concluding Remarks and the Future of Graph-Theoretic Methods

The present paper has drawn attention to various graph theory methods that have been developed for vulnerability analysis of INs. To better follow the existing graph-theoretic methods, this article has proposed a new classification scheme that categorizes different methods based on different topological characteristics. Care is needed in applying graph theory tools to the field of INs. These tools have been criticized on several grounds.

The key contribution of this study is to facilitate future research in applying the graph theory methods to the field of INs by offering a new classification scheme of various graph theory methods. Further, this research provides a thorough understanding of the adoption of various vulnerability-averse indicators of INs.

Based on the emerging trends in the development of graph theory methods for vulnerability analysis of INs, a few directions for the future evolution of these methods is forecast. First, it seems clear that an integrated method that combines graph theory tools with other scientific approaches will be necessary in order to strengthen the ability of these tools. Currently, there is a small body of literature in this context. This trend is likely to intensify in the literature in the future. Second, the current exclusive focus on the topology of networks may be avoided by the development of domain-specific graph-based metrics that take into account both topological and functional attributes of INs. Third, while the overarching aim of applying graph theory to infrastructure asset management is to evaluate the reliability of networks, there is very little recourse to reliability theory. Future research might usefully seek to incorporate the concepts from reliability theory into topological analysis of INs. Finally, the development of dynamic graph theory metrics, which can be automatically updated in response to changes in the topology of a network, is foreseen.

References

Agathokleous A, Christodoulou C, Christodoulou SE (2017) Topological robustness and vulnerability assessment of water distribution networks. Water Resource Manage 31(12):4007–4021

Bloomfield RE, Popov P, Salako K, Stankovic V, Wright D (2017) Preliminary interdependency analysis: an approach to support critical-infrastructure risk-assessment. Reliab Eng Syst Saf 167:198–217

Carlson JM, Doyle J (2002) Complexity and robustness. Proc Natl Acad Sci USA 99(1):2538–2545

Chang SE, Shinozuka M (2004) Measuring improvements in the disaster resilience of communities. Earthquake Spectra 20(3):739–755

Cimellaro GP, Tinebra A, Renschler C, Fragiadakis M (2016) New resilience index for urban water distribution networks. J Struct Eng 142(8):1–13 C4015014

Diao K, Farmani R, Fu G, Astaraei-Imani M, Ward S, Butler D (2014) Clustering analysis of water distribution systems. Water Sci Tech J Int Assoc Water Pollut Res 70(11):1764–1773

Di Nardo A, Di Natale M, Giudicianni C, Musmarra D, Rodriguez Varela JM, Santonastaso GF, Simone A, Tzatchkov V (2017) Redundancy features of water distribution systems. Procedia Eng 186:412–419

Giustolisi O, Ridolfi L, Simone A (2019) Tailoring centrality metrics for water distribution networks. Water Resour Res 55(3):2348–2369

Gunawan I, Schultmann F, Zarghami SA (2017) The four Rs performance indicators of water distribution networks: a review of research literature. Int J Qual Reliab Manage 34(5):720–732

Hashimoto T, Stedinger JR, Locks DP (1982) Reliability, resiliency and vulnerability criteria for water resource system performance evaluation. Water Resour Res 18(1):14–20

Han B, Liu L, Omiecinsky E (2017) A systematic approach to clustering whole trajectories of mobile objects in road networks. IEEE Trans Knowl Data Eng 29(5):936–949

Herrera M, Abraham E, Stoianov I (2016) A graph-theoretic framework for assessing the resilience of sectorised water distribution networks. Water Resour Manage 30(5):1685–1699

Honglu L, Zhihong T, Anqiang H, Zaili Y (2018) Analysis of vulnerabilities in maritime supply chains. Reliab Eng Syst Saf 169(2018):475–484

Jain A (2016) Betweenness centrality connectivity aware routing algorithm for prolonging network lifetime in wireless sensor networks. Wirel Netw 22(5):1605–1624

Johansson J, Hassel H, Zio E (2013) Reliability and vulnerability analyses of critical infrastructures: comparing two approaches in the context of power systems. Reliab Eng Syst Saf 120(2013):27–38

Li J, Duenas-Osorio L, Chen C, Berryhill B, Yazdani A (2016) Characterizing the topological and controllability features of U.S. power transmission networks. Phys A 453:84–98

Li P, Wang BH, Sun H, Gao P, Zhou T (2008) A limited resource model of fault-tolerant capability against cascading failure of complex networks. Eur Phys J B 62(1):101–104

Liu W, Li X, Liu T, Liu B (2019) Approximating betweenness centrality to identify key nodes in a weighted urban complex transportation network. J Adv Transp 2019:1–8 ID: 9024745

Liu W, Song Z (2020) Review of studies on the resilience of urban critical infrastructure networks. Reliab Eng Syst Saf 193:1–16 106617

Naudé A, Santos-Paulino AU, McGillivray M (2009) Measuring vulnerability: an overview and introdution. Oxford Develop Stud 37(3):183–191

Neumann T (2016) Routing planning as an application of graph theory with fuzzy logic. TransNav, Int J Mar Navig Saf Sea Transp 10(4):661–664

Ouyang W, Cheng H, Zhang X, Yao L, Bazargan M (2010) Distribution network planning considering distributed generation by genetic algorithm combined with graph theory. Electric Power Compon Syst 38(3):325–339

Wei P, Sun D (2011) Weighted algebraic connectivity: an application to airport transportation network. J Adv Transp IFAC Proc 44:13864–13869

Wilson R (2013) Early graph theory. In: Wilson R, Watkins JJ (eds) Combinatorics: ancient and modern. Oxford University Press, Oxford, pp 1–21

Yazdani A, Jeffrey P (2012) Applying network theory to quantify the redundancy and structural robustness of water distribution systems. J Water Resour Plan Manage 138(2):153–161

Zarghami SA., Gunawan IF (2020a) A domain-specific measure of centrality for water distribution networks. Eng Constr Archit Manag 27(2), 341–355

Zarghami SA, Gunawan IF (2020b) A fuzzy-based vulnerability assessment model for infrastructure networks incorporating reliability and centrality. Eng Constr Archit Manag 27(3), 725–744

Zarghami SA, Gunawan I, Schultmann F (2019a) Entropy of centrality values for topological vulnerability analysis of water distribution networks. Built Environ Project Asset Manage 9(3):412–425

Zarghami SA, Gunawan I, Schultmann F (2020) Exact reliability evaluation of infrastructure networks using graph theory. Qual Reliab Int J 36(2):498–510

Zarghami SA, Gunawan I, Schultmann F (2018a) Integrating entropy theory and cospanning tree technique for redundancy analysis of water distribution networks. Reliab Eng Syst Saf 176(2018):102–112

Zarghami SA, Gunawan I, Schultmann F (2019b) Integrating topological and hydraulic attributes for robustness analysis of water distribution networks. Int J Ind Eng Oper Manag 1(1):1–12

Zarghami SA, Gunawan I, Schultmann F (2018b) System dynamics modelling process in water sector: a review or research literature. Syst Res Behav Sci 35(6):776–790

Ziha K (2000) Redundancy and robustness of systems of events. Probab Eng Mech 15(4):347–357

Risk-Based Asset Replacement Policy for an Electrical Infrastructure

Jos Wetzer[1](✉), Sungin Cho[2], Sanuri Ishak[3], Yogendra Balsubramaniam[3], Wan Khairuzzaman Wan Nordin[4], and Alisadikin Muhammad Zainal[4]

[1] DNV GL Energy, Utrechtseweg 310-B50, 6812 AR Arnhem, The Netherlands
Jos.Wetzer@dnvgl.com

[2] DNV GL Energy (Presently with the SP Group), 16 Science Park Dr, Singapore 118227, Singapore

[3] TNB Research Sdn Bhd, No. 1, Lorong Ayer Itam Kawasan Institusi Penyelidikan, 43000 Kajang, Selangor, Malaysia

[4] TNB Distribution Division, Tingkat 11, Wisma TNB No. 19, Jalan Timur, 46200 Petaling Jaya, Selangor De, Malaysia

Abstract. With the increased demands regarding the reliability, availability, safety and affordability of key public infrastructures, it becomes increasingly challenging for asset owners and managers to achieve a proper balance amongst performance, risk and cost. As a response, operators of critical infrastructures need sound solutions to maximize the value of their assets, meanwhile realizing their business values and objectives. This requires well-defined decision-making processes with an adequate risk management concept embedded.

This is particularly true for managing ageing infrastructures. Considering the large amount of assets, this requires a balance between managing reliability/availability, budgets and resources. A traditional replacement approach often involves fixed replacement criteria, which are normally being defined for each specific equipment type. The main drawback of this approach is that no account is made for the differences in stress history or in the impact and thereby the risk of failure: within a specific asset class, critical assets follow the same criteria as non-critical assets, leading to (too) high replacement volumes and (too) early replacements.

In this paper, a risk-based replacement process is presented for prioritization, replacement planning and portfolio management, staying away from fixed replacement criteria. Although the risk-based approach itself is well described in literature, it remains a challenge for asset managers to deal with the large amount of assets. This is here solved by using a two-step approach. First, by using predefined triggers, the assets with a potential risk are identified. Rather than simply replacing these assets, these assets (and only these assets) are subjected to a risk assessment by estimating the probability and impact, using the organization's risk matrix. Next, the optimum solution for mitigating the risk is defined. This may consist of replacement, but it may just as well be refurbishment, intensification of maintenance, or even accepting the risk.

A replacement decision is only one of the types of decisions to be made by an asset manager based on a common asset management policy and a unified approach towards decision making. This paper presents a real-life case study at a

A. Crespo Márquez et al. (Eds.): WCEAM 2019, LNME, pp. 102–112, 2021.
https://doi.org/10.1007/978-3-030-64228-0_10

large electrical infrastructure operator, and discusses the opportunities and risks encountered in implementing such a policy revision.

Keywords: Risk-based decision-making · Asset management · Replacement policy · Public infrastructure

1 Introduction

1.1 Background

As the electrical infrastructure is aging, a main concern for the electrical utility is how to manage such assets in a technically and financially balanced manner. To manage assets, the asset management policy defines the approach, whereas the strategy defines the strategic objectives. A smart approach recognizes which group of assets can exceed their design lifetime, and which groups of assets may fail early, depending on the exposure to operational stresses and intensity of utilization. A balanced approach requires a dedicated assessment of the risk, and asset management activities should be planned and prioritized such as to mitigate those identified risks. For asset intensive organizations such as electrical utilities, timely replacement of aged assets is one of the mitigation options to lower the risk. Since replacement involves high costs and requires meticulous planning, in a highly regulated environment, proper justification is needed to avoid failures due to late replacement, and to avoid capital costs due to early replacement. In this paper, we address the following questions;

- What method and what criteria can be used to identify assets with a "potential" risk (age just being one of the criteria)?
- What method and criteria can be used to assess the "real" risk of such assets (age not being a criterion)?
- How do we determine the optimum solution for mitigating the risk?
- How can we translate this to a replacement strategy?
- How can we incorporate this in a corporate risk-based decision-making process?

This paper presents a stepwise approach and process to address these issues and to set up a unified combination of policy and guidelines. To conclude with, the paper discusses the key ingredients for successful implementation of such a decision-making process.

1.2 Risk-Based Asset Management

International standards such as the ISO 55000 series urge utilities to have a structured approach to AM related processes and risk-based decision making. Such a framework provides the platform for the AM system policy and methodologies. On the other hand, utilities are required to empower this platform with technical expertise and experience [1–3]. The asset replacement policy and guidelines, therefore, should describe how to translate the corporate strategy into an asset replacement plan. The replacement policy and guidelines define the framework and process for risk-based decision taking by addressing the following phases:

- Defining the business values served (reliability, safety, environment, cost effectiveness, customer satisfaction, regulatory compliance) and the KPI's the business values are measured by, aligned to the regulations and laws to be complied with
- Organizing the line of sight and the working principles of replacement guidelines/procedures as part of the Asset Management processes
- Detailing the overall Asset Management philosophy and approach towards (risk based) Asset Management including, for example:

  a. the approach towards maintenance – refurbishment – replacement
  b. the choice for corrective, predictive, condition or risk-based maintenance
  c. the preference for proven or innovative practices
  d. the ambition to follow or lead the business

2 Asset Replacement Policy Document

2.1 Asset Management Document Structure

To have adequate asset replacement strategies, it is considered essential to first define the asset management context. Structuring asset management and developing risk-based replacement policies and guidelines requires a set of documents showing the organizational objectives and the asset management objectives, which are eventually translated into replacement objectives. In addition, replacement aspects need to be addressed at a strategic level, tactical level and operational level. Lastly, replacement aspects are related to, or part of, other business processes, such as risk analysis, assessment and management, maintenance processes and programs. Table 1 shows a hierarchy of documents and a short description of the asset management documents that build a replacement policy.

2.2 Building an Asset Replacement Policy Document

A document structure, with a well-defined scope and content per document type (replacement policy, replacement program, replacement guideline) should be based on a clear set of definitions and aim at smart decision criteria. The development of a set of replacement policy documents requires to carefully select the right amount of detail, with enough information to make the right choices without entering into detailed technical solutions. Terms such as replacement decision criteria, replacement assessment criteria and replacement drivers should be well-defined and used consistently. The policy document should not be restricted to high level considerations but provide a clear blueprint for developing a replacement process. The policy should involve the following chapters:

1. **Business values and KPI's.** One should start mentioning the business values. This provides the link with the Asset Management policy where these business values ought to be described. Examples are: Safety, Reliability, Costs, Compliance, Environment, Customer satisfaction and Reputation. Also, the policy should mention the KPI's that Business Values are measured by

Table 1. Document hierarchy for an asset management system.

Document	Short description
Corporate mission, vision and values	Describes how the company positions itself, now and in the future, defining its reason of existence, its position in the surrounding society, and the values cherished.
Corporate strategy & objectives	High level plan, usually on a 3 to 5 years basis, with clear (SMART) objectives, based on stakeholder requirements, external and internal developments (SWOT), including the planning of large or non-routine activities and new developments
Corporate policies	Describe the rules of the game, the high-level approach towards e.g. corporate risk management, safety, environment, decision making and the like
Asset management system	Part of the organization's management system, including the AM policy, AM objectives, SAMP, AM plans and supporting processes and systems
Asset management policy	Describes the rules of the game for the asset management domain, the high-level approach and the methodologies and criteria for managing risks, maintenance, replacement, decision making. It includes the risk management approach, risk matrices and acceptance criteria (appetite).
Strategic asset management plan (SAMP), or asset management strategy	High level asset management plan, usually on a 3-5 years basis, with clear (SMART) objectives, based on stakeholder requirements, external and internal developments (SWOT), planning of non-routine activities (extension, reinforcement, new technology, new processes), referring to processes and decision-making criteria; The strategy includes sub-strategies, e.g. for maintenance development or improvement, replacement, grid reinforcement, data quality and IT, and asset management system development
Asset management plan	(S)AMP worked out in more detail for a shorter timeframe, usually 1 year, describing activities, SMART objectives and planning
Asset management decision making	The process of decision making is the core of the asset management system, and usually involves risk identification, risk analysis, generating solutions, determining urgency and priority, deciding on the portfolio
Asset management risk analysis	Analysis of existing risks (with input from e.g. performance and failure reports) as well as forecasted risks (based on evaluating trends and developments). Risk analysis is based on the probability of an event and the impact of that event, the risks being weighed according to the defined risk appetite for each business value
Process description guidelines and manuals	Describes how the subsequent steps in a process are executed, and the use of methodologies, including the interactions and responsibilities, and criteria, boundary conditions, tools, resources and information sources

2. **Replacement decision process**. Secondly, one should describe the replacement decision process. This process needs to describe how the risks are identified, analyzed and ranked, how the solutions are developed, weighed and prioritized, and how the

replacement portfolio is assembled and decided upon. This is a dedicated application of the overall decision process to be described in the Asset Management policy
3. **Generic replacement triggers**. As described above, replacement decisions in a modern asset management organization require a risk-based decision process rather than fixed replacement criteria. However, subjecting all assets to this decision process is not efficient. In order to select the assets for which this decision process is required, one needs to develop generic replacement triggers as a first filter

3 Risk-Based Decision-Making Process for Asset Replacement

3.1 Development of a Risk-Based Replacement Decision Process

In the past, TNB, being a large electricity distribution company, based their replacement decisions on a fixed set of replacement criteria. TNB experienced the following issues with this approach:

- The approach did not clearly define the amount of risk with respect to specific business values, and did not sufficiently differentiate between impact levels (from negligible to unacceptable)
- The approach did not provide clear inputs for prioritizing planning.

In order to overcome these issues, TNB switched-over to a two-step risk-based approach:

1. First, assets with a possible risk are identified by means of triggers; triggers (as shown in Table 2) are defined per business value and reflect the KPI's that are used to measure the threat to a business value;
2. Second, the identified assets are subjected to a risk-based decision-making process.

The business values define what the most valuable company "assets" are. The performance for each business value is measured against Key Performance Indicators (KPI's), and companies formulate targets (objectives) for each business value. Business values are based on internal and external stakeholders' needs and expectations and are the foundation of the asset management policy (and thereby the replacement policy). Examples of often used business values and corresponding KPI's are shown in Table 2.

The replacement policy and the corresponding guidelines are part of a family of policies, guidelines and processes. Figure 1 provides an overview of how policies, guidelines and decisions are connected to the ISO 55000 line of sight which describes the chain of decision making from corporate strategy down to the portfolio of activities. To accomplish the (strategic) asset management goals the company needs policies, to describe how those goals are being achieved, guidelines to translate the policy to day-to-day practice, and processes to define the day-to-day activities.

3.2 Setting Up Triggers for Risk Analysis and Assessment Process

In order to identify the asset that require a full risk assessment, for each asset type suitable triggers should be designed for the business values defined by in the organization's policy.

Table 2. Typical examples of electricity distribution company business values and short description.

Business value	Examples of key performance indicators
Safety	Exposure to high voltages, current arcs, fire, explosion, hazardous (toxic) materials
Reliability	Condition, integrity, failure frequency, repair history, obsolescence, availability of service and spare parts, age distribution
Cost	Cost of maintenance, refurbishment, repair, replacement, service, spare parts, Total cost of ownership (TCO)
Compliance	Legal and/or regulatory requirements, inability to connect
Environment	Environmental impact such as oil spills, gas leaks, polychlorinated biphenyls (PCB), decommissioning concerns
Customer	Customer satisfaction survey framework
Reputation	Impact on society and industry

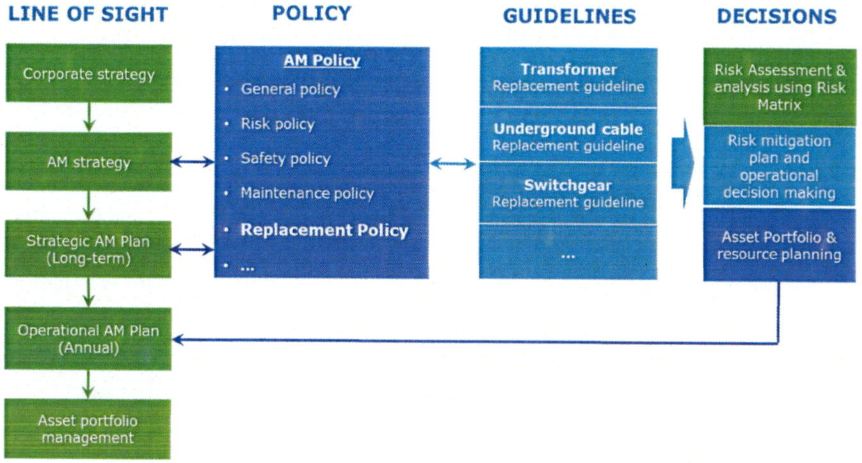

Fig. 1. Linking replacement practice to the asset management line of sight

These triggers may, for example, depend on the design, the materials used, the role of the electrical equipment and the failure history. These triggers serve as a filter, in order to limit the amount of assets that need to be analyzed in more detail. An example of possible triggers for distribution transformers is shown in below Table 3.

Unlike absolute replacement criteria, the triggers defined are based on various business values and account for the risk appetite and KPI's. They reflect the various driving forces for replacement, and allow flexibility for defining the optimum solutions, prioritization, and planning.

Table 3. Examples of the triggers for Distribution Transformers

Business value	Examples of triggers
Safety	The possibility of a failure resulting in a fire or explosion cannot be neglected. Possible symptoms are oil leakages, elevated temperature, and the occurrence of an abnormal sound or smell
Reliability	– The years of operation of the transformer (based on statistical modeling of the failures and estimated useful remaining useful life) – Health index (based on various condition and diagnostic markers to assess the health of the equipment) – Obsoleteness
Cost	– Indirect outage costs – A suspicion of increased or high losses based on type, age, model – Increase of Total Cost of Ownership compared to average value of peer
Compliance	The existing transformer cannot satisfy the load demand
Environment	Trace of polychlorinated biphenyls (PCB) in the insulation oil

3.3 Designing a Detailed Risk-Based Decision-Making Process

The decision-making process proposed is a risk-based process involving the following phases as shows in Fig. 2;

1. **Asset analysis**: The purpose of the asset analysis process is to identify the list of assets with a possible risk to one or more business values, by means of triggers such as shown in Table 2.
2. **Assessment of the risk**: The risk is analyzed against the business values specified in the policy and the criteria specified in the guidelines for individual equipment types. The risk analysis includes business values, probability and impact, and organization's risk appetite. The outcome is a list of assets with categorized risks from negligible to unacceptable.
3. **Assessing the optimum solution**: The risk identified shall be mitigated by refurbishment, replacement, maintenance adjustment or it may be accepted by continuing business as usual. Here the optimum solution is chosen, based on criteria defined in the guidelines (e.g. maximum risk reduction per monetary value). If none of the mitigation options can sufficiently reduce the risk, measures are formulated to define alternatives (adjust specification or design).
4. **Prioritization of the solutions**: The required information and data shall be collected in a standardized way to provide a clear and unambiguous justification of the replacement (or other mitigation option). Prioritization criteria (from the guideline) are used to provide a list of proposed replacements which contains information such as the type of asset, the asset ID number, the reason for replacement, the replacement cost and the expected completion date.
5. **Replacement Planning**: Based on the prioritized list of replacements, a proposal is developed for a mid-to-short term replacement plan, taking into consideration the

available budget and resources, and the impact on the network. This proposal is then submitted for approval and incorporated in the operational planning process.

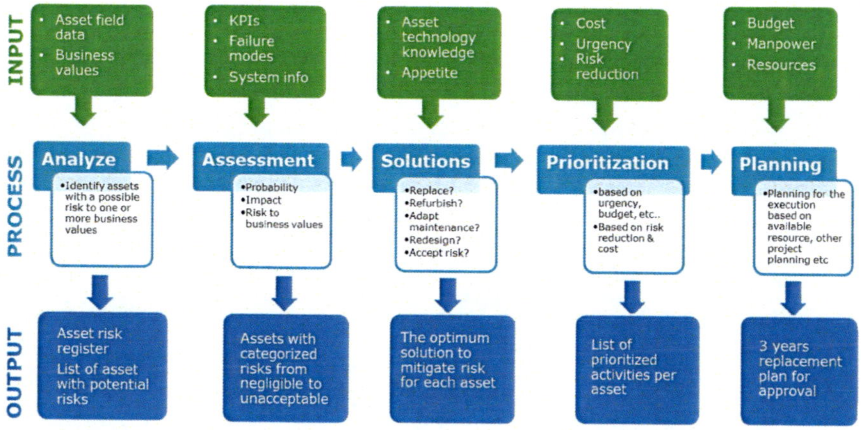

Fig. 2. Designing a risk-based decision-making process for asset replacement

3.4 KPI Setting and Risk Matrix: Capturing Risks in Terms of Probability and Consequence

The performance for each business value is measured against Key Performance Indicators (KPI's) defined by the company. Commonly, for each business value a company formulate targets (objectives). Well-designed KPIs serve as the foundation of the asset management policy and risk-based decision making. The challenge is to define one unique set of business values which is unambiguous and non-debatable. The business values are usually calibrated to one another such that one set of probability and impact classes serves all business values.

A risk matrix is a standardized risk management tool to understand the combined effect of the probability of a hazard (failure) and its impact on one or more business values. For asset-related risks, several methods allow to quantify probability and impact, one of them being the so-called "Asset Health and Risk indexing" [4–6]. In a risk matrix with probability and impact axes, every cell in the matrix corresponds with a combination of probability and impact and is given a so-called risk appetite (e.g. unacceptable, very high risk, high risk, medium risk, low risk, negligible). In [7], various utility risk matrices are shown.

3.5 Selecting the Optimum Solution

The first three (3) steps in the decision-making process (asset analysis, risk assessment and solution assessment) lead to the optimum solution for each specific risk. Before entering this process, it needs to identify assets with a possible risk that need to enter the process. This is done by replacement triggers, which are defined per asset type and per business value. The optimum solution may be any of the following alternatives as shown in Fig. 3:

1. Do nothing (accept the risk and operation as usual)
2. Adjustment of the maintenance
3. Refurbishment
4. Replacement
5. Redesign (note that this option cannot be applied to existing asset, only for the future assets)

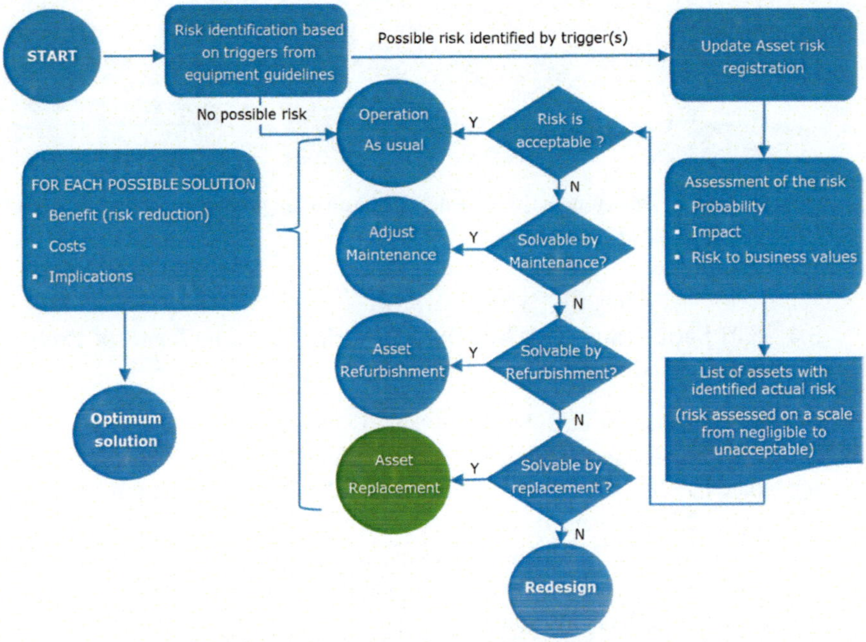

Fig. 3. Example of a process flow chart for asset analysis, risk assessment and solution selection

Based on the above flowchart, assets in the fleet are subjected to the stage of risk identification in order to identify the assets that exhibit a potential risk to business values. This identification is based on pre-set triggers. In many cases the assignment of triggers may be performed based on inspection and maintenance data on top of general asset information such as type/make/model/age etc.

After these potential risks have been identified and registered, a full risk assessment is performed by estimating the probability and impact of asset failure using a well-defined risk matrix.

The risk assessment requires all assets at risk to be positioned in this risk matrix based on the estimated probability and impact. Using the risk appetite (risk acceptance criteria), the asset can be categorized and prioritized for selecting the optimum solution.

For each asset at risk, the optimum solution is derived. The optimum solution depends on the amount of risk reduction, estimated costs and benefit of the solution (the cost effectiveness). The outcome will be: 1) the level of urgency and 2) the optimized solution for one specific risk.

When it comes to selection of replacement as an optimum solution, a proposal is developed for execution of the replacement with various internal and external stakeholders considering available resources, availability of the asset, and time to replacement.

This entire process can be repeated on a yearly basis, and commonly the replacement plan is revisited yearly to maximize the risk reduction with balanced cost and performance, and to account for delays in execution and for new risks arising during the year.

3.6 Prioritization and Investment Planning Optimization

After deciding on the list of assets for which replacement turns out to be the most effective solution, this list needs to be prioritized. The prioritization process is decided by the company on beforehand and is often governed by urgency, cost and effectiveness of the replacement (risk reduction per monetary unit). However, also other aspects may come in such as the availability of resources and grid innovation projects. The ultimate question is how to spend the available resources such that maximum risk reduction is achieved given the constraints.

For the long term, different investment scenarios may be analyzed and compared to arrive at an optimum long-term replacement strategy. In the course of time such a strategy needs to be monitored and improved to account for new information.

3.7 Improvement of Asset Management Performance

For TNB, the newly developed approach has led to a gain in quality and effectiveness. This improvement may be summarized as follows:

- The change from replacement based on fixed triggers to a risk-based replacement decision process has resulted in a smaller replacement volume and better prioritization. This will eventually result in improved performance at lower costs.
- The introduction of a two-stage replacement decision process has significantly enhanced the efficiency of the decision process. It results in a decrease of the number of risk assessments needed to arrive at an adequate replacement portfolio.
- The newly designed decision process has further resulted in a better trade-off between asset management actions (replacement, refurbishment, maintenance). As a result, the company resources are better utilized to achieve the desired grid quality.

4 Conclusion

In this paper we have presented a risk-based replacement process that is consistent with ISO 55000, and that allows to develop justified replacement programs. The approach involves a two-step approach; (a) Identifying assets with a potential risk, using triggers per business value, (b) A risk-based decision-making process to develop the replacement portfolio. When it comes to implementation, this replacement program requires various key ingredients such as high-quality asset data and well-defined organization's risk appetite, risk matrix, and estimated risk reduction per monetary value. Lastly, to deal with large amount of equipment, optimization of the portfolio for resource planning should be performed. The main benefits of this approach are:

1. The decision-making process is aligned with modern asset management standards, such as the ISO 55000 series, embracing and adopting risk-based decision making.
2. Risk based decision making allows to substantiate replacement to the regulator and other stakeholders, by not only showing the technical advantages but also the business benefits. In this way the regulator may be convinced of the necessity and may better understand the impact when not accepting the replacement budget.
3. Risk based decision is based on likelihood, impact and risk, and allows to prioritize replacements such that the overall risks to business values are at a minimum.
4. For TNB, the newly developed approach has led to a gain in quality and effectiveness.

References

1. International Organization for Standards (ISO), "ISO 55000: Asset Management – Overview, principles and terminology (2014), ISO 55001: Management Systems – Requirements (2014), ISO 55002: Management Systems–Guidelines for the application of ISO 55001 (2018)
2. Cho S, Nguyen K, Wetzer JM (2015) Conceptual design for asset management system under the framework of ISO 55000. In: CIRED, 23rd international conference on electricity distribution, Lyon
3. Cho S, Seow, KS, Wetzer JM (2016) Seven fundamental technical quality chapter for asset management system under the frame of ISO 55000. In: Proceedings 21st conference of the electric power supply industry (CEPSI)
4. Vermeer M, Wetzer J, van der Wielen P, de Haan, E, de Meulemeester E (2015) Asset management decision support modeling, using a health and risk model. In: PowerTech. IEEE, Eindhoven, pp. 1–6
5. Wetzer J (2018) Transformer health and risk indexing. Transform Mag 5(2):74–79
6. UK DNO Common Network Asset Indices Methodology. Health and Criticality. Version 1.1., January 2017
7. WG C1.34. ISO Series 55000 standards: Implementation and information guidelines for utility, CIGRE TB 787, December 2019

A Common Risk Framework for Road and Rail Infrastructures

H. Patrício[1(✉)] and N. Almeia[2]

[1] Infraestruturas de Portugal, Lisbon, Portugal
Hugo.patricio@infraestruturasdeportugal.pt
[2] Instituto Superior Técnico (IST), University of Lisbon, Lisbon, Portugal
nunomarquesalmeida@tecnico.ulisboa.pt

Abstract. Infrastructure consumes significant financial resources from public budgets, which are often restricted. It is thus increasingly important to optimize and justify long-term investments needs and operational expenses and set prioritization criteria for deteriorating infrastructures competing for funding. In this regard, public infrastructure organizations have been progressively adopting asset management principles and maturing techniques to demonstrate that their asset-related decisions, plans and activities are transparent and efficient. Those principles and techniques involve the balancing of cost, risk and performance and the demonstration that value is being derived from the infrastructure asset base. Risk-based approaches are popular in this regard as they translate in plain terms the societal impacts deriving from infrastructure failures. While some infrastructure organizations exhibit signs of excellence pushing the boundaries of sophistication and innovation, others use less formal approaches while applying and infusing risk-based thinking within the asset management functions of the organization. This paper deals with the application of a common infrastructure risk assessment process in a Portuguese infrastructure organization managing both the national road and railway transportation networks. It includes the presentation of a maturity model and the harmonization efforts towards a common language for the operational risks of bridges, tunnels, pavements, track, signaling and power stations assets. The challenges of establishing a common risk management framework for two very different asset networks are discussed, namely within a pilot-case study covering the maintenance function. The paper also covers the integration of the proposed infrastructure risk assessment process within the overall corporate risk management strategy and the asset management system of the organization.

Keywords: Transport infrastructures · Risk management · Asset management

1 Introduction

The increasing challenges related with aging infrastructures reaching the end of their useful lives are widely covered in the literature. Several authors refer to concerns arising from the deferral of critical investment and the subsequent accelerated decay of the infrastructures' condition compromising the levels of service [1]. Another common

A. Crespo Márquez et al. (Eds.): WCEAM 2019, LNME, pp. 113–125, 2021.
https://doi.org/10.1007/978-3-030-64228-0_11

theme is that capital investment rates must increase in the forthcoming decades [2, 3]. The key aspects leading to better decisions in optimum maintenance and operation of infrastructure under uncertainty environments have also been widely debated [4, 5].

Independently of the type of infrastructure and the asset management function at stake, it is consensual within the infrastructure asset management community that it is crucial to adopt a systematic approach to optimize and balance infrastructure performance, risk and cost, both in the short and long-terms [2, 6, 7]. The aim of such systematic approaches is to maximize the value generated from infrastructure assets, ensuring that resources are used where the benefits are greater, assisting infrastructure organizations to balance the economic, social and political impacts of their asset-intensive business [8, 9].

This paper describes an empirical case of a public infrastructure organization managing both the Portuguese national road and railway networks. This infrastructure organization involves a wide range of stakeholders with competing interests [10] while seeking ways of providing reliable, safe and sustainable transportation systems. It presents a proposal to optimize and justify investments needs and operational expenses and set prioritization criteria for two different types of transportation networks competing for funding. The proposal is inspired in asset management principles and techniques developed to demonstrate that asset-related decisions, plans and activities are transparent and efficient. Without losing site of the importance of the balancing of cost, risk and performance and the need to demonstrate that value is being derived from the infrastructure asset base, it focuses on the application of a common infrastructure risk assessment process for the two transportation networks. The challenges of establishing a common risk management framework for two very different asset networks are discussed, namely within a pilot-case study covering the maintenance function of bridges, tunnels, pavements, track, signaling and power stations assets, among others. The paper also covers the integration of the proposed infrastructure risk assessment process within the overall corporate risk management strategy and the asset management system of the organization.

2 Method

2.1 Maturity Model for the Risk Management Process

Organizations often use maturity models that are designed to act as drivers of continual improvement. The authors use the guidance of the ISO 55000 series and ISO 31000 to proposes a maturity model based on qualitative indicators with the purpose of enhancing risk management within asset management activities of large-scale infrastructure organizations with diverse asset portfolios. The underlying reasoning is that the more rapidly effective infrastructure risk management is achieved, the more efficiently will the asset management goals be realized. This is in line with the recognition that risk management facilitates continual improvement of the organization (ISO/TR 31004).

The maturity model uses a 6-level scale that maps the progressive implementation of a complete ISO 31000-based risk management process:

- 0 | Innocent: The organisation has not recognised the need for this requirement and/or there is no evidence of commitment to put it in place;

- 1 | Aware: The organisation has identified the need for this requirement, and there is evidence of intent to progress it.
- 2 | Developing: The organisation has identified the means of systematically and consistently achieving the requirements and can demonstrate that these are being progressed with credible and resourced plans in place.
- 3 | Competent: The organisation can demonstrate that it systematically and consistently applies the principles set out in ISO 31000.
- 4 | Optimizing: The organisation can demonstrate that it is systematically and consistently optimizing its Risk management within Asset Management practice, in line with the organisation's objectives and operating context.
- 5 | Excellent: The organisation can demonstrate that it employs the leading practices and achieves maximum value from the management of its assets, in line with the organisation's objectives and operating context.

2.2 Components of the Risk Management Process

The proposed process includes an adapted version of all the core elements of the ISO 31000 risk management process (see Fig. 1): establishing the context, risk assessment (risk identification, risk analysis and risk evaluation) and risk treatment. It also covers the non-core element the ISO 31000 risk management process, namely communication and consultation, monitoring and review and recording and reporting.

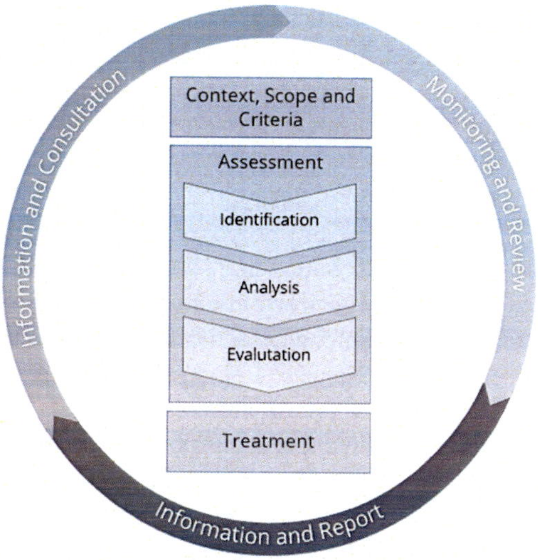

Fig. 1. Core and non-core components of ISO 3100 risk management process

In ISO 31000-based risk management processes, risk can be expressed in terms of risk sources, potential events, their consequences and likelihood. Such a process can be used to support asset management decisions and inform the effects of uncertainties on the organizational, asset management and asset portfolio objectives. ISO 31000-based risk management processes are particularly suitable in operational contexts where the ISO 55000 series of standards on asset management are also in place. The authors have found that this is particularly the case when road and rail networks with multiple types of asset are involved [11].

3 Empirical Case

This proposed components of the risk management process and the monitoring of its implementation through a maturity model is currently being tested in Infraestruturas de Portugal (IP). IP is the infrastructure management organization of both road and railway networks with different asset classes (e.g. bridges both rail and road, rail track, rail signalling, rail power, road) that compete for funding, giving rise to the need of having a common language and framework to support decision making.

Infraestruturas de Portugal (IP) is a public organization that resulted in 2015 from the merger of road and railway administrations. It is responsible for the construction, operation and maintenance of the national rail and road networks. The assessment of the organization's performance corresponds to the level of IP's response to the needs and expectations of various relevant stakeholders. An integrated management system is in place to ensure that these are met. This integrated management system adheres to the requirements established in international standards on quality management, environmental management, occupational safety and health management and asset management.

IP has a corporate risk management process in place, which is allocated to a shared corporate service, as highlighted in Fig. 2 ("Risk and Compliance"). This corporate risk management process was initially designed to cover 5 major risk categories: business, reputational, technological, financial and regulatory & compliance. But it originally excluded operational risks arising from the infrastructure, the management of which is currently allocated to a (Physical) Asset Management department, as highlighted in Fig. 3.

The maintenance function of IP is used as empirical case for testing the maturity model and the components of the proposed risk management process. Asset-related information (e.g. operational context and asset knowledge) is the starting point for the yearly reviewed maintenance plan with two time-frames i) short term (5-year horizon), in which the most critical maintenance activities are identified and ii) long term (20-year horizon), in which the financial resource levels are estimated, taking into account the current asset condition and degradation models (see Fig. 3).

The prioritization of maintenance actions in IP is based on multi-criteria analysis and techniques that implicitly reflect the concerns and objectives of the infrastructure organization. However due to operational and financial constraints, it is not possible to deliver all maintenance activities according to the needs, and thus the resulting backlog is also monitored.

4 Results and Discussion

4.1 Continuous Improvement of Infrastructure Risk Management

The actions taken in IP to address the risks (and opportunities) associated with the combined management of road and rail infrastructure assets include the monitoring

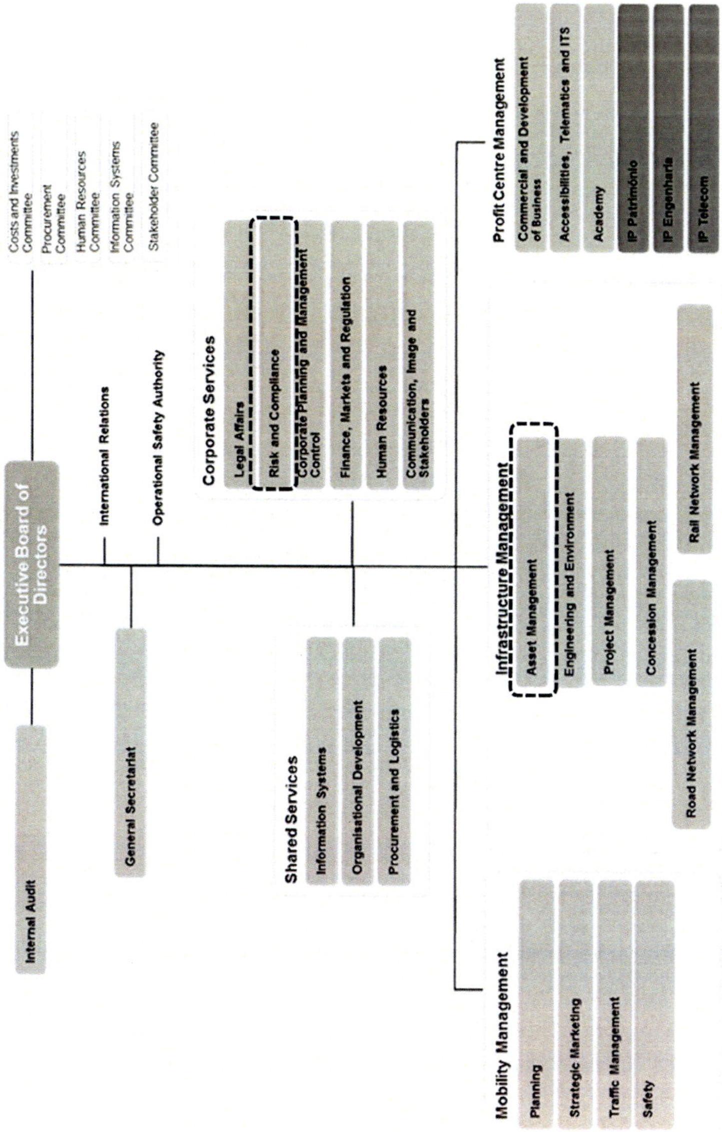

Fig. 2. Allocation of corporate risk management and infrastructure risk management

of the progressive implementation of the ISO-31000 based risk management process. Figure 4 presents the intended target levels of maturity achieved in a period of four successive years.

Table 1 presents the main efforts that were planned and implemented in each of those years.

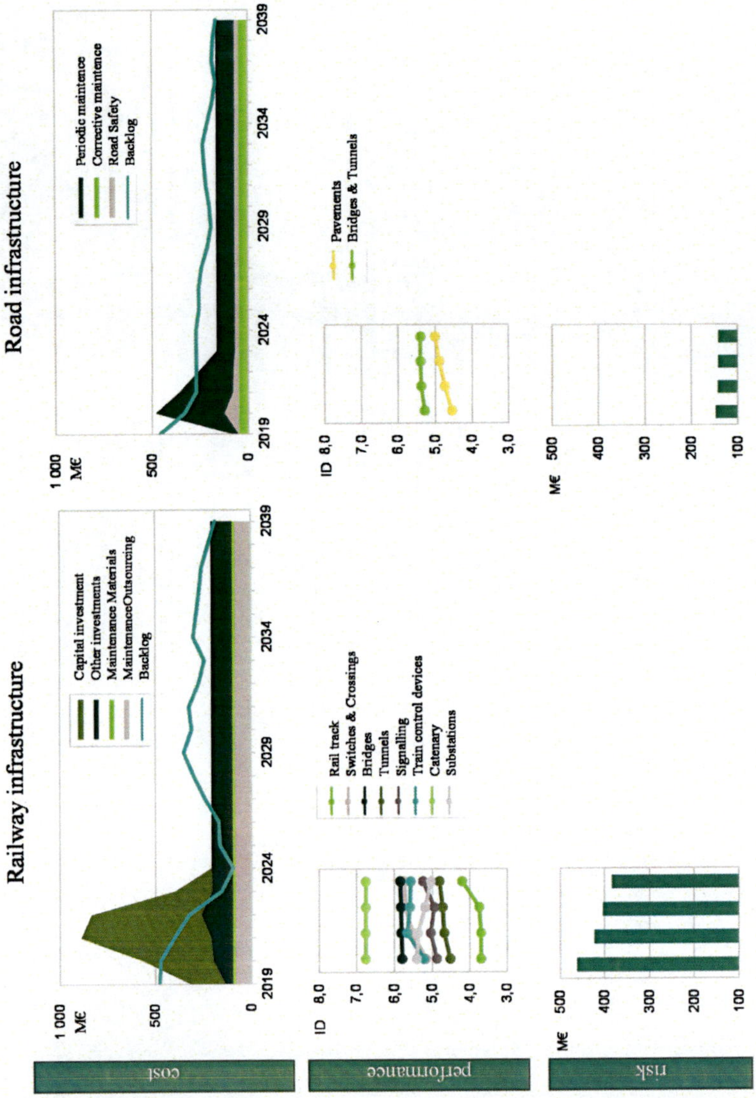

Fig. 3. Executive summary of the infrastructure maintenance plan.

4.2 Core Outputs of the Risk Management Process

The (infrastructure) risk management process described in this study is complementary to the corporate risk management process that is already in place in the organization. Both processes follow the ISO 31000 format. The scope of application of the proposed (infrastructure) risk management process is the road and rail transportation networks and its infrastructure elements that work as systems to perform as required (Fig. 5).

The various asset classes covered by the process differ from each other. Constructed assets (e.g. buildings and retaining walls) are technologically very different from train control software (e.g. electronic interlocking and signalling systems) and mechanical

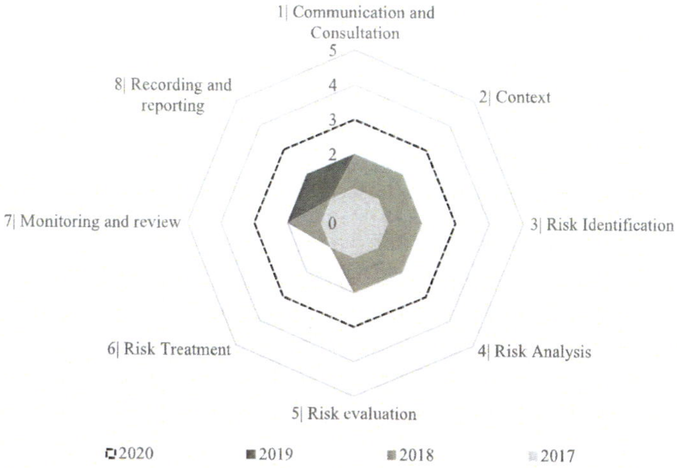

Fig. 4. Yearly maturity target for the components of the risk management process.

Table 1. Main developments for managing the risk of road and rail infrastructure assets.

Year	Components addressed	Main developments and applications
2017	Context Identification Analysis Evaluation Treatment	Infrastructure risk categories Scope (asset classes): bridges and tunnels (road and rail), road pavements, rail track, rail switches & crossings Risk treatment for assets flagged for corrective maintenance Risk level metrics (probability and consequences) Scenario analysis
2018	Context Identification Analysis Evaluation Monitoring and review Communication	Revised scope (asset classes added): rail signaling, rail power supply, road safety Enhanced risk analysis combining pavement and road safety data Emphasis on assets flagged for corrective maintenance
2019	Context Identification Analysis Evaluation Recording and reporting	Monitoring and review of full asset portfolio risk In-depth discussion with top management about risk communicating (e.g. risk levels vs risk cost)

systems (e.g. control of rail track devices). Likewise, spatially speaking, discrete assets (e.g. bridges, power stations) differ from continuous assets (e.g. road pavements, rail track).

Asset management translates organization's objectives into asset-related decisions, plans and activities using a risk-based approach [12]. This study describes the application of the risk management process to maintenance decisions, plans and activities (see Fig. 1).

Historically, road and rail maintenance activities have been delivered by dedicated teams, grouped by asset classes with particular cultures. Table 2. exemplifies the challenge of comparing technical information deriving form siloed activities into a common language, plus translating it to non-technical audiences. It is worth noting that the condition indicators used in the organization, for different asset classes, are fully aligned with internationally recognized best practices and technical standards.

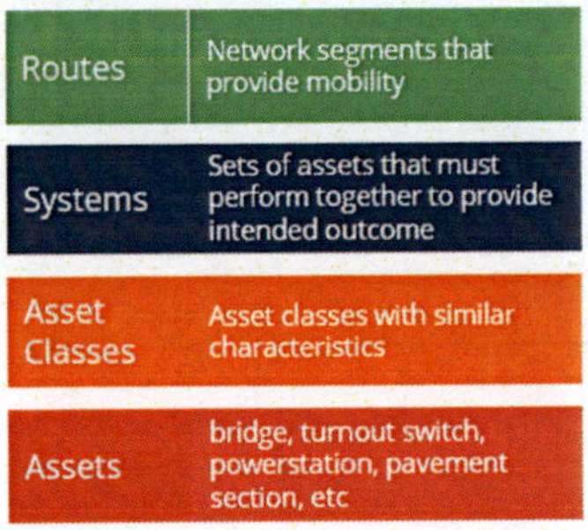

Fig. 5. Asset hierarchy levels and whole life-cycle phases

Table 2. Examples of condition indicators used across asset classes

	Road bridges	Rail bridges	Rail tunnels	Rail track	Road pavements
Best	EC0	IA0	Level I	QN1	5
Worse	EC5	IA4	Level IV	QN3	0
Scale type	Discrete	Discrete	Discrete	Discrete	Continuous

Although the infrastructure organization is experienced with the ISO 31000 risk management process at the corporate level, there was a gap to be filled regarding risks that can arise from operational activities, namely those related to infrastructure assets (see Fig. 2).

A panel of experts identified five standard risks to be considered under the scope of operational risk, as relevant to the asset management objectives, namely (Fig. 6): i) major accident (ACID), ii) unavailability (DISP), iii) level of service reduction (LOS), iv) increasing OPEX, v) compliance (COMP).

The risk analysis currently covers the complete portfolio of the most relevant asset classes for both road and railway infrastructures (bridges and tunnels, rail track, rail signalling, rail power supply, road sections). In 2019, more than 15500 assets were considered, and more than 70 000 risks analysed.

The current situation (Table 3) is that nearly 2000 individual assets constitute a significant source for more than 5000 risks.

The results of the risk analysis are summarized in Fig. 7. and Fig. 8 provides insights into the risk levels resulting from different maintenance budgets in the forthcoming 5-year period, namely by considering that asset condition degrades over time.

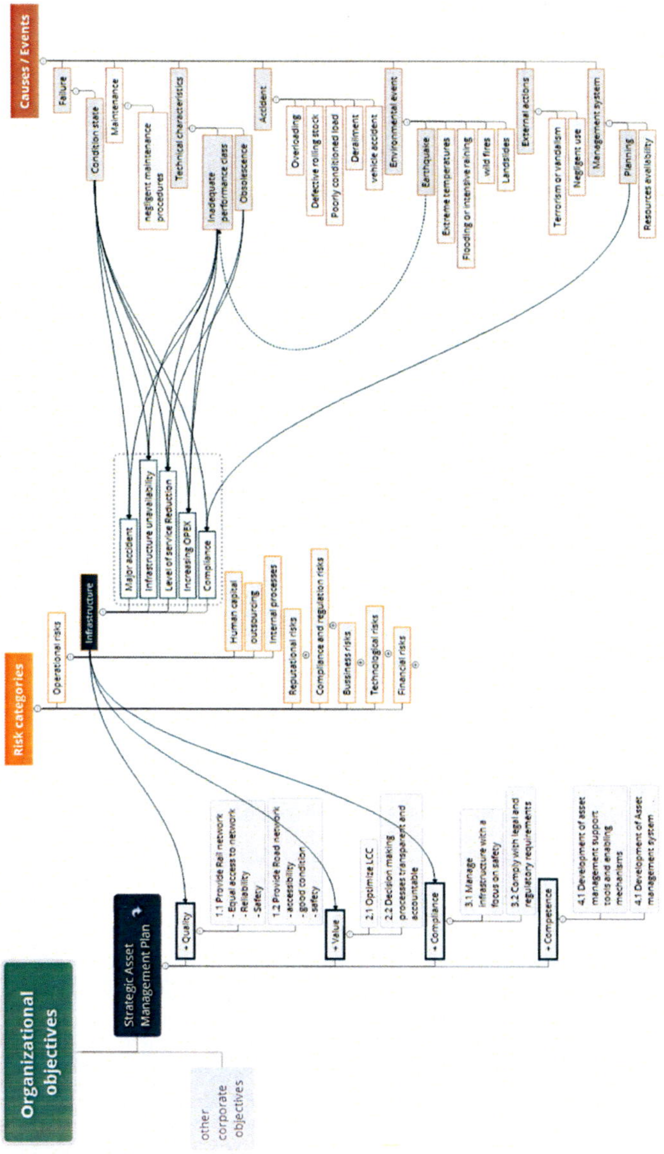

Fig. 6. Linking operational risks categories to organizational and asset management objectives.

Table 3. Asset risk sources having at least one moderate risk (2019)

	Rail	Road	Total
Risk sources	639	1 230	1 869
Risks	2 767	2 665	5 432

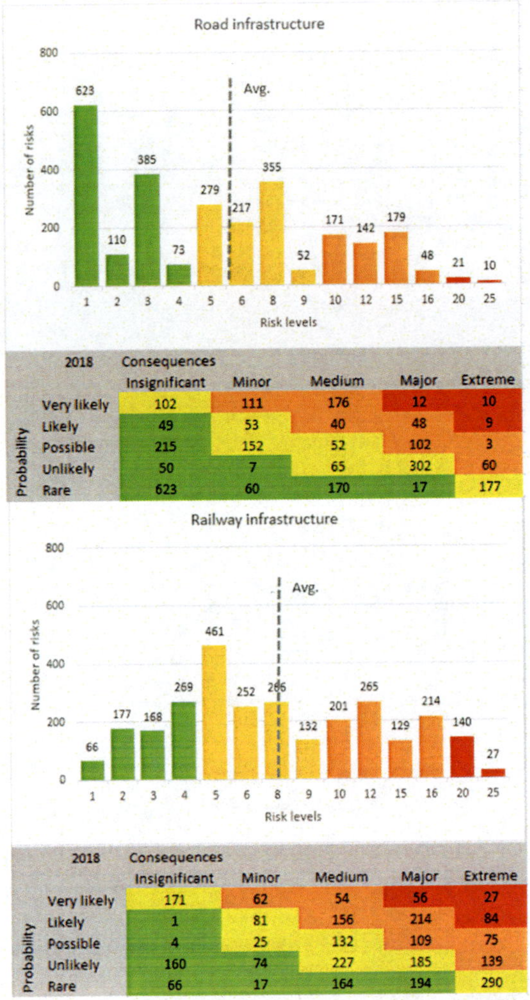

Road infrastructure

2018	Consequences Insignificant	Minor	Medium	Major	Extreme
Probability: Very likely	102	111	176	12	10
Likely	49	53	40	48	9
Possible	215	152	52	102	3
Unlikely	50	7	65	302	60
Rare	623	60	170	17	177

Railway infrastructure

2018	Consequences Insignificant	Minor	Medium	Major	Extreme
Probability: Very likely	171	62	54	56	27
Likely	1	81	156	214	84
Possible	4	25	132	109	75
Unlikely	160	74	227	185	139
Rare	66	17	164	194	290

Fig. 7. Summary output of the risk analysis.

The expected cost for each risk is estimated using the relationship between impact levels and financial consequences and the likelihood of occurrence. Estimations consider both assets in a good (baseline) condition and in a deteriorated (real) condition (see Fig. 9). These estimates allow the calculation of a risk indicator at the network level, conveying the idea that even for an infrastructure in good condition there is always an inherent risk that cannot be totally eliminated. It is degraded asset that add risk aggravation factors that must be controlled and modified as needed.

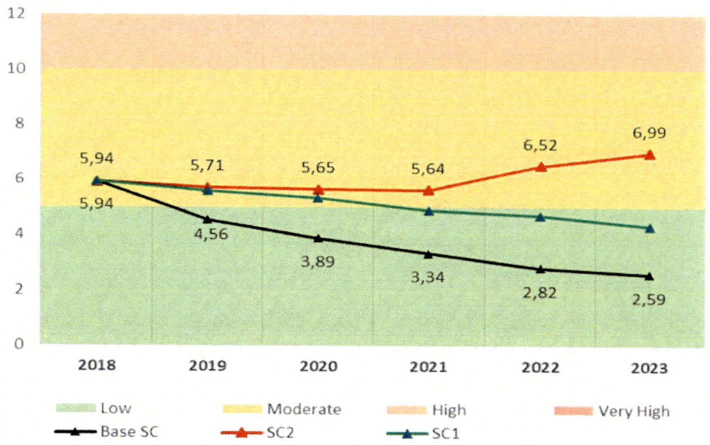

Fig. 8. Average infrastructure risk level (low to very high) for different budget scenarios (base SC, SC1 and SC2).

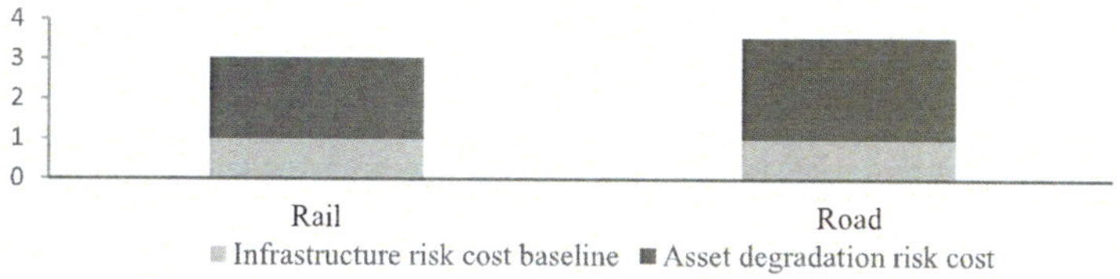

Fig. 9. Current expected risk cost due to asset degradation compared with risk cost baseline (infrastructure in good condition)

4.3 Non-core Outputs of the Risk Management Process

The core-elements of the risk management process (establishing the context, risk identification, risk analysis, risk evaluation and risk treatment) are to be complemented with the components of communication and consultation, monitoring and review and recording and reporting. Each of these non-core components also plays a significant role. It is worth stressing that, for example, the consultation of a panel of experts was needed to develop the risk map presented in Fig. 6 and that a strategy had to be discussed to communicate risks to a non-technical audience of external stakeholders. Figure 10 provides an example of the latter. It was also found that effective risk communication should take into account such factors as the target audience, social perception and other human psychological aspects that might arise the need to deliver the results accordingly albeit based on the same data.

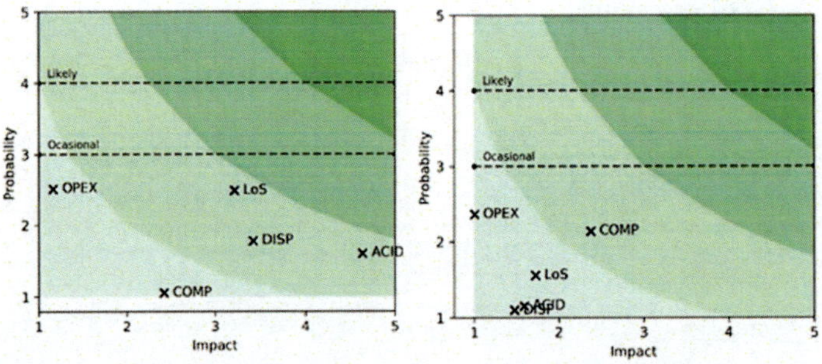

Fig. 10. Average risk level aggregated for each risk subcategories for rail (left) and road (right) networks.

5 Final Remarks

In transport infrastructures such as road and rail, large asset portfolios and asset systems must perform as required in order to secure network and route objectives and deliver value to the organization and its stakeholders. The types of assets owned by road and railway infrastructure organizations can range from real estate property (land, buildings and other facilities), to structures (bridges, tunnels and retaining walls), rail superstructure (rails and sleepers), pavements, electrical signalling, level crossings, etc. These assets are expected to work as a system and deliver an infrastructure with the required levels of service. But the performance of assets varies throughout time and depends greatly from use conditions. Moreover, most asset classes in engineering infrastructures are critical from the operations point of view, meaning that if a failure occurs, the network route becomes unavailable until further actions are accomplished. Understanding the operational risks of these infrastructures are key issues for asset management decision-making, as they may influence life cycle management decisions such as renewing life-expired assets or defining maintenance strategies, amongst others.

This paper presents a risk-based approach established in view of setting a common language across networks and asset classes to support decision making processes, namely when scenario analysis is needed due to financial constraints. These approach links corporate risk management with 5 operational infrastructure risk categories: i) major accident, ii) unavailability, iii) level of service reduction, iv) increasing OPEX, v) compliance.

A maturity model was used to map the implementation progress of all the components of a risk management process based on ISO 31000. The process enabled a high-level language that was useful to bridge different cultures in a large and complex organization and to support asset management decision for both road and rail infrastructure assets. Furthermore, it was found that effective risk communication should take into account such factors as the target audience, social perception and other human psychological aspects that might arise the need to deliver the results accordingly albeit based on the same data.

The proposed process needs further embedment in decision making processes of the various asset management decisions at different levels of the organization and further involvement from multiple organization units is needed to taking into account how risks (and opportunities) can vary in time. The process currently supports the maintenance decisions, plans and activities in IP, but it has been noted that the maintenance function is limited in its capacity to modify significantly a considerable amount of infrastructure risks.

Finally, the scope of this process might contribute to applications beyond maintenance, e.g. the risk evaluation process might result recommend to reconsider asset management objectives, negotiate stakeholders' requirements to deal with compliance risks or to address some risks within research and innovation plans.

References

1. Francisque A, Tesfamariam S, Kabir G et al (2017) Water mains renewal planning framework for small to medium sized water utilities: a life cycle cost analysis approach. Urban Water J. https://doi.org/10.1080/1573062X.2016.1223321
2. Rokstad MM, Ugarelli RM (2015) Minimising the total cost of renewal and risk of water infrastructure assets by grouping renewal interventions. Reliab Eng Syst Saf. https://doi.org/10.1016/j.ress.2015.05.014
3. World Economic Forum (2014) Strategic Infrastructure Steps to Operate and Maintain Infrastructure Efficiently and Effectively
4. Sánchez-Silva M, Frangopol DM, Padgett J, Soliman M (2016) Maintenance and operation of infrastructure systems: review. J Struct Eng. https://doi.org/10.1061/(ASCE)ST.1943-541X.0001543
5. Wijnia YC (2016) Processing Risk in Asset Management. TU Delft
6. PIARC Technical Committee D.1 Asset Management (2019) Innovative Approaches To Asset Management. World Road Association (PIARC)
7. UIC (2016) Railway Application Guide Practical implementation of Asset Management through ISO 55001
8. Van Der Lei T, Herder P, Wijnia Y (2012) Asset management: the state of the art in Europe from a life cycle perspective. Springer, Cham
9. IPWEA (2015) International infrastructure management manual 2015, 5th ed. Institute of Public Works Engineering Australasia (IPWEA)
10. Hugo P, Nuno A (2016) A framework for evaluating the performance of infrastructure assets. Application to the life-cycle of road and railway bridges. In: Life-cycle of engineering systems. CRC Press, Taylor & Francis Group, Boca Raton, pp 1743–1748
11. Patricio H (2019) Risk based approach in maintenance planning in the context of road and railway infrastructure. In: IABSE Symposium of Guimaraes 2019 Towar a Resilient Built Environ Risk and Asset Management, pp 1418–1427. https://doi.org/10.1080/10168664.2019.1624344
12. ISO 55001 (2014) Asset Management - Management Systems - Requirements
13. ISO 31000 (2008) Risk management — Guidelines
14. ISO/TR 31004 (20013) Risk management — Guidance for the implementation of ISO 31000

Development of a Resilience Management Framework Adapted to Complex Asset Systems: Hydro-Québec Research Chair on Asset Management

Georges Abdul-Nour[1](✉), François Gauthier[1], Ibrahima Diallo[1], Dragan Komljenovic[2], Raynald Vaillancourt[2], and Alain Côté[2]

[1] University of Quebec at Trois-Rivieres, Trois-Rivieres, Canada
{georges.abdulnour,farncois.gauthier,Ibrahima.diallo2}@uqtr.ca

[2] Institut de recherche d'Hydro-Québec (IREQ), Montreal, Canada
{komljenovic.dragan,cote.Alain7}@ireq.ca,
Vaillancourt.raynald@hydro.qc.ca

Abstract. Complex asset systems are essential to the functioning of our society. They include but are not limited to: oil platforms, transportation networks, water and gas distribution systems, nuclear power plants, the stock exchange system, communication networks and electrical power utilities. Electrical power generation, transport and distribution utilities such as Hydro-Québec (HQ) have significant asset portfolios. In every step of the chain of production, electrical utilities must know the condition, location and availability of their assets to maximize productivity, reduce service interruptions and ensure the safety of operations and users. When their assets are managed effectively, they can conduct monitoring, obtain vital information and stimulate performance to improve it. This integrated management, commonly known as asset management, allows them to fully exploit the potential of their assets and to convert asset management into a powerful competitive advantage. There are many asset management challenges for electrical utilities: infrastructure maintenance and updating, the obligation to adapt and prepare for all types of changes (standards, types of client, environmental, etc.), new market development, etc. Asset management is considered an efficient approach available to electrical utilities to make better investments in their assets. In this regard, R&D and innovation are crucial. For this reason, HQ, in partnership with UQTR and NSERC, developed and launched a research chair on asset management. The Chair's research will be primarily focused on developing asset management concepts, optimization algorithms and generic models, as well as adapting and transferring results to HQ and other major Canadian electrical utilities. One of the unique features of this Chair is the development of a new integrated, up-to-date and comprehensive asset management paradigm. This paper introduces the background, the key issues and the detailed research proposal for this Chair. One of the leading research projects for this Chair, regarding the development of a resilience management framework, is presented. A resilience case study is presented as an example at the end of this paper.

A. Crespo Márquez et al. (Eds.): WCEAM 2019, LNME, pp. 126–136, 2021.
https://doi.org/10.1007/978-3-030-64228-0_12

Keywords: Asset management · Complexity · Modelling · Resilience · Risk · Uncertainty · Decision making · Complex system · Extreme and rare events · Electrical grid · Climate change

1 Background and Key Issues

Electrical utilities such as HQ must constantly find strategies for remaining competitive in a sector characterized by transitions, uncertainties, turbulence and worldwide competition. Furthermore, achieving the objectives imposed by the challenges of sustainable development and increasingly stringent environmental regulations cannot be neglected. HQ has power generation and distribution infrastructure in increasingly complex technological systems. Its production business is primarily hydropower; it produces more than 37,000 MW. Its transmission network totals close to 35,000 km and the medium-voltage distribution system is approximately 120,000 km. Its assets include modern technological elements that are difficult to integrate into their lifecycle planning and optimization framework. It is therefore essential to distinguish the physical life from the useful life or asset utilization and also to take into consideration the technological obsolescence that may result from software or material incompatibility between generations of equipment, etc. In this context, it is essential for the partner to ensure optimal asset sustainability; asset management will be an advantage.

However, asset management is traditionally performed in silos in different departments in the company. For example, during the acquisition phase, the focus is on implementing a technology within available budget limits and timeframes, while ensuring compliance with technical specifications. The primary goal is to ensure demand is met, while guaranteeing asset availability and reliability through maintenance strategies [1], etc. However, this approach does not capture the complexity of technology-driven companies such as HQ. To that end, asset management requires a paradigm shift to improve and maintain asset value. This leads to a global strategic approach that covers the entire asset lifecycle and integrates other concepts. This is the objective of the research programming presented herein.

In an increasingly complex world, achieving companies' reliability, resiliency and robustness criteria requires considering and integrating numerous interacting factors that result from different layers: technical, technological, environmental, social, organizational, etc. Furthermore, the asset management of electricity companies is particular; these systems are very complex because they must simultaneously consider the dynamics of market demand and production technology [2]. This Research Chair aims at developing a new paradigm based on a set of interrelated concepts. This is necessary to achieve the objective of this project. Thus, our reference framework is broken down based on the following concepts.

- The concept of the asset lifecycle: Ouertani, Parlikad and McFarlane [3] define the asset lifecycle as the succession of five phases: acquire, deploy, operate, maintain and retire at the end of life.
- The concept of the impact of external factors and extreme events: Classic approaches cannot sufficiently analyze these risks. The literature recommends addressing these

types of events and developing appropriate approaches to define them, including resilience management [4–8]. According to Komljenovic et al., given the complexity of the asset management business and operational context, there is a growing risk of extreme and rare events [8].

- The concept of risk management: It is crucial that risk management is considered when making a decision on asset management [9]. Standard ISO 55001 [10] covers the requirement of handling risks. Standard ISO 55000 [11] refers to standard ISO 31000 [12] (Risk Management Standard) on risk management for analyzing and managing risks in asset management.
- The concept of performance measurement: Most research concentrates on performance measurement in asset management systems from a maintenance perspective [13, 14] and rarely from a global perspective. Maletic et al. [15] examined the role of maintenance with a view to improving corporate competitiveness and profitability.
- The concept of decision making: Theron [16] recently proposed a multi-criteria decision-making tool in asset management, particularly decisions to repair/replace physical assets. Maletic et al. [15] examined the role of two contingency factors, namely uncertainty and competitiveness, in asset management practices. The authors concluded that organizations faced with a high level of uncertainty and competitiveness are more engaged in deploying asset management practices. Khuntia et al. [17] moreover presented a literature review on the asset management of electricity distribution systems. In the same vein, Reddy [18] addressed the main challenges of and opportunities for improving decision-making in asset management of electricity distribution networks.

The literature review reveals the need for a global model of asset management that takes into account the complexity of electrical utilities, which are modern technological companies with a business and operational context that simultaneously considers the dynamics of market demand and production technology. This research Chair aims at filling up the gap in the field of Asset management. Section two will present the detailed research project proposal. A HQ based case study is presented in section three. Section four presents a conclusion and expected results.

2 Detailed Research Project Proposal

2.1 Project Objectives

Chair activities are planned over three years and renewable for two years. These activities consist in modelling and optimizing the global asset management process, optimizing the asset lifecycle and modelling the impact of rare and extreme events and external factors on asset management. The long-term objective is to make a significant contribution to the development and integration of a new integrated, up-to-date and global asset management paradigm.

2.2 Themes of the Asset Management Chair

This project is structured around three main themes:

Theme 1. Modelling and optimizing the global asset management process: Objective. Developing a global and integrated concept of design, optimization and decision- making in the asset management process.

Asset management is an integrated, complex and multidisciplinary activity that involves collaboration and coordination between several departments in the company. Our research is aimed at developing a global and integrated concept of the optimal asset management strategy: acquire, implement, operate, maintain and retire at end of life.

Theme 2. Optimization of the asset lifecycle: Objective. Developing a global methodology for modelling and optimizing the asset lifecycle in an asset management context.

The lifecycle is usually analyzed based on components in which an overall "system" vision/approach is missing. The project is aimed at developing integrated approaches in order to address this weakness, making it possible to optimize the entire lifecycle of technical systems and their constituent components.

Theme 3. Modelling the risks of extreme events and external factors in complex asset management: Objective. Developing a global methodology for modelling the impact of extreme or rare events and external factors on the asset management strategy.

It is crucial to take risk management into consideration in asset management decision-making. In a complex business and operational asset management context, the risks of external factors as well as extreme and rare events have a significant impact on company performance [8]. However, classic risk analysis approaches are not sufficient for the analysis of this type of event in asset management. Electrical power generation, transport and distribution systems represent critical infrastructures that are tightly connected in networks with numerous components that interact with each other in a nonlinear manner, and can evolve in dynamic ways that are neither completely regular nor fully random. In these systems, undesirable events often occur due to lack of understanding of the interactions within their structure. The gaps in risk management approaches under deep uncertainty in the context of complex systems further contribute to overall uncertainties. Furthermore, external factors (natural and human) are having an increasingly significant and tangible impact on the performance of the asset management process. This problem has hardly been studied. It is therefore necessary to develop new methods for identifying and including these factors in the decision-making process [2, 5–8].

To overcome these issues, recent scientific publications propose the adoption of resilience management in addition to risk management. Resilience management is recognized as an approach that palliates uncertainties and mitigates the consequences of risks regardless of their unpredictability. It improves a system's ability to recover from a disruption. Thus, one of the leading research projects for this Chair aims at proposing a resilience management framework applicable to complex asset systems, such as an electrical utility, by integrating it within asset management. The objective is to help the decision-making process by choosing from the strategies of (i) traditional risk management, (ii) risk management under deep uncertainty and (iii) resilience management, for coping with disruptions that may affect a complex asset system. The various risk management tools and resilience management approaches applicable to complex asset systems are identified and categorized. The choice of appropriate strategies is based on the characteristics of the system and the uncertainties associated with the potential disruptions. In the next section, a case study of the application of the proposed decision-making framework to a medium voltage overhead electric power distribution network is presented.

3 Case Study

3.1 Background

Problem: Given the uncertainties surrounding climate change and its impacts on electrical utilities, the acceptability of the knowledge base and assumptions, the characterization of asset system behaviour, the needs and expectations of stakeholders as well as the types of interruptions to include in risk analyses create challenges in asset-management decision making. **Research question:** Which strategy traditional risk management, risk management under uncertainty and resilience management is the most appropriate for handling the impacts of climate change on electric power distribution networks? **Objective:** develop a decision-making framework (under risk, uncertainty or resilience) in the context of electrical utilities. **Methodology:** Use information from the literature review to classify the types of systems and their interruptions that influence the choice of the management under risk, uncertainty and resilience strategy. **Result:** A decision-making framework is proposed and its application is taken from a representative case study from the electrical distribution network of Hydro-Québec through a cost/benefit analysis and knowledge bases.

3.2 Moving from Risk Management to Resilience Management Within Complex Systems

Risk is defined most often as the product of a prejudice by its probability of occurrence [19]. Risk management itself provides information for decision making through analyzing and assessing the risk [5].

As a result, risk assessment consists in identifying and quantifying all the possible interruption (danger or threat) scenarios. Within complex systems, interruptions may be in large part unknown [20, 21]. Quantifying the probability is often impossible [20, 22]. In the above-mentioned context, it is appropriate to develop methods for facing the complexity of systems as well as their uncertainties [23]. The next sections are divided as follows: Sect. 3.3 presents the methodology adopted, Sect. 3.4 addresses the design of the decision making under risk, uncertainty and resilience framework and illustrates the application.

3.3 Methodology

The methodology presented in Fig. 1 was followed to develop a resilience management framework integrated into asset management and adapted to complex systems. Literature review helped synthesis the relevant parameters to consider in decision making between traditional risk management, management under uncertainty and resilience management, and lead to the development of a framework. Application of this framework to a case study taken from Hydro-Québec Distribution was done by Modelling costs of major outages related to extreme weather events.

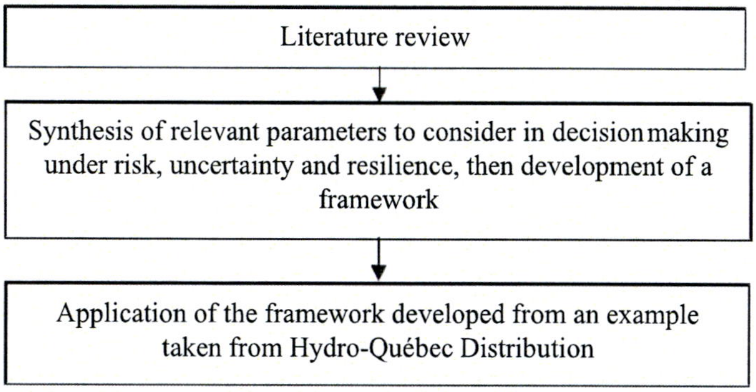

Fig. 1. Methodology structure

In this framework, decision making is divided into five elements, namely: scientific evidence, knowledge bases, global assessment of risks and uncertainty, review and judgment of management, then decision making [24, 25]. On this basis, Fig. 2 shows the steps of the resilience management framework.

3.4 Hydro-Québec Based Case Study

The Quebec government is Hydro-Québec's sole shareholder [24]. It is one of the largest North American electricity generation, transmission and distribution companies. A few facts about the company are [24, 25]: 4.3 million customers; CAD 77 billion in assets including 63 hydro-electric stations and 24 thermal power stations; 34,361 km allocated to the transmission network and 181,130 km to the medium voltage distribution network; CAD 148 million R&D budget; generation capacity of 37,310 MW; more than 99% of its electricity generation is exclusively from renewable energy sources.

Applying the Framework Developed. Assumptions: A1: historical data cannot represent future events in the context of climate events but burying lines is a long-term sustainable solution for handling these events. Transition from resilience to reliability.

Modelling Costs of Major Outages Related to Extreme Weather Events. For the period from 2009 to 2019, the following conditions are verified in the context of outages in Hydro-Québec's overhead electrical distribution service: major outages are continuous (there were continuously costs associated with major outages each year in the period from 2009 to 2019); they are uniformly distributed (annual distribution); the costs represent the maximums recorded during distribution service outages; the causes of these outages are assumed to be independent because they are related to weather events.

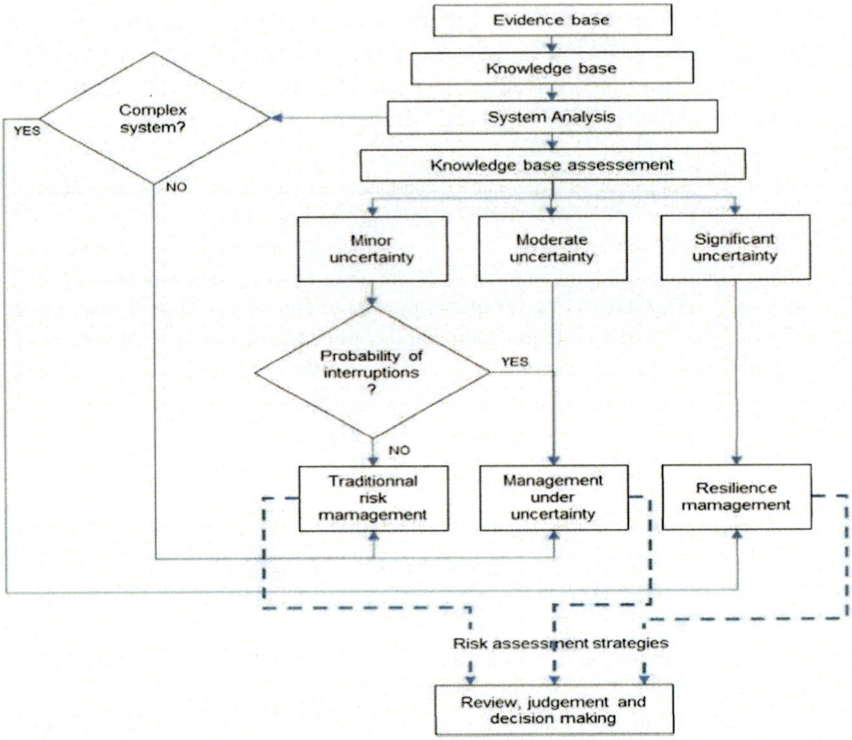

Fig. 2. Steps in the decision-making under risk, uncertainty and resilience framework

Thus, these annual investment costs in major outages are subject to Generalized Extreme Value (GEV) distributions that are used to formulate rare probabilistic events [26].

Namely x_i= random variable that represents the total investment of the electrical utility in major outages for the year *i*; with $i = 2009$ to 2019. The generalized form of the function of the distribution of this type of random variable is as follows [26–28]:

$$F(x_i; \mu, \sigma, \xi) = exp\left\{-\left(1 + \frac{\xi(x_i - \mu)}{\sigma}\right)^{-\frac{1}{\xi}}\right\} with\, 1 + \frac{\xi(x_i - \mu)}{\sigma} > 0; \quad (1)$$

$$\sigma > 0;\; x_i, \mu \text{ and } \xi \text{ defined on } \mathbb{R}$$

Where μ, σ, ξ respectively represents the location, scale and shape parameters.

The greater the number of parameters, the greater the level of uncertainty. Accordingly, by reducing the number of parameters, uncertainties may also diminish.

The analysis of the values of the distribution shape parameter $F(x_i; \mu, \sigma, \xi)$ is summarized as follows [29]: $\xi < 0$ (Weibull's distribution), $\xi = 0$ (Gumbel's distribution) and $\xi > 0$ (Fréchet's distribution). The analysis of the use context of each of these distributions makes it possible to suggest the following assumption $\xi = 0$.

We have:

$$\lim_{\xi \to 0} F(x_i; \mu, \sigma, \xi) = \lim_{\xi \to 0} exp\left\{-\left(1 + \frac{\xi(x_i - \mu)}{\sigma}\right)^{-\frac{1}{\xi}}\right\} \tag{2}$$

$$\text{However}\left(1 + \frac{\xi(x_i - \mu)}{\sigma}\right)^{-\frac{1}{\xi}} = exp\left\{ln\left(1 + \frac{\xi(x_i - \mu)}{\sigma}\right)^{-\frac{1}{\xi}}\right\} \text{because } 1 + \frac{\xi(x_i - \mu)}{\sigma} > 0$$

$$= exp\left\{-\frac{ln\left(1 + \frac{\xi(x_i-\mu)}{\sigma}\right)}{\xi}\right\} \tag{3}$$

$$Accordingly, \lim_{\xi \to 0} exp\left\{-\left(1 + \frac{\xi(x_i - \mu)}{\sigma}\right)^{-\frac{1}{\xi}}\right\} = exp\left\{-exp\left[-\frac{x_i - \mu}{\sigma}\right]\right\} \tag{4}$$

$$\text{Thus: } F(x_i; \mu, \sigma, \xi) = exp\left\{-exp\left[-\frac{x_i - \mu}{\sigma}\right]\right\} When\, \xi = 0 \tag{5}$$

We find Gumbel's allocation of the function of the distribution [30].

Gumbel's distribution was largely used in the literature to characterize several rare and extreme phenomena [30, 31]. For example, Bell [28] applied it to verify the return on the European Union's traded option prices with regard to extreme events affecting financial markets. Park, Park and Lee [32] used it to estimate the distribution of extreme wind speed values in three sites with different meteorological conditions. Yue [33] used it to determine the marginal distribution of peaks and quantities of storms as well as their return periods. Komljenovic et al. [8] used it to determine the risks of extreme interruptions of Hydro-Québec's service based on maximum CHI (customer hours of interruption) recorded per year. Moreover, in the case studied here, the costs related to major outages are strongly correlated ($R2 = 0.9408$) to CHI (customer hours of interruption). Thus, Gumbel's distribution will be used to model the cost distribution of major service outages of Hydro-Québec's overhead electrical distribution service.

There are several existing methods by which to find the parameters of this distribution, including the graphical method, as follows [34, 35]:

- classify the x_i in ascending order;
- assign a rank i (r_i) to each value of x_i
- calculate empirical frequencies F(x_i) = (r_i − 0.44)/(n + 0.12)) (Gringorten's estimate) with n = number of years of observation = 11 in this case.
- place $z_i = \frac{x_i - \mu}{\sigma}$, it becomes $-\ln(F(x_i)) = \exp(-z_i) \Rightarrow z_i = -\ln(-\ln(F(x_i)))$ with $F(x_i; \mu, \sigma, \xi) = F(x_i)$ and $x_i = z_i\sigma + \mu$
- draw graphically z_i(X − xs) x_i in ascending order (Y − xs) and determine a = σ and b = μ as well as the correlation coefficient.

From HQ's data, we deduct that these parameters are:

$\sigma = 13.87$; $\mu = 16.021$ and the correlation coefficient ($R2$) = 0.965.

Thus, the distribution of investments in major overhead outages could be written as follows:

$$F(x_i; \mu, \sigma) = exp\left\{-exp\left[-\frac{x_i - 16.021}{13.87}\right]\right\} = \text{F(xi)}$$

The costs of extreme events are chiefly related to weather phenomena. Accordingly, they affect only overhead lines. The use of this distribution is to find at least the circumstances of the frequency of extreme events even if in the context of climate change, historical data are not representative of future events.

Assumption A1 Sensitivity Analysis. Studies have shown that major climate changes will increase the estimated advantages of burying lines [36]. Undergrounding lines cannot be justified unless the willingness of customers to pay more is included in study parameters [37]. A practical approach for estimating the number of outages on an underground line consists in subtracting outages related to weather conditions, vegetation and outages related to accidents caused by wildlife from the average number of outages on an overhead line [37].

The advantages of an underground electrical grid compared to overhead lines [38] are: the transition from a state of resilience to a state of reliability; reduced frequency of interruptions due to climate change; avoidance of vegetation costs, avoidance of legal liability for damage and injuries; aesthetic advantages; better customer service. In addition, underground lines have the following features in the context of Hydro-Québec Distribution: higher repair costs compared to overhead; a lower outage rate with a ratio of 2.91 compared to overhead for medium voltage and 16.6055 for low voltage per 100 km; a higher planned interruption rate with a ratio of 0.6117 compared to overhead for medium voltage and lower planned interruption rate with a ratio of 4.0435 for low voltage per 100 km.

The analyses in this study were conducted through a cost/benefit analysis of having overhead or underground (buried) distribution lines. They are limited because the costs and technical characteristics assigned to these two strategies are not used for complete overhead or underground line structures in a given area. In this sense, transparent assessments of the costs and advantages of undergrounding and strengthening and reserve budget planning programs are useful for decision makers to make it possible to decide on the most appropriate approaches for the long-term resilience of electrical systems [36]. These assessments are justified by the observation that several studies on burying lines or leaving them overhead are limited to the technical and quantitative aspects, particularly the economic aspect. In this sense, Renn [39] underlines that these assessments should never be the sole criteria for a solid decision-making basis in complex systems.

Thus, in this time of deep long-term uncertainty concerning the development of the impact of extreme weather events on the electrical distribution network, it is appropriate to deepen reflections to launch new solid bases for decision making on how to address resilience management using broader mechanisms that are more inclusive of stakeholders through asset management strategies. This asset management should be contextualized for each organization by taking into account its technical, operational, social, economic and environmental constraints as well as their interconnections. It should also integrate the opinions and involvement of all its stakeholders by incorporating the evolution of uncertainties, knowledge, social values and time horizons.

4 Conclusion and Expected Impact

The expected impacts of this project in the short-and medium-term are to: i) address the industrial partner's problems, ii) update and improve the concept of asset management by developing new scientific approaches, and lastly iii) contribute to training highly qualified personnel on the cutting edge in activity sectors increasingly significant for Canadian companies in this sector. This research chair will have a significant impact on HQ's performance and productivity and on Canadian electrical utilities. It will furthermore help define new asset management standards for them.

References

1. Amadi-Echendu J (2004) Managing physical assets is a paradigm shift from maintenance. In: IEEE s
2. Komonen K, Kortelainen H, Räikkönen M (2012) Corporate asset management for industrial companies: an integrated business-driven approach. In: Van der Lei T, Herder P, Wijnia Y (eds) Asset management, pp 47-63. Springer, Dordrecht
3. Ouertani M-Z, Parlikad A, McFarlane D (2008) Towards an approach to select an asset information management strategy. IJCSA 5(13b):25–44
4. Aven T (2015) Implications of black swans to the foundations and practice of risk assessment and management. Reliab Eng Syst Saf 134:83–91
5. Aven T (2016) Risk assessment and risk management: review of recent advances on their foundation. Eur J Oper Res 243(11):1–13
6. Cox L Jr (2012) Confronting deep uncertainties in risk analysis. Risk Anal 32(110):1607–1629
7. Komljenovic D, Abdul-Nour G (2015) Impact of rare events on the strategy of asset management. In: 11e Congrès international de génie industriel (CIGI2015), Québec, Canada
8. Komljenovic D, Abdul-Nour G, Gaha M, Langheit C, Bourgeois M (2016) Risks of extreme and rare events in asset management. Saf Sci 88:129–145
9. Kumar G, Maiti J (2012) Modeling risk based maintenance using fuzzy analytic network process. Expert Syst Appl 39(111):9946–9954
10. ISO-55001 (2014) Asset management, overviews, principles and terminology, International Organisation for Standardization. ISO
11. ISO-55000 (2014) Asset management, management systems, requirements, International Organisation for Standardization. ISO
12. ISO-31000 (2018) Risk management - guidelines, International Organization for Standardization. ISO
13. Kumar U, Galar D, Parida A, Stenström C, Berges L (2013) Maintenance performance metrics: a state-of-the-art review. J Qual Maintenance Eng 19(13):233–277
14. Simões J, Gomes C, Yasin M (2011) A literature review of maintenance performance measurement: a conceptual framework and directions for future research. J Qual Maintenance Eng 17(12):116–137
15. Maletic D, Maletic M, Al-Najjar B, Gomiscek B (2014) The role of maintenance in improving company's competitiveness and profitability: a case study in a textile company. J Manufact Tech Manag 25(14):441–456
16. Theron E (2016) An integrated framework for the management of strategic physical asset repair/replace decisions. Stellenbosch University, Stellenbosch
17. Khuntia S, Rueda J, Bouwman S, van der Meijden M (2016) A literature survey on asset management in electrical power (transmission and distribution) system. Int Trans Electric Energy Syst 26(110):2123–2133

18. Reddy S (2016) Asset management and maintenance in transmission and distribution networks. Int J Appl Eng Res 11(119):9735–9739
19. Craig C (2018) Risk management in a policy environment: the particular challenges associated with extreme risks. Futures 102:146–152
20. Park J, Seager T, Rao P, Convertino M, Linkov I (2013) Integrating risk and resilience approaches to catastrophe management in engineering systems. Risk Anal 33(13):356–367
21. Aven T (2018) The call for a shift from risk to resilience: what does it mean? Risk Anal 39(16):1196–1203
22. Bostick T, Connelly J, Lambert J, Linkov I (2018) Resilience science, policy and investment for civil infrastructure. Reliab Eng Syst Saf 175:19–23
23. Zio E (2016) Challenges in the vulnerability and risk analysis of critical infrastructures. Reliab Eng Syst Saf 152:137–150
24. Hydro-Québec (2017) Rapport sur le développement durable 2017. Hydro-Québec
25. Hydro-Québec (2018) Rapport sur le développement durable 2018. Hydro-Québec
26. Chen H, Zhao T (2020) Modeling power loss during blackouts in China using non-stationary generalized extreme value distribution. Energy 195:117044
27. Zheng L, Sayed T (2019) A full Bayes approach for traffic conflict-based before-after safety evaluation using extreme value theory. Accid Anal Prev 131:308–315
28. Bell R (2006) Option princing with the extreme value distributions. University of London, London (2006)
29. Mathwave: Extreme Value Distribution. http://www.mathwave.com/articles/extreme-value-distributions.html. Accessed 06 Jan 2020
30. Gómez Y, Bolfarine H, Gómez H (2019) Gumbel distribution with heavy tails and applications to environmental data. Math Comput Simul 157:115–129
31. Miladinovic B, Tsokos C (2009) Ordinary, Bayes, empirical Bayes, and non-parametric reliability analysis for the modified Gumbel failure model. Nonlinear Anal Theory Methods Appl 71(112):e1426–e1436
32. Park S, Park S, Lee S-H (2016) Strategy on sustainable infrastructure asset management: focus on Korea's future policy directivity. Renew Sustain Energy Rev 62:710–722
33. Yue S (2000) The Gumbel logistic model for representing a multivariate storm event. Adv Water Resour 24(12):179–185
34. Saksena, S (2020) Plotting the flood frequency curve using Gumbel distribution, https://serc.carleton.edu/hydromodules/steps/166250.html. Accessed 06 Jan 2020
35. NIST: Extreme wind speeds software: Excel https://www.itl.nist.gov/div898/winds/excel.htm. Accessed 06 Jan 2020
36. Larsen P (2016) A method to estimate the costs and benefits of undergrounding electricity transmission and distribution lines. Energy Econ 60:47–61
37. Glass E, Glass V (2019) Underground power lines can be the least cost option when study biases are corrected. Electr J 32(12):7–12
38. Zamuda C, Wall T, Guzowski L, Bergerson J, Ford J, Lewis L, Jeffers R, DeRosa S (2019) Monetization methods for evaluating investments in electricity system resilience to extreme weather and climate change. Electr J 32(19):106641
39. Renn O (1998) Three decades of risk research: accomplishments and new challenges. J Risk Res 1(11):49–71

Resilience Enhancement of Critical Infrastructure – Graph-Theoretical Resilience Analysis of the Water Distribution System in the German City of Darmstadt

Imke-Sophie Lorenz[1(✉)], Lena C. Altherr[2], and Peter F. Pelz[1]

[1] Chair of Fluid Systems, Technische Universität Darmstadt, Darmstadt, Germany
imke.lorenz@fst.tu-darmstadt.de

[2] Faculty of Energy, Buildings, Environment, Münster University of Applied Sciences, Stegerwaldstraße 39, 48565 Steinfurt, Germany

Abstract. Water suppliers are faced with the great challenge of achieving high-quality and, at the same time, low-cost water supply. Since climatic and demographic influences will pose further challenges in the future, the resilience enhancement of water distribution systems (WDS), i.e. the enhancement of their capability to withstand and recover from disturbances, has been in particular focus recently. To assess the resilience of WDS, graph-theoretical metrics have been proposed. In this study, a promising approach is first physically derived analytically and then applied to assess the resilience of the WDS for a district in a major German City. The topology based resilience index computed for every consumer node takes into consideration the resistance of the best supply path as well as alternative supply paths. This resistance of a supply path is derived to be the dimensionless pressure loss in the pipes making up the path. The conducted analysis of a present WDS provides insight into the process of actively influencing the resilience of WDS locally and globally by adding pipes. The study shows that especially pipes added close to the reservoirs and main branching points in the WDS result in a high resilience enhancement of the overall WDS.

Keywords: Resilient infrastructure · Resilience assessment · Resilience metric graph theory · Water distribution system · Case study · Pressure loss

1 Introduction

Water is one of the most important resources for human survival and well-being and crucial to many sectors of the economy. Water distribution systems (WDS) are therefore considered as critical infrastructures. A failure or massive restriction of water supply can have a massive impact on residents and even more on critical water users in economy and health care [1, 2]. Possible failures or restrictions in water supply can result from pipe breakage, water contamination, or water shortage. Especially the impact of the first two scenarios can be reduced by closing proximate valves and thereby changing

A. Crespo Márquez et al. (Eds.): WCEAM 2019, LNME, pp. 137–149, 2021.
https://doi.org/10.1007/978-3-030-64228-0_13

the typical water flow in the pipes. These scenarios can be induced by natural or man-made hazards. Furthermore, the demand pattern is prone to short or long-term changes. Asset management deals with both, changes in the demand pattern and the availability of the resource water [3]. In this mostly robustness analysis are conducted and used for future WDS design. The conducted research classifies as a proactive asset management strategy allowing for decisions where to adapt the WDS and furthermore leading the way to advanced optimization tools for predictive asset management [4].

One cornerstone to protect and ensure water supply is the planning of resilient infrastructures, which offer increased capabilities "[...] to provide and maintain an acceptable level of service when normal operations are challenged", cf. [5], as well as, "given the occurrence of a particular disruptive event (or set of events) [...]" are able to "efficiently reduce both the magnitude and duration of the deviation from targeted system levels", cf. [6].

In order to improve the resilience of water distribution systems so that a predefined required minimum function can still be guaranteed even if individual subcomponents fail, suitable methods for the analysis and optimization of water networks are required: In a first step, the status quo of the WDS has to be assessed. Based on this assessment, in a second step, optimal enhancements of the network can be found.

In this paper, we first present a literature review of different proposed resilience metrics for WDS. For our further investigations, we focus on a popular graph-theoretical approach presented by Herrera et al. [7], and show its practical application within a case study for the German city Darmstadt. Instead of solely assessing the present resilience of the WDS considered, this work studies which adaptations to the WDS lead to the largest resilience enhancement compared to its cost. Prior to this research empirical studies on the resilience for different relative pipe failure scenarios where assessed, which showed that the graph-theoretical resilience index behaves similarly to hydraulically-defined resilience measures. To further improve the practical applicability of this approach, we add physical considerations to the proposed metric, so that not only topological but also physical properties such as pressure loss and geodetic height are included in the resilience analysis.

Based on this enhanced assessment, we are able to identify suitable additions to the current network structure that help to increase the resilience of the overall WDS, as well as help to ensure the water supply of a specifically critical social infrastructure within the network, a hospital.

2 Related Works

The resilience of a technical system describes its ability to maintain a minimum level of functionality in case of disturbances, and a subsequent possibility of recovery from these disturbances [8]. To assess the resilience of water distribution systems, several approaches and metrics have been proposed in literature [9].

Since the structure of critical infrastructures, in this study WDS, can be represented by a mathematical graph, especially graph-theoretical resilience metrics have been considered. Recently it has been shown that for a broad range of exemplary WDS, selected graph-theoretical metrics show a clear correlation with the system's resilience [10].

For the graph-theoretical approach, the WDS is converted into a graph whose nodes can be distinguished as consumer nodes and source nodes, while its edges connecting these nodes represent existing pipes, cf. Figure 1. The graph itself is assumed to be planar, i.e. it can be drawn such that its edges intersect only at their endpoints and no under or over crossing of pipes occurs.

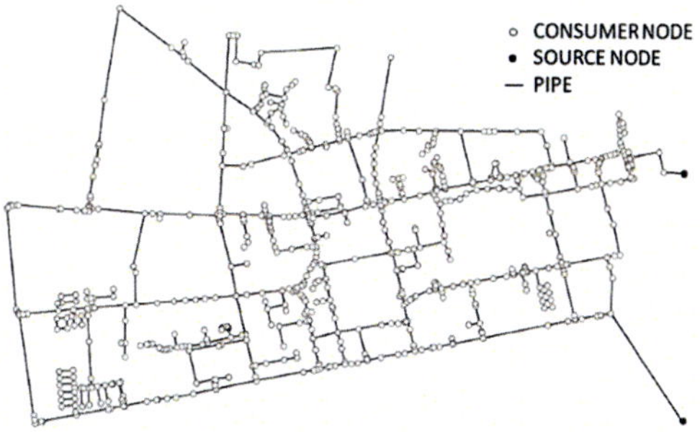

Fig. 1. Planar graph of a WDS. Source and consumer nodes are connected by edges which represent pipes.

Graph-theoretical metrics proposed in the literature that are relying on this graph representation can be distinguished according to the approach of examining the WDS: *First*, there are metrics that focus on assessing the network's topology and can be used in a general manner independent of a specific failure scenario. *Second*, there are metrics tailored to specific failure scenarios. Their focus is on determining what proportion of water demand can still be covered in the respective scenarios. *Thirdly*, there are metrics that cover dynamical aspects and assess the functional restoration following a failure. In the latter two cases, different failure and disturbance scenarios are conceivable, such as component failure, increased water demand or contamination of the water in individual areas.

Within the first approach, the network topology can be classified by several metrics. These include simpler metrics such as statistical or spectral metrics. For example, the graph density compares the number of existing edges in the graph to the number of all edges of the complete graph, in which all nodes are linked with each other, cf. [11]. Other metrics used for analysis are the maximum expansion of the graph along the existing paths (graph diameter) (cf. e.g. [12]), and the meshedness coefficient (cf. e.g. [13]). While these metrics per se are not considered specific resilience metrics, often a combination of these is used for assessing the network resilience, cf. e.g. Yazdani et al. [14]. It is pointed out that in this case, resilience properties of the WDS are a result of the combination of a robust *and* a redundant network [15]. The resilience index of Herrera, cf. [7], also follows this topological approach.

Resilience metrics of the second type, which evaluate the functional performance with respect to specific failures, usually quantify which proportion of water demand can still be covered in the respective failure scenario. Todini has defined a resilience index that can be used for different scenarios, since it is independent of the specific failure type [16]. It has therefore been extended by various researchers, and multiple adjustments were proposed, cf. [9]. Amarasinghe et al. in particular investigates supply problems subject to a strongly increased demand or insufficient precipitation [17]. An investigation of the WDS resilience depending on the specific reason for failure is conducted by Diao et al. cf. [18].

Thirdly, there are resilience metrics that cover dynamic aspects and evaluate the time-varying fulfilment of the water demand after a specific failure event. For assessing resilience, Hashimoto uses the inverse of the time during which the system does not fully meet the water demand [19]. Japan, as a region particularly confronted with natural disasters such as earthquakes and heavy rainfall, has laid legal foundations for the design of resilient water networks. One goal is to minimize the downtime of the water network during natural disasters [20]. Meng et al. examine the entire recovery process after a failure event until the complete functionality is restored, cf. [10]. The result is therefore not a single resilience measure, but rather various evaluation criteria for an analysis of the recovery process.

In this work, we aim to optimally adapt the topology of an existing WDS with respect to its general resilience. We therefore build our works on an approach from Herrera, cf. [7], that is solely based on topological considerations and therefore independent of a specific failure scenario. As such, Herrera's approach does not cover dynamic restoration aspects, which strongly depend on the specific disturbances. Instead he considers two aspects which are crucial for finding optimal topology enhancements: On the one hand, the best possible existing supply path is assessed for each consumer node. On the other hand, in order to cover the case of the best supply path failing, alternative supply paths within the network are assessed. We therefore tested Herrera's resilience index for its physical and technical applicability for the water network in our case study, and used the result as a basis for the subsequent network adaptation. Our goal in this work is to critically examine Herrera's index for its physical-technical significance and to enhance it such that not only topological features, but also the physical feature of pipe resistance making up the different supply paths pressure are covered. This is done by a thorough order of magnitude analysis for the pressure loss in the pipes of the WDS as well as a physical feasibility study of the supply paths. This yields a topological and even more physically motivated resilience metric.

3 Graph-Theoretical Approach

Following the proposed approach in [7], we compute a resilience index for each node of the water network. For this purpose we model the WDS as planar mathematical graph $G = (V, E)$ with node set V and edge set E. The node set consists of the set of the consumer nodes, C, and the set of the source nodes, S, i.e. $V = C \cup S$.

Herrera's resilience index I_{GT} is based on the following assumptions:

The water supply of a consumer node becomes more resilient if there are many paths from the water sources (e.g. tanks, reservoirs) to this specific node. In effect, the additional paths to provide water to a consumer node in case of pipe failure are taken into consideration.

Furthermore, not only the pure network topology, but also the hydraulic properties of the network are considered. High pressure losses in pipes making up a path to a consumer node could make it impracticable or even impossible to supply the node under consideration.

Combining both considerations, Herrera's resilience index for a node $i \in C$ is defined as

$$I_{GT}(i) = \sum_{s=1}^{|S|} \left(\frac{1}{K} \sum_{k=1}^{K} \frac{1}{r(k,s)} \right), \tag{1}$$

where $|S|$ is the total number of sources of the WDS, and $r(k, s)$, is a dimensionless factor of the pressure drop across the k-th path from source s to node i. Since the calculation of all paths linking a node to the water sources is computationally very expensive for larger networks, instead only the K shortest paths are considered. The minimum number of K shortest paths to identify a valid resilience index depends on the WDS itself. Therefore, it is determined for a critical transfer node as proposed in [7], see Sect. 6.

The dimensionless factor $r(k)$ of the pressure drop Δp_l for one path, i.e. the paths resistance to water supply, is given as

$$r(k) = \frac{\Delta p_l}{\varrho/2\, u_0^2} = \sum_{m=1}^{M} \left(\frac{u_m}{u_0} \right)^2 \left(f(m) \frac{L_m}{D_m} + C_d \right) \approx \sum_{m=1}^{M} f(m) \frac{L_m}{D_m}, \tag{2}$$

where M is the number of pipes making up the k-th shortest path. Length and diameter of each pipe m are denoted by L_m and D_m, respectively. Due to the branching of the WDS towards the consumers, in WDS a narrowing of the pipe diameters in flow direction is common. Therefore, throttling losses C_d in the WDS are assumed to be neglectable. Furthermore, the proportion of flow velocities in the system is close to one, due to the narrow range of velocities in which WDSs operate, $u_m/u_0 \approx 1$. A lower limit is given due to the risk of biological build-up for stagnant water. In turn, an upper limit is chosen to reduce the pressure drop along the pipes and therefore to operate efficiently. In WDS turbulent flow is the case, as it can be observed considering the Reynolds number $\mathrm{Re} = u^* D^* / \nu$ of the system. An order of magnitude analysis shows that for the characteristic velocity $u^* = O(10^0)$ m/s [21], the range of pipe diameters $D^* \geq O(10^{-1})$ m and the kinematic viscosity of water $\nu = O(10^{-6})$ m^2/s turbulent flow in pipes exists $\mathrm{Re} = O(10^5) > Re_{\mathrm{crit}} = 2000$. To determine whether flow in a rough or smooth pipe has to be assumed, the pipe roughness in WDS of order $a^* = O(10^{-4})$ m [21] is decisive. Resulting, the critical product of flow velocity, pipe roughness and reciprocal kinematic viscosity is in the order of $u^* a^* \nu^{-1} = O(10^2) > 70$. Therefore, turbulent flow in hydraulic rough pipes can be assumed and the pipe's friction factor $f(m)$ is given by Prandtl-Kármán's law as $f(m) = (2\lg(D_m/a_m) + 1.74)^{-2}$ [22]. The given physical discussion of the parameter r shows that the presented resilience index depends on the WDS's design only, i.e. its topology instead of its current load scenario.

Similarly, the present derivation of the dimensionless pressure losses can be realized on the basis of the Hazen-William equation where the head loss of a certain pipe segment is given by an empirical relationship. In this case the friction induced pressure losses are estimated by an empirically determined roughness coefficient.

4 Physical Feasibility

To study the physical feasibility of the supply paths considered for the graph-theoretical resilience index, energy conservation has to be examined. Therefore, Bernoulli's equation considering the previously discussed friction losses Δp_l is derived from the Navier-Stokes equation. In this, the elevation profile as well as the static and dynamic pressure difference between source and consumer node of the WDS is of importance to consider whether the pressure losses in the WDS are covered by the existing head in the water reservoirs H_s. Therefore, an order of magnitude analysis as well as an assessment of the friction losses is conducted.

$$\frac{\varrho u_s^2}{2} + p_s + \varrho g h_s = \frac{\varrho u_i^2}{2} + p_i + \varrho g h_i + \Delta p_l \tag{3}$$

As derived in Sect. 3 the friction losses for turbulent flow in a hydraulic rough pipe are given as follows.

$$\Delta p_l = \frac{\varrho u_0^2}{2} \sum_{m=1}^{M} \frac{1}{(2\log_{10}(D_m/a_m) + 1.74)^2} \frac{L_m}{D_m} = \frac{1}{2d} O\left(10^5\right) \frac{\text{Pa}}{\text{km}} L_{s,i,k} \tag{4}$$

In this, the friction factor $f(m)$ is derived as order of magnitude $O\left(10^{-2}\right)$. For water the density is of order of magnitude $O\left(10^3\right)\text{kg/m}^3$. As this order of magnitude analysis has an accuracy of one decimal potency, further the range of the diameter plays a significant role, which is considered by $D = d * O\left(10^{-1}\right)$. Therefore, the significance of the pipe diameter for the pressure losses is adequately considered. The length of the k-th supply path $L_{s,i,k}$ is the other parameter playing a significant role on the pressure losses. Therefore, especially notably longer supply paths compared to the shortest supply path have to be validated in respect to their feasibility.

To bring this in relation to the conservation of energy and therefore the existing energy in the system, two cases can be distinguished. On the one hand, the source can be an open tank and therefore have ambient pressure. On the other hand, the source could be pressurized. The pipeline pressure at the consumer node should be at minimum $p_i = 3 * 10^5$ Pa [21]. As stated before, the velocities in the system are in a very close range and therefore the outlet velocity of the source u_s and the flow velocity at a node i are of similar size. As a result, the dynamic pressure terms cancel out. The geodetic pressure of the system is in turn dependent on the specific network. In general static, dynamic and geodetic pressure can be reformulated as a pressure head $\Delta H_{s,i}$ of the system for which the following has to hold true.

$$\Delta H_{s,i} = \Delta p_{s,i}/\varrho g + \Delta h_{s,i} = \Delta p_l/\varrho g = \frac{1}{2d} O\left(10^1\right) \frac{\text{m}}{\text{km}} L \tag{5}$$

To validate that the existing pressure head for any consumer node in the system is high enough to appropriately feed the consumer node, a worst case scenario estimation can be made instead of assessing every single consumer node. Therefore, the least pressure head has to fulfill the longest considered supply path for the specific use case.

5 Case Study

In this paper we conduct a WDS assessment considering the introduced graph-theoretical resilience index I_{GT} defined by Herrera (cf. Eq. (1)) for a district of the German city Darmstadt. The polygon in Fig. 2a) shows this district. Using the DynaVIBe tool [23], a virtual WDS for this district is generated based on OpenStreetMap data and a digital elevation map [24]. The generated WDS consists of two source nodes as well as 729 demand nodes which are linked by 763 edges modelling pipes of diameters ranging from 50 mm to 250 mm. This classifies the network to be of medium size, according to [7]. As stated within for a resilience assessment of every demand node medium sized networks allow for the testing of possible resilience measures while at the same time having feasible computational costs. The total water demand of this network is estimated to be 100 l/s and expected to be uniformly consumed in the considered district. The source nodes model the existing pressure vessels outside the city, depicted by basins, which are simulated as reservoirs with a constant head. In the considered district of Darmstadt, a hospital is situated, which is marked by a cross in Fig. 2. The generated WDS is mapped in Fig. 2b and the colormap of the nodes shows their graph-theoretical resilience index I_{GT}. In this study, the influence of adaptations to the WDS on the resilience index is investigated. These changes are the addition of five different pipes (i – v), cf. Fig. 2b). The edges to add pipes were selected due to the low resilience index of the nodes being connected (i – iv) and their anticipated socio-economic influence (v). The latter is chosen to increase the resilience index at the hospital. All edges chosen are below an existing street or path and therefore below common ground. This is of importance as supply infrastructures are enjoined to run in parallel to streets and paths in urban areas [25]. This case study gives the possibility to apply the derived resilience metric to a real-world-like WDS generated by an established tool and post processed to meet the known standards of the region. It also shows the already well developed WDS in industrial urban areas. As these areas and therefore their supply infrastructures are under changing conditions regarding demand pattern as well as ageing pipes, the resilience improvement is of importance and current subject for asset management strategies [3].

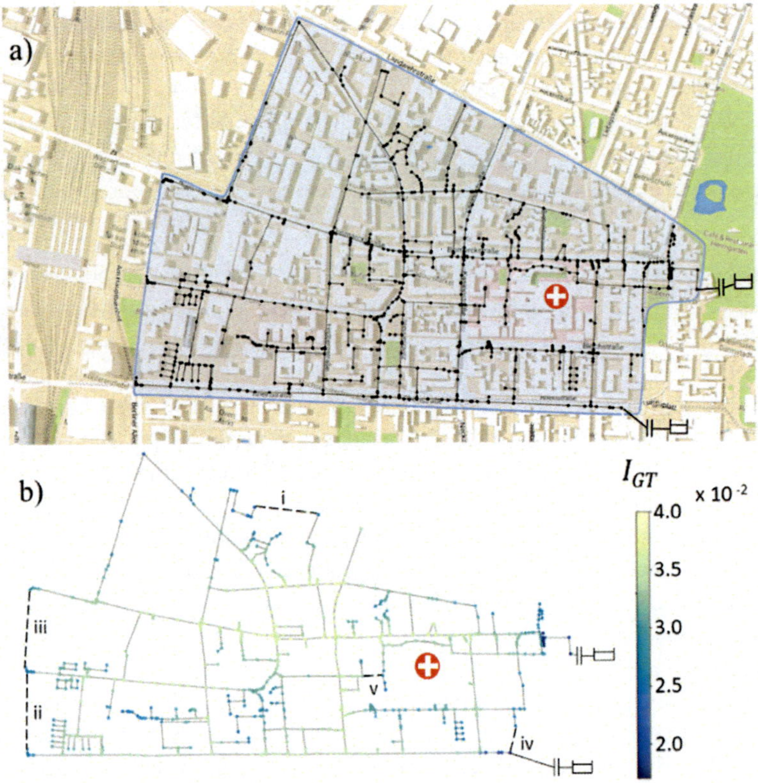

Fig. 2. a) Water distribution system of a district in a German City, Darmstadt, and b) its water distribution system created using DynaVIBe. a) The blue polygon shows the considered district of Darmstadt for which the WDS is analyzed. b) The planar graph of the current WDS generated using DynaVIBe is pictured. The WDS is fed by two water reservoirs outside the city depicted by basins. The color code of the nodes gives Herrera's resilience index. Dotted lines with roman numbering show the pipes which are added in the case study. The hospital's location is marked by the cross icon.

6 Technical Implementation and Preliminary Considerations

To process and tailor the generated Network data as well as to calculate its resilience index, the python packages WNTR [26] in combination with NetworkX [27] prove useful. As described by Herrera, cf. [7], first the number K of shortest paths to be considered in order to obtain a sufficiently good approximation of the resilience index is determined. Conform with Yazdani et al. [14], who showed that especially nodes with a high betweenness centrality index are decisive for resilience. The influence of the number K of paths considered to determine the resilience index is studied for such a critical transfer node. Consistent with [7] the relative difference of the mean conductance $g(K)$ between $K+1$ and K is calculated as

$$d_r(K,s) = \frac{g(K+1,s) - g(K,s)}{g(K_{max},s)}, \quad g(K) = \frac{1}{K}\sum_{k=1}^{K}\frac{1}{r(k,s)}. \tag{6}$$

For varying values of K, the influence is depicted in Fig. 3a). It shows that this difference converges for large values of K. The critical number of $K = 17$ shortest paths allows for a deviation of $d_r = 1\%$ for the relative change of the resilience index.

In this work the influence of the addition of pipes on the WDS's resilience is studied. Therefore, the five different pipes are added in every possible combination, this results in 325 different networks. Furthermore, a relative number of 2.5% of pipes is added to compare to the removal of this many pipes as studied in [7].

As stated in Sect. 4 an estimation of the pressure losses in the longest considered supply path has to be estimated and compared with the minimum available pressure head of the system to assure that any considered supply path is feasible. Each supply path is made up of the supply line, which is 4 km long and at maximum double the range of two consumer nodes farthest apart in the network, i.e. 2 km. For the supply line the diameter is known to be $d_s = 250\,\text{mm}$ while in the network for a worst case scenario assessment a mean diameter of $d_i = 100\,\text{mm}$ is assumed. Considering Eq. 4, the maximum friction losses Δp_l of the highest resistance k-th path can be estimated to have an upper limit of $\Delta p_\text{l} = 1/2 * \mathcal{O}(10^5)\,\text{Pam/km} * (4\,\text{km}/0.25\,\text{m} + 2\,\text{km}/0.1\,\text{m}) = 1.8 * \mathcal{O}(10^5)\,\text{Pa}$ (cf. Eq. (4)). Following the energy conservation, the difference in dynamic pressure, static pressure, and geodetic pressure have to account for these losses so that this worst scenario supply path is feasible.

The minimum geodetic height difference between the lower source and the highest consumer node is $\min(\Delta h_{s,i}) = 45\,\text{m}$. In this case study both sources are open tanks resulting in ambient pressure at the source node, $p_s = 10^5\,\text{Pa}$. Since at the consumer nodes a static pressure of $p_i = 3 * 10^5\,\text{Pa}$ is required, the static pressure difference results in $\Delta p_{s,i} = -2 * 10^5\,\text{Pa}$. Following Eq. (5), the minimum pressure head of the present water distribution system can be determined as $\varrho g \min(\Delta H_{s,i}) = \Delta p_{s,i} + \varrho g \min(\Delta h_{s,i}) = 2,41 * 10^5\,\text{Pa} > \Delta p_\text{l}$. This shows that the lower limit of pressure head meets the demand of the worst scenario pressure losses well and even with a small gap to account for higher friction losses in the smallest pipes of 50 mm diameter. With this positive assessment of the physical feasibility the calculations of the present resilience as well as the influence of the addition of pipes to the existing water distribution system can be conducted.

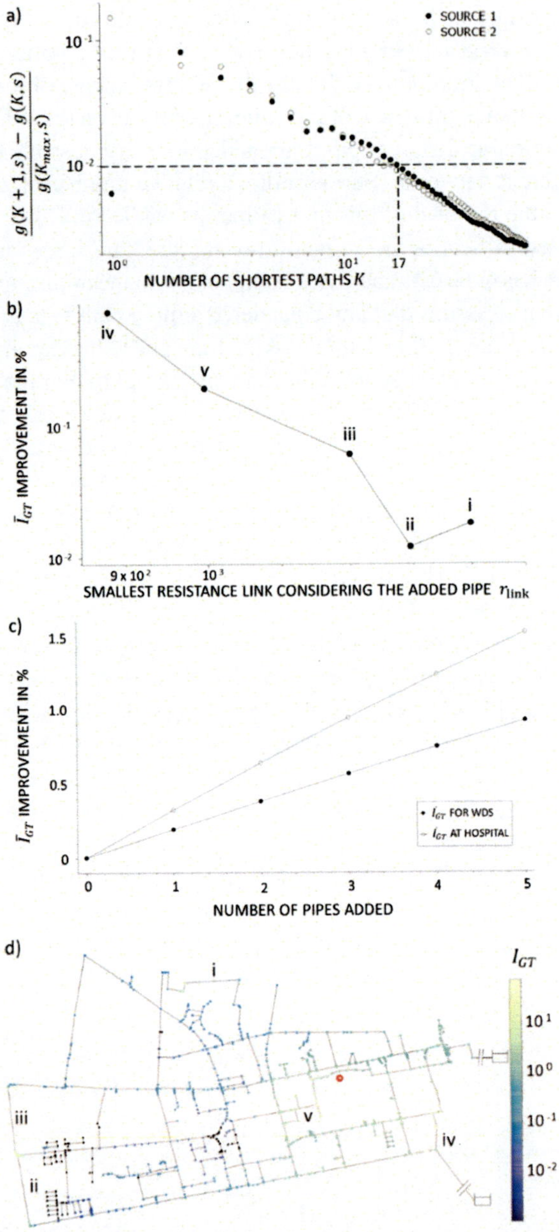

Fig. 3. a) Validation of the number of shortest paths having to be considered for the change of resilience index to be less than 1%, b) difference in the resilience index for different pipes additions depending on the dimensionless pressure drop of the link connecting both sources, c) average of the mean resilience improvement for the WDS and the consumer node at the hospital for a given number of added pipes, d) the WDS's change in resilience index with all five pipes added.

7 Results and Conclusion

First, the random successive addition of 5 different pipes, introduced in Sect. 6, is studied. The addition of all 5 pipes leads to a resilience increase of 0.93%. Its distribution is pictured in Fig. 3d). The largest improvements are made in the area where pipe iv is added. Therefore, the influence of the different pipes added singly to the WDS is studied closer. The influence of the addition of single pipes on the improvement of the averaged WDS resilience depends strongly on the dimensionless pressure drop across the path with the smallest resistance connecting both reservoirs considering the added pipe. This dependency is pictured in Fig. 3b), where the addition of pipe iv allows for the minimum weight link of the reservoirs and results in the highest increase of resilience averaged over the WDS. As the course of the graph shows, the improvement in resilience for a pipe does not only depend on its resulting smallest link. Since the resilience index considers the number of shortest paths possible as well as the resistance of these paths, possible additional factors are the degree of branching of the nodes connected and the pipe's diameter, respectively.

As pictured in Fig. 3c), the randomised addition of pipes steadily increases the average of all possible WDS' mean resilience for a given number of added pipes. The average resilience at the hospital node improves also with the number of added pipes. Strong influence results from the addition of pipe iv, whose influence on the WDS's resilience impacts the hospital node, also, and even more from the intended improvement by the addition of pipe v. Outcome is that the resilience of socio-economic important areas is influenced positively by a higher meshedness in this area. Additionally, an improvement of the general WDS is likely to positively influence the resilience of the considered area.

Furthermore, the addition of pipes to an existing WDS compared to its elimination as done in [7] provided interesting insights. These findings trace back to the structure of the WDS. While the elimination of 2.5% of the WDS's pipes leads to a relative reduction of 9.4% for C-town, considered in [3], an addition of the same percentage of pipes to the present WDS leads to a 7.4% increase in resilience. Though both WDS are of similar size, the number of sources varies widely. The WDS of C-town has one reservoir and seven additional tanks within its system, all modelled as sources. Still, the influence of the addition or removal of pipes seems to be similar.

Moreover, these results show that the addition of pipes influences the resilience increase of WDS stronger than linearly and therefore is a valid measure to increase the resilience of WDS. This shows the resilience increase by making use of functional redundancy in a network as also studied by Altherr et al. [28]. Within this study it is also shown that the adaptation of a network is not trivial but instead the combinatorial size quickly exceeds human brute force computational capabilities. This also explains the small amount of number pipes chosen to assess within this study. Therefore, the results can be considered as directive for future studies but need to be tested on larger scale. To do so it would be of interest to solely consider the main-line WDS, which consists of the subset of pipes with diameters larger than 100 mm and has more possibilities to be appended by pipes considering the strict constraint of the urban transportation system as limiting factor of where pipes can be added. To master the resulting combinatorial explosion of possibilities to add pipes when considering a larger number of pipes to add, in further work a mathematical optimisation model is set up. In a first approach, the resilience index is linearized to further study the behaviour of the network resilience depending on the addition of pipes. Future work covers the optimisation of water distribution systems' resilience for the nonlinearized resilience index introduced in this work. Moreover, further physical considerations to adapt the resilience index can be included or different resilience indices can be considered.

The application of the derived resilience metric solely based on the topology of the WDS and no longer considering the current hydraulic state within the system is based on the assumption of similar flow velocities within the pipes. The applicability of this assumption is given in literature and has been confirmed by local water providers but needs to be testified for transfer studies. If this assumption is not applicable the calculation of the current hydraulic state within the network is essential for the resilience assessment with the proposed graph-theoretical resilience metric.

Acknowledgements. The authors thank the KSB Stiftung Stuttgart, Germany for funding this research. Moreover, we thank the German Research Foundation, DFG, for partly funding this research within the Collaborative Research Center SFB 805 "Control of Uncertainties in Load-Carrying Structures in Mechanical Engineering" within the subproject "Resilient Design".

References

1. USEPA (US Environmental Protection Agency) (2004) Emergency Response Plan Guidance for Small and Medium Community Water Systems to Comply with the Public Health Security and Bioterrorism Preparedness and Response Act of 2002
2. Welter G, Socher M, Needham P, Bieber S, Bonnaffon H (2013) Cross-sector emergency planning for water supply utilities and healthcare facilities. J Healthcare Risk Manag 32(4):5–14
3. Ugarelli R, Venkatesh G, Brattebø H, Di Federico V, Sægrov S (2010) Asset management for urban wastewater pipeline networks. J Infrastruct Syst 16(2):112–121
4. Lorenz I-S, Pelz P (2020) Optimal resilience enhancement of water distribution systems. Water 12(9):2602
5. Croope SV, McNeil S (2011) Improving resilience of critical infrastructure systems postdisaster. Transp Res Rec J Transp Res Board 2234(1):3–13
6. Vugrin ED, Warren DE, Ehlen MA, Camphouse RC (2010) A framework for assessing the resilience of infrastructure and economic systems. In: Sustainable and Resilient Critical Infrastructure Systems, Springer, Heidelberg, pp 77–116
7. Herrera M, Abraham E, Stoianov I (2016) A graph-theoretic framework for assessing the resilience of sectorised water distribution networks. Water Resour Manage 30(5):1685–1699. https://doi.org/10.1007/s11269-016-1245-6
8. Altherr LC et al (2018) Resilience in mechanical engineering - a concept for controlling uncertainty during design, production and usage phase of load-carrying structures. Appl Mech Mater 885:187–198
9. Shin S et al (2018) A systematic review of quantitative resilience measures for water infrastructure systems. Water 10(2):164
10. Meng F, Fu G, Farmani R, Sweetapple C, Butler D (2018) Topological attributes of network resilience: a study in water distribution systems. Water Res 143:376–386
11. Chen G, Wang X, Li X (2014) Fundamentals of Complex Networks. John Wiley, New York

12. Meng KK, Fengming D, Guan TE (2007). Introduction to graph theory - H3 mathematics. World Scientific
13. Cardillo A, Scellato S, Latora V, Porta S (2006) Structural properties of planar graphs of urban street patterns. Phys Rev E 73(6):066107
14. Yazdani A, Otoo RA, Jeffrey P (2011) Resilience enhancing expansion strategies for water distribution systems: a network theory approach. Environ Model Softw 26(12):1574–1582
15. Yazdani A, Jeffrey P (2011) Complex network analysis of water distribution systems. Chaos Interdiscip. J Nonlinear Sci 21(1):016111
16. Todini E (2000) Looped water distribution networks design using a resilience index based heuristic approach. Urban Water 2(2):115–122
17. Amarasinghe P (2014) Resilience of water supply systems in meeting the challenges posed by climate change and population growth. Queensland University of Technology, Brisbane, Australia
18. Diao K, Sweetapple C, Farmani R, Fu G, Ward S, Butler D (2016) Global resilience analysis of water distribution systems. Water Res 106:383–393
19. Hashimoto T, Stedinger JR, Loucks DP (1982) Reliability, resiliency, and vulnerability criteria for water resource system performance evaluation. Water Resour Res 18(1):14–20
20. The World Bank (2018) Resilient water supply and sanitation services: the case of Japan
21. Baur A, et al. (2019) Mutschmann/Stimmelmayr Taschenbuch der Wasserversorgung. Springer Fachmedien Wiesbaden, Wiesbaden
22. Spurk J, Aksel N (2019) Strömungslehre, 9th ed. Heidelberg: Springer
23. Mair M, Rauch W, Sitzenfrei R (2011) Spanning tree-based algorithm for generating water distribution network sets by using street network data sets. World Environ Water Resour Congress 2014(2014):465–474
24. Jarvis A, Reuter HI, Nelson A, Guevara E (2008) Hole-filled seamless SRTM data V4, International Centre for Tropical Agriculture (CIAT)
25. DIN-Normausschuss Bauwesen (2018) DIN 1998 - Placement of service conduits in public circulation areas – Guideline for planning, ICS 93.025
26. Klise KA, Bynum M, Moriarty D, Murray R (2017) A software framework for assessing the resilience of drinking water systems to disasters with an example earthquake case study. Environ Model Softw 95:420–431
27. Hagberg AA, Schult DA, Swart PJ (2008) Exploring network structure, dynamics, and function using NetworkX. In: Proceedings of the 7th Python in Science Conference (SciPy), vol 836, pp 11–15
28. Altherr LC, Leise P, Pfetsch ME, Schmitt A (2019) Resilient layout, design and operation of energy-efficient water distribution networks for high-rise buildings using MINLP, October 2019

Asset Data and Information

Aligning IIoT and ISA-95 to Improve Asset Management in Process Industries

Yong-Lip The[1](✉) and Anastasia Govan Kuusk[2](✉)

[1] OSIsoft LLC, San Leandro, CA, USA
ythe@osisoft.com
[2] University of South Australia, Mawson Lakes, SA, Australia
Anastasia.kuusk@unisa.edu.au

Abstract. Despite standards such as International Standards Association (ISA-95) and Manufacturing Enterprise Solutions Association (MESA) International guiding operational process, asset automation and control for some time, the integration challenge remains for Industrial Internet of Things (IIoT) that the strategies of converging Information Technology (IT) and Operational Technology (OT) domains remain disparate. IT systems prioritize data governance and reliability to ensure the needs of enterprise users are met while OT systems prioritize human safety and asset reliability to ensure that the production process remain safe and effective. These differing priorities lead to greater complexities in the design of convergent systems necessary to progress Industrial Internet of Things, coined more recently as Industry 4.0.

The gaps are between IoT computing layers and alignment to the ISA-95 automation layers plus the evolving constructs of Industry 4.0 and how in the future un/trusted data moves seamlessly between converged OT and IT domains. In doing so, the information governance links between technology, organizational processes, and people (TOP) across the ISA-95 and Manufacturing Operations Management (MOM) System stacks must be considered to provide true secure convergence of the OT and IT domains. This paper explores and builds upon the growing industrial manufacturing OT/IT/IoT convergence literature [1, 31], extending exploration of the role of TOP governance factors to RAMI4.0, suggesting a single OT/IT strategy for organizations to achieve Industry 4.0 initiatives.

The case study research method provides insights into emerging industry and standards convergence practices. Applying these standards and aligning the interdependence between OT and IT strategies with TOP constructs, are explored in the context of achieving Industry 4.0 objectives robustly and resiliently through applying a holistic Data Infrastructure Platform (DIP) in process-centric organizations.

Keywords: Industry 4.0 · ISA-95 · IIoT · Edge · Fog · Cloud · OT/IT · RAMI4.0

1 Introduction

This paper provides insights into technologically advancing considerations of east to west as well as the traditional north to south view of facilitating data from the asset to the

A. Crespo Márquez et al. (Eds.): WCEAM 2019, LNME, pp. 153–163, 2021.
https://doi.org/10.1007/978-3-030-64228-0_14

office when organizations optimize their engineering assets. Process-centric industries like Utilities, Oil & Gas, Chemical, Manufacturing and Mining currently assess people, process, and Operational Technology convergence [1] when determining their overall internet, industry, data and process control strategy for managing and optimizing their assets.

The primary challenge facing these industries is how to holistically leverage the IT (Information Technology) centric, knowledge economy into their OT (Operational Technology) centric, process-oriented businesses [1] in a seamless yet maturity-based, profound manner; whilst accounting for advancements in people, process and technology frameworks (including security, hardware and software) when governing real-time and time-series data.

Digital transformation is made up of traditional and next generation concepts, further requiring a truly holistic view of OT and IT convergence [2], such as Industrial Internet of Things (IIoT), sensors, edge and fog computing, PLC (Programmable Logic Controller), historians, SCADA(Supervisory Control And Data Acquisition) systems and Industry 4.0. IIoT or the Industrial Internet of Things is one of the fundamental building blocks to Industry 4.0. Industry 4.0 is suggested as the next generation of manufacturing where "smart" technologies are used to automate entire supply chains, leveraging machine connectivity and ultimately, the power of the Internet and the Cloud [3] with converged management frameworks such as RAMI4.0 [31] requiring organizations to consider significant technology architecture changes.

The paper defines the current digital transformation elements and challenges the traditional extension of people, process and technology in the OT/IT, ISA-95 based convergence (north to south view) with the growing influence of IIoT and converged state of the art frameworks such as RAMI4.0. Advancements in hardware, software, sensors, security threats and network communications warrants revisiting to ensure a holistic view of data governance management, to facilitate organizational maturity progress. This includes next generation concepts, such as the introduction of a Data Infrastructure Platform (DIP) that aligns the OT world of Process Control Systems (PCS) with the IT world of Cyber Physical Systems (CPS). This holistic view is considered the east to west view of OT/IT convergence.

2 Edge-Fog-Cloud for the OT World

There is broad acceptance that the growth of connected devices is accelerating [4] (25 billion by 2021 [5]). There will be ten times more data created from these devices than stored, so Edge and Fog computing are the solutions expected to fill this gap [6]. Logically, this infers aggregation of data and distribution of computation at various "layers" between the sensor and the cloud. It is important to understand the impact of these computation expectations to the world of asset-intensive process industries. To do so, we must first analyze the two perspectives that IIoT can be viewed from – the IT perspective and the industrial automation perspective. This is the east to west view of OT and IT convergence.

Fog nodes can be split to 2 layers, one at the LAN (Local Area Network) level to encapsulate Edge computers (e.g. within a process control systems network) and the other at the enterprise WAN (Wide Area Network) level (e.g. corporate intranet infrastructure) which then acts as the gateway to the Cloud.

Important Fog applications involve real-time interactions rather than batch processing [7, 13]. An Edge-Fog node therefore fulfils the key SCADA to PLC/smart device communication requirement of real-time interactions. Therefore, a SCADA system in the Process Control LAN is not dissimilar to an Edge-Fog node. We will refer to this as the LAN Fog node.

A Fog node can also provide support for on-line and/or streaming analytics and interplay with the Cloud where scalable, enterprise applications and analytics are perfectly placed. A Cloud-Fog node therefore addresses the same ERP-MOM (Enterprise Resource Planning - Manufacturing Operations Management) communications requirements. Therefore, a MOM system connected to the corporate WAN and the Cloud is not dissimilar to a Cloud-Fog node. We will refer to this as the WAN Fog node.

In each of these Fog nodes, computers can validate, aggregate and contextualize data to reduce the volume (bytes) before sending it to the Cloud as well as run any localized applications vital to the operations for decision support, etc. [6].

3 IIoT Dependency on Fog Capability

From an industrial automation perspective, Edge computing is of vital importance and is a means of ensuring that the operational assets are sustained in a safe, effective, and reliable manner [8]. By definition [7], Edge computing in PCS (Process Control Systems) comes in the form of DCS (Distributed Control System), PAC (Process Automation Control), PLC and RTU (Remote Terminal Unit) [9]. These are highly specialized and complex systems that communicates using digital and analogue signals with field devices and sensors. More recent developments from industry also saw the introduction of the ePAC (Ethernet Programmable Automation Controller) from the likes of Schneider Electric [10] that comes with native IP routing capability to eliminate bottlenecks and complexity when connecting with both wired and wireless Ethernet-based intelligent devices, via the process control network. This is all within the OT domain.

Two differing paradigms and cultures separate the OT and IT departments of process-centric industries in the past. OT have separated from IT to the point of having an "air gap" between the corporate WAN and the PCS LANs. This is a safety and process reliability driven decision [11]. Today, with the demands of process data from the enterprise, OT policies have accommodated IT requirements by providing secured access from external networks. This is usually via a DMZ (Demilitarized Zone) within the firewall between the corporate and process control networks wherein typically lies the Process Historian for real-time collection, storage, and retrieval of time-series data from the process [9]. The melding of these two technology paradigms (firewall and more recently, a historian application) is known as the point of OT/IT convergence [12] and is where the Edge-Fog or LAN Fog node is located. (See: Edge Computing ref [6], Fog computing ref [13]). Therefore, the Process Historian typically fits as an application in the LAN Fog Node as a process control data gateway, alongside the SCADA (or DCS) system.

Process Historians today are quite advanced and possess much of the computational and analysis needs of geographically spread, mission-critical operations such as electricity grids [14]. At this point, such Historians can take on more of a DIP (Data Infrastructure Platform) role, connecting to numerous edge devices and controllers. In the case of Dominion Virginia Power [14], the DIP is based on the OSIsoft PI system and performs some of the distributed intelligence capability of Fog computing [7] as well as some of the visual and functional capability of MOM systems [15]. As a DIP, its powerful and vast interfacing capability can be further augmented by other specialized systems within the corporate network and/or in the Cloud. It may also address a significant portion of the technology requirements that could arise in a typical Digital Transformation Journey. It can fulfil the roles of the Operational Backbone and some aspects of the Digital Services Platform [16].

In a pure IoT solution, without the "Industry" prefix to the acronym, a DIP does not necessarily need to consider the sensitivity of industrial safety and process reliability. Think of the ride sharing business, Uber, by definition [17] an IoT platform. The biggest risk to failure is that someone (a rider) cannot get from point A to point B as planned. In the case of IIoT, a DIP must factor in a consistent and reliable connection from the Sensor to the Cloud [3] as the consequences of industrial failure can be catastrophic. Industrial failures can be reduced when digital transformations account for robust, flexible and secure digital data platforms [16, 24] such as applying Process Historians to capture IIoT data, particularly where systems are now expected to collect, cleanse and analyze 500,000 data points every few minutes [30].

Leveraging the access to real-time data from the physical process to aid in operations management adds value to businesses in terms of asset health, process improvements, supply chain visibility and overall decision support [18]. With the rise of IIoT converging with OT and IT data, this level of complexity appears to increase, particularly when securely integrating trusted and untrusted domain data [19]. With the increased convergence of IIoT and PCS (Process Control Systems), establishing a model for aligning these two worlds eases the journey towards Industry 4.0. One clear method to ensure such alignment is through the establishment of a common DIP for consolidating the management of all real-time and time-series data, including state of the art, holistic security policies and data architecture.

Given that we have established the close alignment between IIoT and the OT PCS paradigm, we must first attempt to look at IIoT solutions through a common lens shared by both the IT and OT professional. This aligns the ISA-95/IEC62264 model [20, 21] to the IoT Layers [13] to link IIoT to PCS.

4 Alignment of IIoT Layers to ISA-95/IEC62264

In manufacturing and process industries, where automation is a critical part of their operation, OT/IT convergence occurs where ISA-95 Level 2 and Level 3 meet [20]. The ISA-95 standard defines a model of data exchange between manufacturing control functions with business process functions and from that we can essentially split up which computing functions tend to be deployed at the Edge (where control occurs) and which functional services would be deployed at the different Fog layers. The purpose of the view in Fig. 3 is to align the IT perspective of CPS to the OT perspective of PCS acknowledging that it lacks communication protocols to handle data flows.

IIoT is an IT or Information Technology concept. IT departments are comfortable with the internet and how it works, including security threats. It is therefore not surprising that Cisco has come up with the IT concept of the "Fog" computing layers to claim the space between the asset Edge computing and Cloud computing [7]. The Fog remains a key computing layer for functional services that encompasses applications that will eventually be consumed by users in ISA-95 Levels 2 and 3.

NIST [32] cites that CPS-based Automation model can be mapped to the ISA-95 automation hierarchy through describing the functionalities of the various devices and/or services as shown in Fig. 1. Each colored dot in the CPS-based Automation "network" diagram shown on the right, corresponds functional services that are typically found in the ISA-95 stack shown in the diagram on the left, albeit in a RAMI4.0 converged, reusable microservices data architecture.

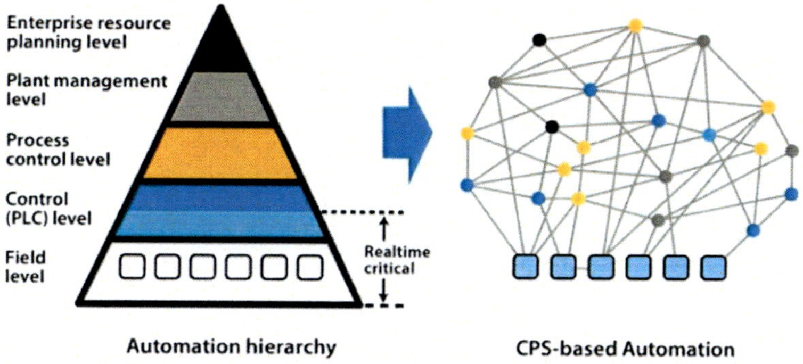

Fig. 1. Decomposition of the automation hierarchy with distributed services

From an interaction perspective, the CPS-based Automation is a more agile, application plug and play federated architecture that may break the typically well-defined ISA-95 hierarchy rules for data exchanges as OT and IT worlds converge. As shown in Fig. 1, functional mapping must persist for CPS to remain aligned with the PCS world. Thus, to ensure this east to west alignment, NIST standards can be extrapolated as state of the art for converged OT/IT services as shown in Fig. 2.

These well-adopted OT and IT standards necessitates the alignment of the CPS computing layers into the ISA-95 stack to accommodate the converged RAMI4.0 technology, process, and people framework. Figure 3 shows the east to west view of how the IoT computing layers also aligns with the Automation layers according to the ISA-95 standards:

This model can be used to understand how data can be prepared and aggregated at each layer between the sensor and the cloud, providing a future state holistic west, east, north, and south view of all operational systems and asset communications and interactions. By doing this, we can ensure that a data lake for analysis in the cloud is populated with contextualized and validated information to avoid the so called "Data Swamp" that is increasingly being reported as an organizational issue [22].

Applying this model means it is possible to deploy a DIP that incorporates a common set of advanced tools to manage, cleanse, synthesize and analyze data for critical business

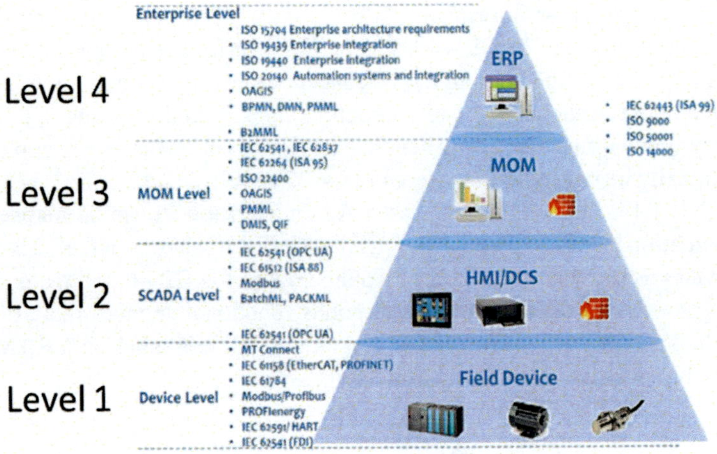

Fig. 2. Standards aligned to the ISA95 model

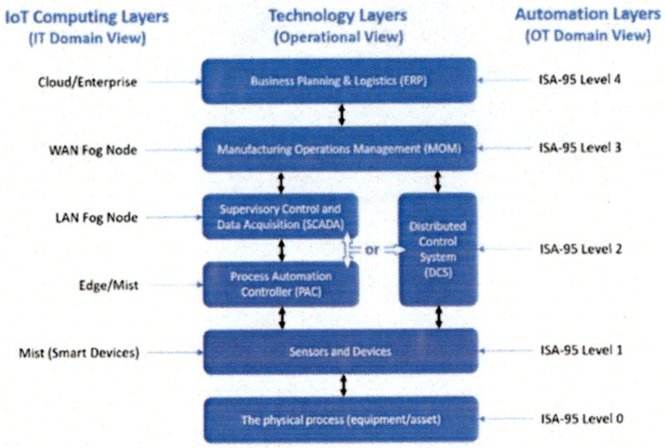

Fig. 3. Holistic data governance alignment of OT and IT domains accounting for IIoT edge, fog, and cloud computing

insights across all of their CPS and PCS managed assets [30]. A common DIP will also resolve the Information layer of the RAMI4.0 model to ensure that common access to data is assured. To fully comply with the intent of the RAMI4.0 model in future, a DIP should also resolve how select data can be securely exchanged across the supply chain or an entire community. Figure 4 below shows how a common DIP across the value stream and hierarchy of assets resolves a key layer of the RAMI4.0 model.

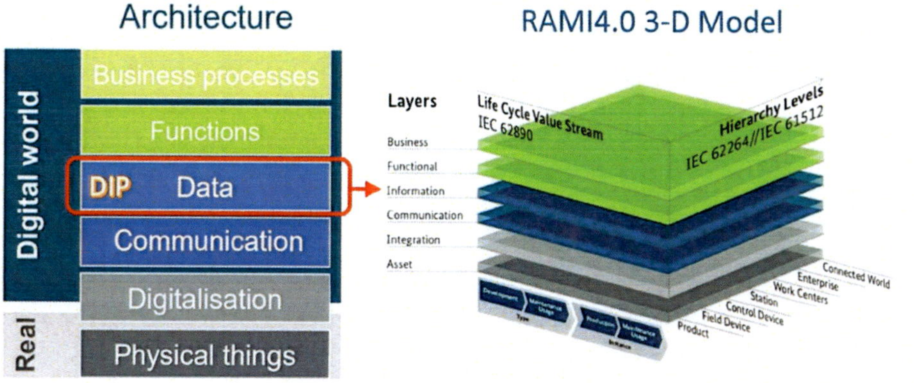

Fig. 4. Information sharing using a DIP to facilitate the RAMI4.0 model

5 Identifying Organizational Maturity in Industry 4.0

Smart Manufacturing or Industry 4.0 is about "Climbing the DIKW (Data, Information, Knowledge, Wisdom) Pyramid" [2] and is a next generation key concept, along with OT/IT convergence and ISA-95 to unlocking the value of IIoT in process industries. Each level of the DIKW Pyramid is a "Tier of Awareness" and we will use its concepts to segment the different computing functions within a business.

This DIKW view simplifies the 5C (Connection, Conversion, Cyber, Cognition, Configuration) reference architecture for Industry 4.0 proposed by Radanliev et al. [23] and in doing so, can guide the future state development of Industry 4.0 [Scanning the industry p. 17].

For the purposes of this paper, the DIKW terms are interpreted as follows:

- **Data:** Digital representation of the state of the process and at a fixed location and time
- **Information:** Contextualization of Data through aggregation and/or associative groupings
- **Knowledge:** Application of information to form cause-effect relationships and used for decision support
- **Wisdom:** Application of Knowledge (system) and enhanced by human interpretation to predict outcomes and to recommend optimized actions and reapplied back to the Knowledge base

This segmentation of Awareness is important because depending on the role of the individual within business, their decision-making needs will differ and therefore they will require different functional and decision support services. Figure 5 extends a traditional ISA-95 OT/IT domain alignment diagram with the dimension of Awareness tiers and the corresponding user requirements (computing functionality) at each layer for optimal business and operational performance, further extending the east to west view of OT/IT convergence.

The Tiers of Awareness (DIKW) are an overarching measure of business maturity when viewed under the lens of Industry 4.0 [2] governed through the OT/IT convergence

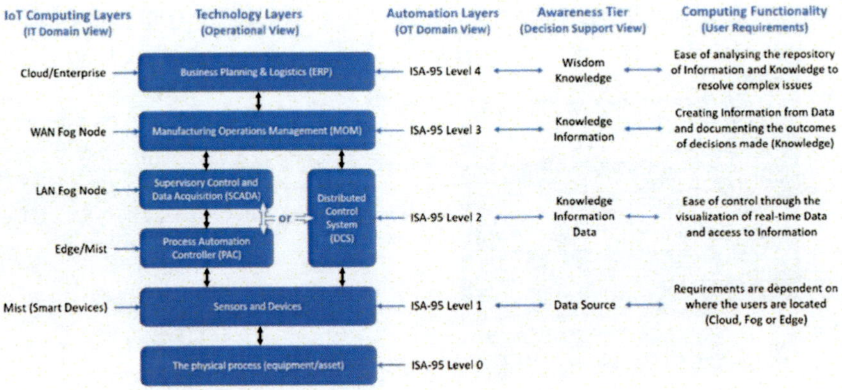

Fig. 5. Mapping of technology layers to awareness and computing functionality

steps recommended of the Australian Plan, Converge, Align and Integrate framework [1]. Most businesses today possess Wisdom, but this usually lies within their people. The goal is to codify and scale this Wisdom beyond the individual by using technology, so entire operations can be optimized using systems, models, and processes.

Table 1. Towards Industry 4.0 - process industry maturity matrix

Tier/Maturity	1	2	3	4	5
DIKW Tiers (Awareness)	Data Silos (no data veracity)	Data visibility	Information availability	Knowledge captured	Wisdom powered
Likely decision support systems	Spreadsheets and reports	Real-time KPI & Dashboards	Situational awareness	Simulations & Digital Twin	AI & Cognitive computing
Production process [24]	Executed processes	Managed processes	Defined processes	Quantitatively managed	Autonomy, optimized processes
Maintenance process [25]	Reactive (Break-fix)	Preventive	Condition based (CBM)	Predictive (PdM)	Prescriptive
Industry 4.0 [26]	Computerized & Connected (Industry 3.0)	Visibility for data-driven decisions	Transparency of context-rich information	Predictability of decisions	Adaptability to exogenous conditions
Digital readiness (DREAMY) [28]	Initial	Managed	Defined	Integrated & Interoperable	Digital oriented

The need for Production process maturity [24] and Maintenance process maturity [25] is well established with industry 4.0 [26] and can be used as a benchmark for future state Industry 4.0 Smart Manufacturing. The business case for IIoT and Industry 4.0 [27] cannot be based on immediate financial returns alone. For example, a DIP is foundational to achieving Industry4.0, however by itself, it will not provide a direct ROI. It will be the secure applications and functional services supported by the DIP that provides the ROI. Digital Readiness Assessment Maturity [28] Model is a tool that provides a method of measuring the overall readiness of the business at each step of the journey while providing guidance on how to move forward in terms of functions (Table 1).

The table above provides maturity stages of mapping a Digital Transformation Journey towards Industry 4.0 for process-centric, asset-intensive businesses aligning to Capability Maturity Model Integration CMMI institute model [29]. Moving up the Tiers from left to right (1–5) is an Industry 4.0 continuous improvement journey. This is a recommended approach to achieve the most successful returns on investment for Digital Transformation: The "step-by-step" approach as recommended by Strother & Ravens [30].

Since 2016 [32], engineering asset management-based organizations experienced gaps in their DIKW maps as shown in Fig. 6. In the four years since, gaps appear to be narrowing, however many are still moving up the DIKW ladder to a new state of the art, converged OT/IT data state. By benchmarking a business using the Process Industry Maturity Matrix, the journey mapping can provide a healthy ROI towards Industry 4.0 achievement.

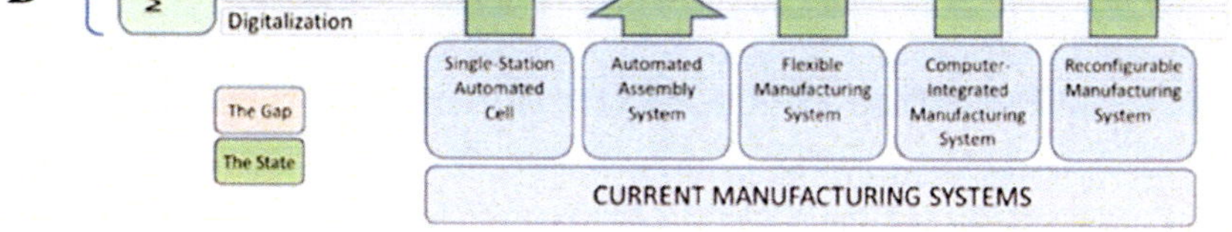

Fig. 6. Gap between manufacturing systems and Industry 4.0 in 2016

6 Conclusion

The paper explores the current digital transformation elements of Industry 4.0 and challenges the extension of the traditional ISA-95 based, OT/IT convergence north to south, with a west to east view, that aligns with the state of the art RAMI4.0 framework, incorporating real and structured secure data posed by the rise of IIoT data management.

The paper explores concepts such as Fog, facilitating data from next generation sensors into historians and increasingly holistic next generation maturity models for organizations to gauge progress of their Industry 4.0 journey. Such extenuating concepts suggests the next generation view of data governance for process control in a continually technologically evolving process centric, engineering asset management-based context that incorporates a holistic north, south, east and west data governance view, leveraging a Data Infrastructure Platform for organizations to achieve commercial success in Industry 4.0.

References

1. Kuusk A, Gao J (2015) Consolidating people, process and technology to bridget the great wall of operational and information technologies. In: Engineering asset management: towards certification, Springer, Cham
2. Lobov A (2018) Smart manufacturing systems: climbing the DIKW pyramid. In: IECON 2018 - 44th annual conference of the IEEE industrial electronics society, Washington DC
3. Wang L (2016) Big data in cyber-physical systems, digital manufacturing and industry 4.0. Int J Eng Manufact 6:1–8
4. Cisco (2018) Cisco global could index: forecast and methodology, 2016–2021 White Paper, 19 November 2018. https://www.cisco.com/c/en/us/solutions/collateral/service-provider/global-cloud-index-gci/white-paper-c11-738085.html. Accessed 7 May 2019
5. Gartner (2018) Gartner identifies top 10 strategic IoT technologies and trends, 7 November. https://www.gartner.com/en/newsroom/press-releases/2018-11-07-gartner-identifies-top-10-strategic-iot-technologies-and-trends. Accessed 7 May 2019
6. Shi W, Cao J, Zhang Q, Li Y, Xu L (2016) Edge computing: vision and challenges. IEEE Internet Things J 3(5):637–646
7. Bonomi F, Milito R, Zhu J, Addepalli S (2012) Fog computing and its role in the internet of things. In: MCC 2012 Proceedings of the first edition of the MCC workshop on mobile cloud computing, Helsinki, Finland
8. Breivold HP, Sandström K (2015) Internet of Things for industrial automation - challenges and technical solutions. In: IEEE international conference on data science and data intensive systems, Sydney, Australia
9. Stouffer K, Falco J, Scarfone K (2011) Guide to industrial control systems (ICS) security. National Institute of Standards & Technology (NIST), Gaithersburg, MD, USA
10. Technical Review Middle East (2016) Schneider Electric introduces world's first ePAC for IIoT in GCC, IT Section, 29 November 2016
11. Johnson RE (2010) Survey of SCADA security challenges and potential attack vectors. In: International Conference for Internet Technology and Secured Transactions, London, UK
12. Agarwal N, Brem A (2015) Strategic business transformation through technology convergence: implications from general electric's industrial internet initiative. Int J Tech Manage 67:196–214

13. Lorga M, Goren NS, Mahmoudi C (2018) Fog computing conceptual model, recommendations of the national institute of standards and technology, NIST, Gaithersburg, MD, USA
14. Liu H, Guo J, Yu W, Zhu L, Liu Y, Xia T, Sun R, Gardner RM (2016) The design and implementation of the enterprise level data platform and big data driven applications and analytics. In: IEEE/PES transmission and distribution conference and exposition (T&D), Dallas, TX, USA, 3–5 May 2016
15. Bikfalvi P, Kulcsár G, Forrai MK, Erdélyi F, Tóth T (2013) On some functions of the MES applications supporting production operations management. In: Applied information science, engineering and technology, Springer, Cham, pp. 103–129
16. Beath C, Sebastien I, Ross J, Beath C, Mocker M, Moloney K, Fonstad N (2017) How big old companies navigate digital transformation. MIS Q Executive 197–201 (2017)
17. Malli SS, Vijayalakshmi S, Balaji V (2017) Real time big data analytics to derive actionable intelligence in enterprise applications. In: Internet of Things and big data analytics toward next-generation intelligence. Studies in Big Data, vol. 30, Springer, pp 99–121
18. Ochs T, Riemann U (2018) Smart manufacturing in the Internet of Things era. In: Internet of Things and big data analytics toward next-generation intelligence, Springer, pp 199-217
19. McKinsey & Company (2015) The Internet of Things: mapping the value beyond the hype. McKinsey Global Institute
20. International Society of Automation, ISA95, Enterprise-Control System Integration. https://www.isa.org/isa95/
21. International Organization for Standardization, Enterprise-control system integration. https://www.iso.org/standard/57308.html
22. Koch R (2018) Draining the data swamp: an organization's data lake is only as good as the preparation and maintenance planning that go into creating it, Strategic Finance, 1 June 2018
23. Radanliev P, Nicolescu R, Huth M, De Roure DC (2019) A reference architecture for integrating the Industrial Internet of Things in the Industry 4.0, University of Oxford, Oxford, UK
24. Kosieradzka A (2017) Maturity model for production management. In: 7th international conference on engineering, project, and production management
25. Gill B, Shea T (2018) Industrial IoT brings benefits and opportunities for rotating equipment suppliers and plant operators. In: Asia turbo machinery & pump symposium, Singapore
26. Schuh G, Anderl R, Gausemeier J, ten Hompel M, Wahlster W (2017) Industrie 4.0 Maturity index: managing the digital transformation, acatch, Munich, Germany
27. Hand A (2018) The value proposition of IIoT, Industry 4.0, Automation World, 16 June 2018
28. De Carolis A, Macchi M, Negri E, Terzi S (2017) A maturity model for assessing the digital readiness of manufacturing companies. In: Advances in production management systems, Part 1, Hamburg, Germany, 3–7 September 2017
29. Yamfashije J (2017) Capability maturity model integration, Carnegie Mellon University, Pittsburg, PA, USA, April 2017
30. Strother N, Ravens S (2019) Finding the ROI in digital transformation – case studies and analysis of the OSIsoft PI system, Navigant Research, Navigant Consulting Inc. pp 2–5, 26
31. Alcácer V, Cruz-Machado V (2019) Scanning the Industry 4.0: a literature review on technologies for manufacturing systems. Eng Sci Tech J. www.elsevier.com
32. Lu Y, Morris KC (2016) Current standards landscape for smart manufacturing systems, National Institute of Standards and Technology (NIST), US Dept of Commerce, Feb 2016. http://dx.doi.org/10.6028/NIST.IR.8107

Real Value of Data in Managing Manufacturing Assets

Teuvo Uusitalo(✉), Jyri Hanski, Helena Kortelainen, and Toni Ahonen

VTT Technical Research Centre of Finland, Visiokatu 4, 33101 Tampere, Finland
{teuvo.uusitalo,jyri.hanski,helena.kortelainen,
toni.ahonen}@vtt.fi

Abstract. Manufacturing networks contain vast amounts of data over the whole lifecycle of assets. There is much talk of data based services and business models but limited number of success stories from the industry. Valuation of data is identified as a major obstacle for successful service business. Actors operating in the value network may have different perspectives about the value of data and do not necessarily see the opportunities that the data offers to other actors in the network. Often the concept of data value may be limited to a single dimension such as quality or security. This paper aims at providing a more holistic perspective on the value of data in manufacturing asset management. The paper presents dimensions of data value based on a literature review. Additionally, future research needs regarding value of data in managing manufacturing assets are identified.

Keywords: Data-driven services · Value · Asset management · Digitalization

1 Introduction

Traditionally, manufacturers have focused on producing a physical product and capturing value by transferring ownership of the product to the customer through sales. The asset owner is then responsible for the costs of operating, maintaining and upgrading the product. Digitalization allows for a radical change of this traditional business model. Digital products and services require the rethinking of the value proposition. The entire life cycle of products and services must be reconsidered. The optimization of customers' processes is an area in which digitalization creates new opportunities.

New services that are based on data pose a presumption that the service provider and customer agree on data-sharing principles. Furthermore, the parties need to reach a mutual understanding of the value creation opportunities. One challenge is to identify and communicate the value of the data. This has been recently identified e.g. by the World Manufacturing Forum [1]. In order to generate profitable data based business it is essential to identify the value of the data and its benefits and the potential end users and customers. The value creation is possible by processing the data, but on the other hand, refining requires expertise, analytics and tools. Big data has become of interest also in asset management. It refers to using of data that is larger than what can be utilized in common software applications. It is characterized by great volume, various

A. Crespo Márquez et al. (Eds.): WCEAM 2019, LNME, pp. 164–174, 2021.
https://doi.org/10.1007/978-3-030-64228-0_15

modalities, rapid generation, and huge value but very low density [2]. Big data creates value for companies in two ways: 1) data is used for the incremental improvement and optimization of current business practices, processes and services, and 2) new products and business models can be innovated based on data use [3].

The focus of research in manufacturing sector is usually on a certain dimensions of data value such as data quality, refinement of data, costs, transparency, supporting decisions or new business models [4–7]. There is no consensus on the determinants of data value in manufacturing [8]. Additionally, comprehensive and systematic literature reviews that consider value of data in manufacturing sector are very limited e.g. [4, 9]. None of the identified reviews focused on asset management. Therefore, there is a need for a review and a comprehensive framework that compiles the different aspects of data value that are relevant for supporting the asset management decisions in the manufacturing sector.

This paper aims to research the dimensions of the data value in the manufacturing sector, with a focus being on asset management. The first research questions is as follows: "What are the main elements of data value in the manufacturing sector?" To answer to this question the literature on data value is researched using scoping review as the research method. The second research question is "what are the main dimensions of data value for asset management in the manufacturing sector?" This question is answered by analysing the identified elements of data value through the asset management framework.

1.1 Value of Data in Manufacturing and Asset Management

Asset management can be defined as "coordinated activity of an organisation to realize value from assets" [10]. Assets or manufactured products typically have a life cycle that has the following phases: concept, development, realization, utilization, enhancement and retirement [11]. These life cycle phases produce, and require, large amounts of data that support the asset management related decision-making. In other words, data has value for asset management.

The term "value" has intrinsically a positive connotation, which can be related to the ideas of appreciation and benefit generation. Value as a term is usually related to the usefulness, quality, importance, price and worth of a subject [12]. A further definition reaches towards ecosystem thinking by stating that "Value is the set of benefits derived by a stakeholder from an exchange" [13]. This definition reflects the idea that a value network generates economic, environmental and social value through complex dynamic exchanges between one or more enterprises, customers, suppliers, strategic partners and the community.

Most innovative data-driven business models are showing a wide variety of value creation opportunities, from direct data monetization to access-based valorization of data assets on sharing platforms [14]. In manufacturing, value of data has been extensively discussed in research fields such as servitization and outcome-based business models [4, 6, 15, 16], big data [4, 7], supply chains [5, 17] and sustainability and circular economy [18].

2 Methodology

Scoping review was selected as the research strategy in this study due to the exploratory nature of this research, to reach an overview on the topic and to identify the key characteristics related to the concept of data value in asset management. In a scoping review, literature in a given area is rapidly gathered, aiming to provide an overview of the type, extent and quantity of research available [19, 20].

With search engine eKnowledge, covering a variety of relevant databases such as Scopus and Web of Science and Google search engine, key scientific and grey literature i.e. governmental, NGO, industrial and commercial publications were identified. A combination of keywords including “value, data, information, manufacturing, industry, asset management, internet of things and decision-making” were applied. Papers published between 2000 and 2020 were included in the review to include only the most recent papers. Literature search was complemented with peer-reviewed articles, conference papers and reports that the authors were already familiar with. In addition, new articles were discovered by analysing the key references in the original papers identified in the previous phase. Altogether 31 papers were selected.

In data analysis, the relevant data from individual papers was extracted and synthesized. The data was coded and analysed inductively to identify emerging categories for data value dimensions. A qualitative data analysis software was used in this phase to increase the transparency of chain of reasoning and to support the simultaneous analysis by the authors. The validity and reliability of results were increased by data and investigator triangulation [21]. This iterative process included creation of several versions of the framework based on analysis of the data and feedback received by the authors.

3 Results - Dimensions of Data Value in Manufacturing Asset Management

The focus of research in manufacturing sector is usually on certain dimensions of data value. These include data quality [22], analytics [6, 17], direct and indirect costs [22], transparency to value network and stakeholders [17], quicker product introductions, new business models, supporting customer success and product as part of broader system [6], refinement of data [4], selling data to other entities [4], and data-driven decisions, ecosystem change and management processes [7]. Subjectivity and case-specificity applies also to the value of data that is strongly linked to its use case and context e.g. [23]. The following sections summarizes the attributes, features and other dimensions that are attached to data and that may affect to its (monetizable) value and the ways value can be harnessed from the data.

3.1 Business Models

In general, business dimension includes approaches such as selling data, analyzing data or creating products or services based on data [4, 24]. Hartmann et al. [3] identify a more specific list of data based business models: free data collector and aggregator, analytics as a service, data generation and analysis, free data knowledge discovery, data aggregation as a service, and multi-source data mash-up and analysis.

Value creation models that are highly digitized and leverage data to connect through to the end user are becoming integrated with value delivery and value capture models, allowing manufacturing and supply chains to better scale up [25]. In managing manufacturing assets, the data based business models can be divided according to level of data refinement and the value to the asset owner as data (DaaS), information (IaaS), knowledge (KaaS) and wisdom as a service (WaaS). The theoretical background is derived from the data, information, knowledge and wisdom (DIKW) hierarchy [26] and the 'knowledge pyramid' [27]. At the DaaS level, the asset owner is provided with an opportunity for gathering and storage of asset data or a specific set of data. IaaS requires further refinement of the data. KaaS contains collecting and analysing information by exploiting service provider's own capabilities and experience in co-operation with the asset owner. WaaS requires a deep understanding of asset owner's business and a wide access to asset owner's information system and other data sources. An example of WaaS could be service provider's ability to develop physical products and services needed in the future and thus, to provide competitive edge for asset owners [28].

Different approaches for data valuation have been proposed. These include traditional asset valuation concepts based on cost, fair market value, and future revenue. In addition, indirect approaches to data valuation emphasize risk and cost associated with lack of data quality and data management [29]. High infrastructure and hardware costs need to be considered when assessing the data value [30]. Similarly, value of information should be taken into account relative to its costs [10] and cost-benefit analysis is presented as a method for determining value of information over asset lifecycle [31].

3.2 Data Characteristics

Data is the fuel of any intelligent industrial system. However, it is not yet considered as a factor of production such as raw material, capital, labour, and energy [1]. To understand the value of the data, the concept of data and its attributes need to be discussed first. Data can be seen as a raw material, which companies derive from their activities [32]. Data are symbols that represent the properties of objects and events [26], unrefined and unfiltered information or a set of discrete objective facts about events [33]. Data is non-rivalrous i.e. one single data set can be exploited by multiple actors at the same time [32]. It is non-fungible i.e. different data sets contain different information and hold different value and experience i.e. the value of data is only known after it has been put into use (ibid.). Data can be specific such as data that is specifically collected at a site for an individual machine. Examples of data characterization include real-time such as measurements or sensor data, history or discrete such as event or incident data.

Amount of data has increased drastically over the last decades. Therefore, big data has been strongly in the focus of research and companies. Big data includes 5Vs: Velocity (acquisition sources), Volume (storage), Veracity (analysis), Variety (development sources) and Value-adding [34]. Recent articles focus on processes of creating value from big data [35, 36] and value of big data in industrial companies [37]. Value elements of big data include integrating heterogeneous data sources and combining data in new ways, and transferring and accessing data from one context to be used in other contexts

[38]. Even though the focus of research and business has been on big data, traditional data is still considered more valuable than unstructured big data [37].

3.3 Data Value Chain

Organization should define processes for where and how the data is reworked into usable information and how it will be communicated [10]. Several authors have presented elements of the data value chain (e.g. [28, 30]) Data value chain includes phases such as data collection, data pre-treatment, descriptive data analysis, data modelling, business knowledge and business decisions.

The real added value in managing manufacturing assets arises when the data is exploited for the prediction of future behaviour, for the follow-up of asset performance, for estimating remaining useful life, identifying the cause of underperforming systems and supporting planning and decision-making [28].

3.4 Decision-Making

Due to data being an experience good, there is no inherent value in data to businesses and it becomes valuable only when placed in relevant context [23, 32, 37]. Value of data relates to the value derived from using data in different use-cases and decision contexts. As an example, data is often specific for a certain machine or contexts and may become obsolete if it is replaced.

Organisation's decision context consists of factors such as decision-making level, objectives, constraints and business impact of decisions [39]. Organisation should determine the goal for utilising data [37]. Organizational objectives, constraints and business impact of decisions vary at different organisational levels i.e. strategic, tactical or operational. In addition, decision-making levels and decision situations related to asset management have differing needs of information. Strategic decisions may require information on trends and developments in the operating environment, bottlenecks, and improvement potential of current processes [39]. Aspects covered by strategic asset management include operational and technical, financial, environmental, social, data and information management, organisational factors and quality of processes and systems, strategic management, regulation and external stakeholders, technology and market and competition [40]. In comparison, operational decisions are typically supported by information on asset condition, type, location, and availability (ibid.).

At operational level, the data supports daily operation and maintenance, and organisational constraints may include maintenance strategy and the criticality of equipment [39]. In this kind of decision-making situations, the business impact is seldom crucial, whereas at strategic level the business impact may be major (ibid.). Type of decision-making situation affects the value of data. Asset management related decision situations can be classified into operations performance, maintenance decision-making and investment decision-making [28]. Decision-making situations can be defined based on, for instance, time to use for decision, frequency of decisions and effect of decision [41]. Time perspective i.e. if the decision is based on data on past, present or future affects the value of data [42].

3.5 Management

Management dimension contains factors such as security and privacy, data quality, data governance, data and information sharing and data ownership [30]. Asset management standard ISO 55000-2 gives several guidelines to the management of data that consider governance and sharing perspectives [43]. An organization should consider the complexity of its processes for managing its asset information. Organization should also consider the need to align its information requirements to suit the level of criticality of the asset in the manufacturing system and the risk that an asset, or managing it, poses. Duplication of data unnecessarily should be avoided. Data and information should be easily exchangeable with service providers and use of common terminology is emphasized.

Data management needs to take into account the security and privacy concerns. Organizations can be reluctant to exchange data with their partners due to these concerns. Combination of different data sources can lead to situations where it is not clear what the privacy implications could be [38]. IDS provides guidelines on how the security aspects should be taken into account in data supply chains [44]. For successful data sharing relationships, it is essential to build trust between collaborating partners. To promote trust, companies should regard data as a business asset and build relational contracts that focus on common goals and mutual benefit [45].

Quality of information is the ability of an information collection to meet user requirements [46]. Quality of information should be relative to its use [29, 43]. It includes dimensions such as accuracy, completeness, security, understandability, timeliness and consistency [31, 46]. Collaboration with stakeholders is needed to ensure quality of asset management information [10].

Development of digital technologies creates a need to develop a common understanding of ownership rights regarding the data created by connected devices. A sensor manufactured by one company can operate in a system developed by another, and be deployed in an environment owned by a third. Agreement will be needed on who has which rights to the resulting data [47]. On the other hand, it can be argued that by sharing the data and partially the products, companies have increased informal dialogue, collaboration in monitoring, training and remediation. This suggest that data should not be considered only from the viewpoint of return of investment but also of the amount of reuse when it is compared to the effort of publishing. Therefore, the value of data emerges from use and reuse instead of ownership [28].

3.6 Organisational Perspective

Information needs in different levels of organization are different [10]. The value of data is case-specific: it is created in the use case, for example in improving customers processes. Therefore, it is important to identify the roles involved in creating value from data. IDS has identified the following roles: data owner, data provider, data consumer, data user, broker service provider, clearing house, identity provider, app store, app provider, vocabulary provider, software provider and service provider [44].

Value of information is influenced by the information flow enabled by the business processes and the overall information capacity of the information system i.e., information diffusion and information capacity [8]. Organizations experience organizational learning as data are applied to organizational processes [22]. As data are increasingly utilised to guide future learning and activities, organisations improve their absorptive capacity (ibid.). Increased data collection is argued to improve organisation's ability to integrate the data into its processes and make better decisions [22].

The knowledge pool of organizations consists of tacit and explicit knowledge. Tacit knowledge refers to non-verbalized action models, skills, ideas and thoughts, which give contents to the verbalised company information. Tacit knowledge can be learned only by experience, and communicated only indirectly, through metaphor and analogy. Data needs to be combined with tacit knowledge to be able to create a deeper understanding of the current case and thus form relevant alternatives for decisions [41].

3.7 Enabling Technical Solutions

Data is generated, treated, analyzed and utilized in many IT and manual systems and with help of many technical solutions. Analytics methods can be classified into descriptive, predictive and prescriptive that answer to questions what has happened, what is likely to happen and what is required to do more, respectively [30]. Intelligent assets require creation of a drastically new multi-layered technology infrastructure (technology stack) that consists of platforms, programs, networks, services, processes, and actors [48].

Smart manufacturing consists of the physical world i.e. physical assets, software and people that are connected to internet of things (IoT) through sensors and the cyber world, which consists of virtual mapping of the physical world (digital twin) and digital services [49]. Integration of Information Technology (IT), Operational Technology (OT) and Engineering Technology (ET) is a major factor affecting the value of data. Integration of IT, OT and ET is understood as connecting the data (CAD and PLM) from new product development and design (ET), to the production and resource planning (ERP and MES) (IT), and real-time analysis of manufacturing data from the shop floor (IoT sensor and machine tool data) (OT) [1].

Successful data sharing in manufacturing requires interoperability of sensors, machines, and companies to enable easy aggregation and analysis of data. Interoperability requires several layers of standardization [45]. Semantic models can provide solutions to technical interoperability for specific use cases. In addition, artificial intelligence based solutions are available to manage semantic interoperability [50].

3.8 Sustainability and Business Ethics

Sustainability i.e. economic social and environmental aspects have considerable impact on data value. Sustainability is tied to demands of transparency and conscientious management of organisation's resources [40]. Data have both positive and negative impact on sustainability. Firstly, digital technologies and their widespread utilisation is associated with rising energy consumption in data centers and networks [51, 52]. Secondly, data has huge potential to increase the material and energy efficiency of manufacturing [1]. Economic benefits of data include increase in profit, business growth, and competitive advantage resulting from big data adoption [38]. Utilisation of big data to guide strategic and operative decisions has positive impact on financial performance [38].

Circular economy is a rising trend that aims to increase the sustainability of manufacturing and has an impact on data value. Circular economy aims to keep resources in use as long as possible and eliminate waste [1]. Circular economy value drivers include extending the useful life and maximising the utilisation of assets, looping assets, and regenerating natural capital [18]. Data on location, condition and availability through sensors, data analytics, machine learning and blockchain enable and drive environmental actions in the same way they are driving business actions [1, 18]. To increase the sustainability of manufacturing, five strategies are suggested: 1. redesign products and materials selection for reuse, 2. conserve and recover resources from used products for new products, 3. develop new ways of production, 4. implement service-based business models and 5. shift to renewables and eliminate toxic chemicals [1].

4 Discussion and Conclusions

Multiple data value dimensions have implications for asset management in the manufacturing sector. Traditionally, asset management is focusing on rather limited aspects of data value such as costs and impacts, decision-making and risk management. The emergence of Industry 4.0 and IoT based solutions, has sparked an interest on advanced technologies and ecosystems. This in turn, has accentuated the need for data security, privacy and novel business models based on data. Increased volume and quality of data has enabled more advanced decision support and risk management, life cycle cost and benefit assessments and network level optimization of activities.

Increased implementation of IoT solutions and improved information systems have increased the amount of data available from different systems. However, increased data do not automatically correlate with value for asset management. For example, data from process and automation systems rarely support asset management decisions.

When considering the life cycle of assets, the focus of data value is currently on use phase. There is a need for more emphasis on the earlier phases such as concept and development and later stages such as enhancement and retirement. As an example, novel business models based on sustainability and circular economy set requirements for both concept and retirement phases.

Future research is needed on assessment and realization of value of data for different stakeholders in industrial value networks. Case studies on the topic are needed. Relevant topics include, for instance, business models, contract types, risk management, and measurement and demonstration of the value of data. Specific research needs include:

- **Manufacturing and asset management data typology**. Different perspectives of data and value of data exist. How to build a bridge between the different areas, activities, companies and scientific fields (e.g. design and O&M)? Common typology is needed for data and value of data for different business actors and researchers.
- **Data is the core competence of asset-intensive companies of the future**. Manufacturing companies are currently developing data based services. In the future, their competitiveness may be determined by the data (see e.g. [4]). Is the data the source of value instead of novel technical solutions? Novel technical solutions provide opportunities for data collection, analysis and utilization and highlights the importance of quantification of costs and benefits (see e.g. [31]).

- **Newcomers and changing roles of companies in the data based business networks.** Data based services change the roles of incumbent companies in the business networks. What are the new roles of companies and other actors in the manufacturing networks of the future? (see e.g. [49]). What are their roles in the data value chain? How and for whom value is created?
- **Data-based service business models.** Data based business highlights the role of business models. What kind of business models are needed for data based services in the manufacturing sector in the future? How is the data monetized?

References

1. World manufacturing foundation, world manufacturing forum (2018) The 2018 world manufacturing forum report, recommendations for the future of manufacturing
2. Chen M, Mao S, Liu Y (2014) Big data: a survey. Mob Netw Appl 19(2):171–209. https://doi.org/10.1007/s11036-013-0489-0
3. Hartmann PM, Zaki M, Feldmann N, Neely A (2016) Capturing value from big data – a taxonomy of data-driven business models used by start-up firms. Int J Oper Prod Manag 36:1382–1406. https://doi.org/10.1108/IJOPM-02-2014-0098
4. Opresnik D, Taisch M (2015) The value of big data in servitization. Int J Prod Econ 165:174–184. https://doi.org/10.1016/J.IJPE.2014.12.036
5. Hazen BT, Boone CA, Ezell JD, Jones-Farmer LA (2014) Data quality for data science, predictive analytics, and big data in supply chain management: an introduction to the problem and suggestions for research and applications. Int J Prod Econ 154:72–80. https://doi.org/10.1016/J.IJPE.2014.04.018
6. Rymaszewska A, Helo P, Gunasekaran A (2017) IoT powered servitization of manufacturing – an exploratory case study. Int J Prod Econ 192:92–105. https://doi.org/10.1016/J.IJPE.2017.02.016
7. Sheng J, Amankwah-Amoah J, Wang X (2017) A multidisciplinary perspective of big data in management research. Int J Prod Econ 191:97–112. https://doi.org/10.1016/J.IJPE.2017.06.006
8. Batini C, Castelli M, Viscusi G et al (2018) Digital information asset evaluation: a case study in manufacturing. ACM SIGMIS Database DATABASE Adv Inf Syst 49:19–33. https://doi.org/10.1145/3242734.3242737
9. Cai L, Zhu Y (2015) The challenges of data quality and data quality assessment in the big data era. Data Sci J 14:1–10. https://doi.org/10.5334/dsj-2015-002
10. ISO (2014) ISO 55000-2014 - asset management - overview, principles and terminology
11. IEC (2014) IEC 60300-1: dependability management – part 1: guidance for management and application. international electrotechnical commission (IEC), Geneva
12. Zeithaml VA (1988) Consumer perceptions of price, quality, and value: a means-end model and synthesis of evidence. J Mark 52:2–22. https://doi.org/10.1177/002224298805200302
13. Evans S, Rana P, Short S (2012) State-of-practice in business modelling and value-networks, emphasising potential future models that could deliver sustainable value
14. de Vallejo LI, Scerri S, Tuikka T (2019) Towards a European data sharing space. Brussels
15. Visnjic I, Jovanovic M, Neely A, Engwall M (2017) What brings the value to outcome-based contract providers? Value drivers in outcome business models. Int J Prod Econ 192:169–181. https://doi.org/10.1016/J.IJPE.2016.12.008

16. Garcia Martin PC, Schroeder A, Ziaee Bigdeli A (2019) The value architecture of servitization: expanding the research scope. J Bus Res 104:438–449. https://doi.org/10.1016/J.JBUSRES.2019.04.010
17. Kamble SS, Gunasekaran A, Gawankar SA (2020) Achieving sustainable performance in a data-driven agriculture supply chain: a review for research and applications. Int J Prod Econ 219:179–194. https://doi.org/10.1016/J.IJPE.2019.05.022
18. Ellen MacArthur foundation (2016) Intelligent assets: unlocking the circular economy potential. Ellen MacArthur Found, pp 1–25
19. Curtin University (2019) Systematic Reviews in the health sciences. https://libguides.library.curtin.edu.au/systematic-reviews. Accessed 27 Jan 2020
20. Munn Z, Peters MDJ, Stern C et al (2018) Systematic review or scoping review? Guidance for authors when choosing between a systematic or scoping review approach. BMC Med Res Methodol 18:7. https://doi.org/10.1186/s12874-018-0611-x
21. Patton MQ (1987) How to use qualitative methods in evaluation. Sage, Newbury Park
22. Hazen BT, Weigel FK, Ezell JD et al (2017) Toward understanding outcomes associated with data quality improvement. Int J Prod Econ 193:737–747. https://doi.org/10.1016/J.IJPE.2017.08.027
23. Koutroumpis P, Leiponen A (2013) Understanding the value of (big) data. In: Proceedings - 2013 IEEE international conference on big data, big data. IEEE
24. Thomas LDW, Leiponen A (2016) Big data commercialization. IEEE Eng Manag Rev 44:74–90. https://doi.org/10.1109/EMR.2016.2568798
25. Joglekar N, Parker G, Srai JS (2020) Winning the race for survival: how advanced manufacturing technologies are driving business-model innovation
26. Ackoff R (1989) From data to wisdom. J Appl Syst Anal 16:3–9. https://doi.org/citeulike-article-id:6930744
27. Rowley J (2007) The wisdom hierarchy: representations of the DIKW hierarchy. J Inf Sci 33:163–180. https://doi.org/10.1177/0165551506070706
28. Kortelainen H, Hanski J, Kunttu S, et al (2017) Fleet service creation in business ecosystems – from data to decisions
29. Fleckenstein M, Fellows L (2018) Modern data strategy. Springer, Heidelberg
30. Sivarajah U, Kamal MM, Irani Z, Weerakkody V (2017) Critical analysis of big data challenges and analytical methods. J Bus Res 70:263–286. https://doi.org/10.1016/J.JBUSRES.2016.08.001
31. Ouertani MZ, Parlikad AK, McFarlane D (2008) Asset information management: research challenges. In: Proceedings of the 2nd international conference on research challenges in information science, RCIS 2008, pp 361–370
32. MIT technology review insights and oracle (2016) The rise of data capital
33. Liew A (2007) Understanding data, information, knowledge and their inter-relationships. J Knowl Manag Pract 8:1–16
34. Addo-Tenkorang R, Helo PT (2016) Big data applications in operations/supply-chain management: a literature review. Comput Ind Eng 101:528–543. https://doi.org/10.1016/J.CIE.2016.09.023
35. Ylijoki O, Porras J (2019) A recipe for big data value creation. Bus Process Manag J 25:1085–1100. https://doi.org/10.1108/BPMJ-03-2018-0082
36. Urbinati A, Bogers M, Chiesa V, Frattini F (2019) Creating and capturing value from big data: a multiple-case study analysis of provider companies. Technovation 84–85:21–36
37. Dueholm Müller S, Jensen P (2017) Big data in the Danish industry: application and value creation. Bus Process Manag J 23:645–670. https://doi.org/10.1108/BPMJ-01-2016-0017
38. Günther WA, Rezazade Mehrizi MH, Huysman M, Feldberg F (2017) Debating big data: a literature review on realizing value from big data. J Strateg Inf Syst 26:191–209. https://doi.org/10.1016/J.JSIS.2017.07.003

39. Ahonen T, Hanski J, Uusitalo T (2018) Approach to digital asset management service development. In: Proceedings of 13th world congress on engineering asset management (WCEAM2018). Presented at Stavanger, Norway, 24–26 September 2018
40. Hanski J, Ojanen V (2019) Sustainability in strategic asset management frameworks: a systematic literature review. Int J Strateg Eng Asset Manag 3:263–294
41. Kunttu S, Ahonen T, Kortelainen H, Jantunen E (2016) Data to decision - knowledge-intensive services for asset owners. In: Proceedings of EuroMaintenance 2016, Athens, Greece
42. Erguido A, Marquez AC, Castellano E, et al (2019) Asset management framework and tools for facing challenges in the adoption of product-service systems. IEEE Trans Eng Manag 1–14. https://doi.org/10.1109/TEM.2019.2951438
43. ISO (2014) ISO 55000-2, Asset management. Management systems. Guidelines for the application of ISO 55001
44. IDS (2018) IDS reference architecture model industrial data space
45. World economic forum (2020) Share to gain : unlocking data value in manufacturing
46. Pipino L, Lee Y, Wang R (2002) Data quality assessment. Commun ACM 45:211–219
47. OECD (2017) The next production revolution
48. Juhanko J, Jurvansuu M, Ahlqvist T, et al (2015) Suomalainen teollinen internet – haasteesta mahdollisuudeksi: taustoittava kooste
49. Tao F, Qi Q (2019) New IT driven service-oriented smart manufacturing: framework and characteristics. IEEE Trans Syst Man Cybern Syst 49:81–91. https://doi.org/10.1109/TSMC.2017.2723764
50. World economic forum (2020) A roadmap for cross-border data flows: future-proofing readiness and cooperation in the new data economy
51. Jenkin TA, Webster J, McShane L (2011) An agenda for "green" information technology and systems research. Inf Organ 21:17–40. https://doi.org/10.1016/j.infoandorg.2010.09.003
52. de Vries A (2018) Bitcoin's growing energy problem. Joule 2:801–805. https://doi.org/10.1016/j.joule.2018.04.016

Internet of Things Adoption Challenges in Enterprise Asset Management Organisations

Wei Jye Seo(✉) and Nina Evans

University of South Australia, GPO Box 2471, Adelaide 5001, Australia
wei_jye.seo@mymail.unisa.edu.au

Abstract. Enterprise Asset Management (EAM) is used to support strategic planning and budgeting, maintenance, repair and operations of organisational assets throughout the asset lifecycle. The purpose of EAM is to improve the assets' reliability and performance without compromising health, safety and the environment, where Internet of Things (IoT) devices are useful to sense, monitor, respond to, or report asset conditions in real-time with minimum human intervention. Typically, IoT is adopted or integrated into an existing business process in an EAM organisation without reengineering the overall business process. As IoT technologies are usually not custom made for specific business process, adequate governance is critical to the success of IoT adoption. In this paper, adoption of IoT in EAM organisations and its challenges will be discussed from the perspective of data management. Literature reviews and academic case studies are also included to help identify the research gaps and data governance challenges related to adopting IoT technologies into EAM organisations. The factors influencing the IoT adoption will be presented from a technology, organisational, and people perspective. Different categories of challenges have been identified, including standardisation, security and privacy, data quality, organisational culture, business continuity and resilience. The paper concludes with some recommendations to improve IoT adoption in EAM organisations, as well as suggestions for future research directions.

Keywords: IoT · Internet of Things · Adoption challenges · TOP · Enterprise Asset Management

1 Introduction

When organisations invest in assets that are vital for their business success, they need a system to take care of these asset and extend their operational status as cost effectively as possible Similar to enterprise resource planning (ERP) systems, Enterprise Asset Management is a dedicated enterprise level program to maximise the value of physical assets, while reducing the cost of ownership [1]. Throughout the asset lifecycle, from asset planning to decommission, EAM helps to minimise risks and costs while maximising business value. In order to optimise the capital investment of the organisation, EAM aims to reduce maintenance costs and asset downtime [1]. However, EAM can be

A. Crespo Márquez et al. (Eds.): WCEAM 2019, LNME, pp. 175–186, 2021.
https://doi.org/10.1007/978-3-030-64228-0_16

complicated as it requires substantial information collection from different parts of the organisation.

The concept of managing assets remotely via the internet started more than twenty years ago [2]. The primary goal of developing asset management via the internet, was to monitor and prevent equipment failure that could cause adverse business effects. The definition of IoT is not restricted to a standard or industry and Gubbi et al. [3] define IoT as “the interconnection of sensing devices with the ability to share information over the internet” [3]. With the integration of IoT, the information collection can be done timely and more effectively with internet connectivity. The data collected can be pre-processed to reduce usage of central processing power and communication bandwidth. For example, relating to the maintenance aspect, the physical condition of asset is monitored in real time. This enables the organisation to replace reactive maintenance with proactive (predictive and preventive) maintenance with the support of machine learning.

2 IoT Adoption

In the 1980s, researchers articulated their ambition to embed technology into the background of our everyday life. In the past decade, many organisations have been adopting IoT related technologies. Many have done so without realising that it is IoT, as the definition of IoT is still being developed and adapted as the technology evolves [3]. As time passes, IoT is gaining more coverage and this technology will be adopted in various industries. The next section describes how various sectors are actively adapting IoT as part of their EAM.

2.1 Manufacturing

As more IoT technologies are being adopted into the manufacturing industry, organisations in this sector are experiencing enhanced performance and productivity [4]. The adoption of IoT has supported the optimisation and automation of processes in manufacturing organisations and enabled improved integration of administrative control into their production lines. The analysis of the data entering through the IoT devices (through machine learning and advanced statistics), plays an important role in self-organisation and logistic optimisation has been made possible, as the status of the asset and resources can now be easily tracked [5]. As the machines and processes are monitored through IoT, maintenance or replacement of parts can be automatically scheduled. Some manufacturers and IT companies are developing IoT-based manufacturing platforms that reduce equipment maintenance downtime using predictive maintenance [6]. These capabilities relate to the important features of the so-called *Industrie 4.0*, which was initiated by the German government. The data acquired by IoT devices are integrated with the manufacturing data to support strategic planning and decision making.

2.2 Aerospace and Aviation Industry

As about half of the commercial aviation budget is spent on maintenance, repair and operation (MRO), Integrated Vehicle Health Management (IVHM) is integrating IoT as part of the system [6]. Predictive maintenance in the aviation industry is mainly focused on reducing aircraft downtime. Safety is also part of the objective of condition monitoring systems, even though airplanes have many redundant features that make them reliable. In situations where critical and spontaneous actions are required for safety reasons - such as oxygen leaks or insufficient pressure in hydraulic arms - immediate feedback to the pilot and instant suggestions of alternative solutions are needed to prevent harm to the passengers.

2.3 Supply Chain Integration

The integration of IoT technologies supports improved resource management and management of the state of assets throughout their lifecycle [4]. Sensors are being used to track and eventually automate processes in real-time. For example, items can be located and presented with smart shelves and inventory and shelfing can be scheduled in advance. The logistics of the overall supply chain is thereby optimised [5]. IoT has also been playing an important role in processes such as counterfeit prevention and detection, environmental emission monitoring, hazardous and threat detection, and other industrial or organisational specific processes. In the pharmaceutical industry, sensors are used to label drugs and other sensitive products and to monitor their use, to improve security and safe handling throughout the supply chain [5].

2.4 Healthcare Industry

IoT technologies has been in the health sector to improve the quality of care. These applications range from RFID that tracks patients, medication and equipment, to wireless controlled implants [4]. IoT technologies add intelligence to the system and enable professional clinical staff to improve the overall medication process and deliver better quality of healthcare [7]. IoT technologies are widely adopted to manage assets towards improved energy management, reduction in human errors and real-time patient monitoring. Medical sensors bridge the gap between the IoT servers in the hospital and the patients, e.g. implants with wireless connectivity devices have been used to monitor and store patient information in real-time. For example, medical implants such as pace makers and other electrical simulation systems that restore or simulate dedicated function in the human body, can be remotely controlled with wireless technology [5].

2.5 Automotive Industry

The communication between automotive vehicles and other IoT devices is gaining more attention in the automobile industry [4]. Research refers to the Internet of Vehicles and Car-to-X Communication. IoT technologies can improve road safety and real time road condition, by providing geographical data and broadcasting real time information to relevant agencies or other vehicles via ad hoc networks. Vehicles can therefore act as intelligent items that can potentially become autonomous during hazardous road or driving conditions. Other communication technologies, such as Dedicated Short Range Communication (DSRC) that are used with the Vehicle-to-Vehicle (V2V) and Vehicle-to-Infrastructure (V2I) communication, enhances Intelligent Transport Systems (ITS) integration for better traffic safety [5]. IoT is currently being integrated into on electric cars for more responsive and transparent data sharing. The purpose of the IoT adoption is to improve the efficiency of accessibility of asset information for the electric car manufacturer. IoT data acquisition will therefore support the risk assessment and lifecycle delivery of the asset [8].

2.6 Energy Provider

In environments that are dangerous to humans or where the equipment is costly - such as the seabed drilling for the oil and gas industry - IoT is critical for supporting successful and safe operation. IoT helps to prevent or reduce the occurrence of accidents through the use of smart wireless sensors that transmit real time data in hazardous environment [5]. Furthermore, IoT devices can also be used to monitor pressure, electricity, gas, chemical, and other metrics during the energy generating operation [4]. By capturing and analysing real time data, the machine and equipment health could be monitored. This will prevent unplanned downtime and report potential hazardous risk before they occur. Other examples of the adoption of IoT in this industry include the monitoring of critical assets in an electrical grid to mitigate the risks of asset failures and prioritise maintenance actions on the grid.

2.7 Smart Cities

Smart Cities originated in 2008, to bring better interconnectivity and intelligence to the planet and to resolve some of the critical issues in urban living, including waste management, transportation, resources management, education, unemployment, healthcare, and other civil governing sectors. Ultimately, Smart Cities aim to optimise the usage of limited resource that are required to accommodate growing number of human populations. Smart Cities are designed to have the following six characteristics: a smart economy that sustains the transaction; smart people who use the system; smart governance that regulate the overall operation; smart mobility to promote accessibility; smart environment to allow autonomous and robust climate management; and smart living, which ensures that the habitat is liveable for human [9–11]. Other researchers also included e-government and urban planning as core Smart Cities domains [12].

The scale of the adoption and coverage of IoT in the society - community, cities, government and even the planet - has surpassed expectations. Some countries are beginning to make use of IoT to as the infrastructure for Smart Cities [10–18]. In Dublin, Ireland, data modelling and prototyping of the architecture of the public transport systems have been integrated into the intelligent transportation service. This helps users to plan their urban journey through visualising traffic condition [13]. As part of IBM's Global Business Service (GBS), Smart Cities has been included as one of the long-term Smart Planet themes [10]. SmartSantander is a large scale IoT project that covers Santander, a city in Spain, as a real-world experiment [15]. The project includes environmental monitoring, outdoor parking management and driver guidance, parks and garden precision irrigation, augmented reality of the city scape, and participatory sensing using mobile phones as sensors. SmartSantander is a big step toward the future of Smart Cities and many other areas will require more thorough research and development. Further popular implementations of Smart Cities with the support of IoT, include the Sensing China project in Shanghai, IoT implementation by the Melbourne City Council in Australia [16], and the IoT Island in the city of Padova, Italy [11].

Some of the other sectors that are actively adapting IoT and that are getting IoT technology include environmental monitoring, media and entertainment industry, agriculture industry, and the education sector. It is challenging to find an organisation that does not have IoT in their business process and that does not depend on IoT technology to some extent. IoT is not only found in commonly used internet connected devices, such as laptop, smartphones, tablet, household appliances and other non-commercial devices, but IoT is also included in digital devices such as thermostats, embedded systems, industrial sensors and other devices that are essential for organisations to conduct their business operations [19].

3 IoT Characteristics - TOP

Technology, Organization, and People are also known as the 'triplet' that is used to perceive characteristics and challenges in three different dimensions [20]. Each dimension also has a major influence on changes in the business environment and vice versa. The three dimensions are discussed next.

3.1 Technological Aspects

The nature of technology depends very much on the business processes involved in the organisation. The usage and functionality of technology in business operation is usually driven by a business need. Many of the research in IoT are focusing on technical challenges, such as new communication protocol, advanced machine learning algorithm, secured data cryptography, IoT architecture, cost reduction, improving the amount of operating time, enhancing data transmission, optimising large volume data processing on the individual IoT devices, automation intelligent maturation, and the list goes on [3, 19].

1. New and more complex IoT devices integration is constantly available. Although they have similar features, there can be difference in data standard and quality requirement. There can be multiple providers with different software platforms and measurement units (e.g. pounds versus kilograms) [19, 21, 22]. IoT implementation can be large scale, as with constant sensing and streaming data IoT generate large volume of data. However, most data generated may not be useful [23–26].
2. Limited IoT hardware performance may result from its mobility nature, so advanced pre-processing or analysing is restricted. This also includes additional encryption and other aftermarket security features [27].

Legacy infrastructure may limit the support of massive IoT deployment, from bandwidth, storage, analytical method, security, and others [28]. Due to mobility, power efficiency, sleeping and wakeup features, IoT devices can have dynamic device status, which allows the device to cope with inconsistent connection [29, 30]. Data from multiple sources are processed and integrated to produce better interpretation of the physical environment [24, 31].

3.2 Organisational Aspects

Apart from technical challenges in providing more advanced or comprehensive services from the functionality aspect of the IoT, other challenges may also impact the business operation of an organisation [20]. As more attention and investments are allocated to research and enhance the features that IoT could offer, it is also equally important to address some of the unique characteristic that brings challenges, apart from technical advancement. The organisation dimension defines the roles and relationship of the people both internally and externally, as well as regulating business strategy of the organisation. To achieve its business goal an organisation coordinates and manages the available resources.

1. Due to data retention regulation in some countries, large volumes of data may have to be stored and archive [32].
2. IoT store data or communicate with other devices via internet. It stores data in the cloud or other location, such as vendor or manufacturer server, which may raise other managerial issues [21].
3. Not only IoT is connected to the internet, it is also connected to the organisation's intranet. With weak protection or security configuration, it might expose the organisation to security vulnerabilities.
4. Existing analysis methods may not be effective or efficient to process and handle data generated by IoT [33].
5. Data sharing with external parties results in more interested stakeholders that are demanding access to data, including customer, supplier, government, and even competitor [34, 35].
6. Emerging new IoT policies and regulations that aim to better govern IoT devices can backfire, as conflicting policies and regulations can cause differences in the devices and their data [36, 37].

3.3 People Aspects

People are the core of internal integration of an organisation, and their impact on the organisation should not be underestimated. Although all organisations include people, these people are dynamic and individual in the way they react and behave.

1. The ability and skillset of new personnel need to be acquired in order to develop or apply adequate analysis methods [33].
2. Different ways of handling IoT devices include installing, configuring, calibrating, disposal, and other operational process that affect the data quality, which could be different from other operators [38].

3. Awareness and knowledge of handling IoT can be new to the organisation personnel. Other than technical personnel, there can be other personnel that may not have enough understanding of IoT and how to handle them.

Many scholars adopt the STOPE framework, which refers to Strategy, Technology, Organization, People, and Environment [39–41]. However, the context of STOPE adoption may not be fully applicable for the IoT. From the IoT perspective, the Strategy can be considered as part of the Organization dimension, while the Environment is part of both Technology (environmental issues) and Organization (external parties). Therefore, it is more relevant to adopt TOP as the overall three dimensions in the context of IoT.

4 Challenges of IoT Adoption

4.1 Standardisation

Standards are designed to ensure that services and devices will be produced according to the same specification, procedure or guideline. Standardisation is crucial to facilitate compatibility between different technologies embedded in IoT. Interoperability, i.e. the ability to work with other devices or systems, is another key aspect of standardisation. With standardisation, the quality of output from the IoT will be manageable and controlled by the users or other interested parties. The same or similar requirements of the target community can also be consistently fulfilled and regulated, if necessary. Rapid adoption of IoT in the society increases the need for different architectures and standards, due to different requirements. IoT are being used in a new area where analogue was previously used, or where IT has never been involved before. For example, a thermostat usually monitors temperature, but more features such as humidity monitoring, air quality monitoring to use the thermostat as an environment monitoring device are gradually added. When calculations and algorithms are added to the device, it supports simple automation or communication with the users in the same environment, which makes it a true IoT device. However, to support more comprehensive environmental monitoring, the device might need to communicate with other devices. Doing so may not be a simple task, especially if there is a difference in data quality control and communication protocol. This has resulted in a lack of unity when there are too many standards in the industry [42]. Technically, excessive standards may not be fully compatible with other standards [42] and some of the standards are not backward compatible with older standards [3].

4.2 Security and Privacy

Connecting to the internet may create an open channel to expose IoT to the external world [43]. With vast volume of data being collected by sensors, the manufacturer are accountable not only during the data collection and stored, but more critically for the use of the data [44]. Whist in some cases it is crucial to utilise sensitive data for analysis of more comprehensive services provision, the same data could violate the privacy of the people. Mismanagement of sensitive data can be intrusive and pose a threat of information leakage, especially where IoT devices are found in an environment where confidentiality is important e.g. military, government offices, health industry, and critical infrastructures. Most importantly, many of the IoT devices are continually collecting data in real time. For example, the mobile phone that people carry around can listen to voice command to access the mobile device and other devices it is connected to. The mobile phone is constantly listening to the environment and to the conversations and noise in the environment, while waiting for a command. The next issue is the accessibility of the data store in electronic devices, which has always been a debatable area. The issues are how much data is stored in the devices and how long the data should be kept. These are some of the important administrative or managerial issues. Without proper handling of the security and privacy of IoT, the behaviour in the IoT environment or trade secrets can be accessed by unauthorised personnel, and the data can also be used or changed for crime activities that may have serious and irreversible consequences [45]. Without proper handling knowledge of the IoT, data collected on the enterprise asset may expose trade secrets or operational detail to unauthorised party.

4.3 Data Quality

Data quality refers to accuracy, timeliness, compatibility, usability, relevance, completeness, conformity, and consistency [38], which has a direct social and economic impact on organisations [46]. The definition of quality is perceived differently by different users. Some users may require more detail than others, while some may require different formatting for different purposes. When similar data are used in an area for which it was not intentionally designed, it may cause data quality issues. One extreme example could be using a surveillance camera for quality control vision inspection in a manufacturing production line. Without adequate control of the data quality, negative consequences may result, such as deflated employee morale, mistrust between organisations, data integrity issues, and others. Data quality has a significant impact on organisational operations, as well as the strategic level and tactical level [47].

The data quality in IoT can make a huge difference, especially in this business intelligence age, where big data and data mining are common tools used in organisations. Capturing real-time multi-dimensional IoT data with improved data quality, will reduce the cost and other resources required to process and mine the data. Improved data quality reduces the cost to resolve the data quality issues, to comply with legal and regulation requirements, and to support managerial decision making. Data quality is affected by resource constraints (power, storage, network and etc.) and extreme environmental conditions or terrains [48]. It is crucial to link the purpose of the IoT data and the data being used when data quality is defined.

4.4 Organisation Culture

Every organisation is formed with people who have their own way of doing things. Organisation culture is defined as "a pattern of basic assumptions that has worked well enough to be considered as valid and therefore, to be taught to new members as the correct way to perceive, think and feel in relation to those problems" [49]. By sharing assumptions, values, and beliefs, organisational culture reflects the behaviour of the managers in the organisation. In this section, the impact of organisational culture as a challenge for IoT adoption and utilisation is presented. When IoT is gradually adopted into organisations, it affects the way the organisation conducts its business and interacts with people, including its own employees, clients, and the wider community. Depending on the scale of adoption, there will be changes in roles and responsibilities. Some people may view IoT as a threat, rather than a benefit, especially when changes push people out of their comfort zone. The acceptance of such changes introduced by IoT can be challenging. If some people in an organisation are not familiar with the IoT technologies, the lack of understanding and subsequent uncertainty may result in negativity. With more robust and automated technologies with IoT, the productivity and performance of the organisation should be more assessable than ever. However, from the organisational perspective, the success of the transition depends to a great extent on people acceptance, a strong organisation culture that accepts change and embraces innovation, as well as motivated leaders who are supportive of IoT adoption.

4.5 Business Continuity and Resilience

One of the unique features of IoT is the ability to automate simple decision making with less human intervention. However, if the organisation is overly dependent on automation, it would be fatal if threats and disruptive events strike that may disturb or disable the technological functionality of IoT. A life may even be lost due to a technical fault, for example in the case of the self-driven car in Arizona [50]. Therefore, business continuity planning is critical to recover from disastrous events. Whenever there is a dynamic change to a business operation, it could have negative consequences such as supply chain interruption, loss of data, damage to critical infrastructure. In a situation where there is insufficient knowledge and experience is insufficient, it can be a challenge to ensure business continuity and resilience after an interruption event.

5 Conclusion and Recommendation

Inevitably, industries that are already having EAM will continue to adopt IoT as part of their business operation. Such transformation occurs in more industries than listed above. Disruptive innovation is an innovation that significantly change the existing way of doing things both on a strategic and operational level. IoT causes changes and challenges with its own unique characteristics that do not exist in traditional IT. Most notably would be its constant data acquisition that results in overwhelmingly huge volume of data. For instance, there are more than ten thousand sensors on one wing of Airbus 350 generating in excess of 2.5 Tb of data every day [51]. Collectively this is beyond the capability of many organisation to own and process. However, the data acquisition is essential for EAM activities, e.g. planning for future organisational strategy, decision making, risk assessment, and asset lifecycle analysis.

This study stresses that IoT is beneficial to the EAM, yet it brings unique challenges that require extensive attention. Organisations should be concern with the IoT adoption and usage challenges mentioned. As IoT technology evolved, the list of unique characteristics will also continue to grow, thereby introducing more features but also challenges to existing EAM. To overcome these challenges, organisations will have to list their IoT devices and their characteristics.

References

1. Gao J, Koronios AP (2006) Key data quality issues for enterprise asset management in engineering organisations. IJEBM 4(1):98–110
2. Royal Jr WC, Watkins RO (1999) Internet asset management system for a fuel dispensing environment. Google patents
3. Gubbi J, Buyya R, Marusic S, Palaniswami M (2013) Internet of Things (IoT): a vision, architectural elements, and future directions. Future Gener Comput Syst 29(7):1645–1660. https://doi.org/10.1016/j.future.2013.01.010
4. Haller S, Karnouskos S, Schroth C (2008) The internet of things in an enterprise context. In: Future internet symposium. Springer, Heidelberg, pp 14–28
5. Bandyopadhyay D, Sen J (2011) Internet of things: applications and challenges in technology and standardization. Wireless Pers Commun 58(1):49–69. https://doi.org/10.1007/s11277-011-0288-5
6. Kwon D, Hodkiewicz MR, Fan J, Shibutani T, Pecht MG (2016) IoT-based prognostics and systems health management for industrial applications. IEEE Access 4:3659–3670. https://doi.org/10.1109/ACCESS.2016.2587754
7. Rahmani AM, Thanigaivelan NK, Tuan Nguyen G, Granados J, Negash B, Liljeberg P, Tenhunen H (2015) Smart e-Health gateway: bringing intelligence to internet-of-things based ubiquitous healthcare systems. In: 2015 12th Annual IEEE consumer communications and networking conference (CCNC), 9–12 Jan 2015, pp 826–834
8. Metso L, Happonen A, Rissanen M, Efvengren K, Ojanen V, Kärri T (2020) Data openness based data sharing concept for future electric car maintenance services. In: Advances in asset management and condition monitoring. Springer, Cham, pp 429–436
9. Perera C, Zaslavsky A, Christen P, Georgakopoulos D (2014) Sensing as a service model for smart cities supported by Internet of Things. Trans Emerg Telecommun Technol 25(1):81–93. https://doi.org/10.1002/ett.2704
10. Paroutis S, Bennett M, Heracleous L (2014) A strategic view on smart city technology: the case of IBM smarter cities during a recession. Technol Forecast Soc Chang 89:262–272. https://doi.org/10.1016/j.techfore.2013.08.041
11. Zanella A, Bui N, Castellani A, Vangelista L, Zorzi M (2014) Internet of things for smart cities. IEEE Internet Things J 1(1):22–32. https://doi.org/10.1109/jiot.2014.2306328
12. Piro G, Cianci I, Grieco LA, Boggia G, Camarda P (2014) Information centric services in smart cities. J Syst Softw 88:169–188. https://doi.org/10.1016/j.jss.2013.10.029
13. René M, Deirdre L (2011) Context-aware pervasive services for smart cities. In: Ubiquitous developments in ambient computing and intelligence: human-centered applications. IGI Global, Hershey, pp 1–16
14. Perera C, Zaslavsky A, Christen P, Georgakopoulos D (2014) Context aware computing for the internet of things: a survey. IEEE Commun Surv Tutorials 16(1):414–454. https://doi.org/10.1109/surv.2013.042313.00197
15. Sanchez L, Muñoz L, Galache JA, Sotres P, Santana JR, Gutierrez V, Ramdhany R, Gluhak A, Krco S, Theodoridis E, Pfisterer D (2014) SmartSantander: IoT experimentation over a smart city testbed. Comput Netw 61:217–238. https://doi.org/10.1016/j.bjp.2013.12.020
16. Jin J, Gubbi J, Marusic S, Palaniswami M (2014) An information framework for creating a smart city through internet of things. IEEE Internet Things J 1(2):112–121. https://doi.org/10.1109/jiot.2013.2296516

17. Li X, Lu R, Liang X, Shen X, Chen J, Lin X (2011) Smart community: an internet of things application. IEEE Commun Mag 49(11):68–75
18. Vlacheas P, Giaffreda R, Stavroulaki V, Kelaidonis D, Foteinos V, Poulios G, Demestichas P, Somov A, Biswas AR, Moessner K (2013) Enabling smart cities through a cognitive management framework for the internet of things. IEEE Commun Mag 51(6):102–111
19. Swan M (2012) Sensor mania! the internet of things, wearable computing, objective metrics, and the quantified self 2.0. J Sens Actuator Netw 1(3):217–253. https://doi.org/10.3390/jsan1030217
20. Khalid H, Kay HK, Robert H (2006) E-business capabilities model: validation and comparison between adopter and non-adopter of e-business companies in UK. J Manuf Technol Manage 17(6):806–828. https://doi.org/10.1108/17410380610678819
21. Ježek J (2012) Data architecture for sensor network. Geoinf. FCE CTU 7:31–38. https://doi.org/10.14311/gi.7.3
22. Botta A, de Donato W, Persico V, Pescape A (2014) On the integration of cloud computing and internet of things, pp 23–30. https://doi.org/10.1109/ficloud.2014.14
23. Soma B, Munmun S, Souvik M, Subhajit D (2011) Role of middleware for internet of things: a study. Int J Comput Sci Eng Surv 2(3):94–105. https://doi.org/10.5121/ijcses.2011.2307
24. Barnaghi P, Bermudez-Edo M, Tönjes R (2015) Challenges for quality of data in smart cities. J. Data Inf Qual 6(2–3):1–4 (2015). https://doi.org/10.1145/2747881
25. Kitchin R (2013) The real-time city? Big data and smart urbanism. GeoJournal 79(1):1–14. https://doi.org/10.1007/s10708-013-9516-8
26. Tönjes R, Barnaghi P, Ali M, Mileo A, Hauswirth M, Ganz F, Ganea S, Kjærgaard B, Kuemper D, Nechifor S (2014) Real time IoT stream processing and large-scale data analytics for smart city applications. In: Poster session, European conference on networks and communications
27. Oriwoh E, Al-Khateeb H, Conrad M (2015) Responsibility and non-repudiation in resource-constrained internet of things scenarios
28. Zhang Z, Cho MCY, Wang C, Hsu C, Chen C, Shieh S (2014) IoT security: ongoing challenges and research opportunities. In: 2014 IEEE 7th international conference on service-oriented computing and applications, 17–19 November 2014, pp 230–234
29. Ma M, Wang P, Chu C-H (2013) Data management for internet of things: challenges, approaches and opportunities. In: 2013 IEEE International conference on green computing and communications and IEEE internet of things and IEEE cyber, physical and social computing, 20–23 August 2013, pp 1144–1151
30. Tsai C-W, Lai C-F, Chiang M-C, Yang LT (2014) Data mining for internet of things: a survey. IEEE Commun Surv Tutorials 16(1):77–97. https://doi.org/10.1109/surv.2013.103013.00206
31. Ishaq I, Carels D, Teklemariam G, Hoebeke J, Abeele F, Poorter E, Moerman I, Demeester P (2013) IETF standardization in the field of the internet of things (IoT): a survey. J Sens Actuator Netw 2(2):235–287. https://doi.org/10.3390/jsan2020235
32. Rose K, Eldridge S, Chapin L (2015) The internet of things: an overview. The internet society (ISOC), pp 1–50
33. Lee I, Lee K (2015) The internet of things (IoT): applications, investments, and challenges for enterprises. Bus Horiz 58(4):431–440. https://doi.org/10.1016/j.bushor.2015.03.008
34. Jernigan S, Kiron D, Ransbotham S (2016) Data sharing and analytics are driving success with IoT. MIT Sloan Manage Rev 58(1):0
35. Yang L, Yang SH, Plotnick L (2013) How the internet of things technology enhances emergency response operations. Technol Forecast Soc Chang 80(9):1854–1867. https://doi.org/10.1016/j.techfore.2012.07.011
36. Borgia E (2014) The internet of things vision: key features, applications and open issues. Comput Commun 54:1–31. https://doi.org/10.1016/j.comcom.2014.09.008
37. Vermesan O, Friess P (2014) Internet of things-from research and innovation to market deployment. River Publishers Aalborg

38. Strome TL (2013) Data quality and governance. In: Healthcare analytics for quality and performance improvement. Wiley, Hoboken, pp 75–90
39. Bin-Abbas H, Bakry SH (2014) Assessment of IT governance in organizations: a simple integrated approach. Comput Hum Behav 32:261–267. https://doi.org/10.1016/j.chb.2013.12.019
40. Saleh MS, Alfantookh A (2011) A new comprehensive framework for enterprise information security risk management. Appl Comput Inf 9(2):107–118. https://doi.org/10.1016/j.aci.2011.05.002
41. Bakry SH (2003) Toward the development of a standard e-readiness assessment policy. Int J Netw Manage 13(2):129–137. https://doi.org/10.1002/nem.466
42. Baskarada S, Koronios A, Gao J (2007) IQM-CMM: a framework for assessing organizational information quality management capability maturity
43. Weber RH (2010) Internet of things – new security and privacy challenges. Comput. Law Secur Rev 26(1):23–30. https://doi.org/10.1016/j.clsr.2009.11.008
44. Michael PP, Chris DN, Hui W, Liming C (2011) Smart home research: projects and issues. In: Ubiquitous developments in ambient computing and intelligence: human-centered applications. IGI Global, Hershey, pp 259–272
45. Pishva D (2017) Internet of things: security and privacy issues and possible solution. In: 2017 19th international conference on advanced communication technology (ICACT). IEEE, pp 797–808
46. Wang RY, Strong DM (1996) Beyond accuracy: what data quality means to data consumers. J Manage Inf Syst 12(4):5–33
47. Redman TC (1998) The impact of poor data quality on the typical enterprise. Commun ACM 41(2):79–82. https://doi.org/10.1145/269012.269025
48. Karkouch A, Mousannif H, Al Moatassime H, Noel T (2016) Data quality in internet of things: a state-of-the-art survey. J Netw Comput Appl 73:57–81. https://doi.org/10.1016/j.jnca.2016.08.002
49. McDermott CM, Stock GN (1999) Organizational culture and advanced manufacturing technology implementation. J Oper Manage 17(5):521–533. https://doi.org/10.1016/S0272-6963(99)00008-X
50. Wakabayashi D (2018) Self-driving uber car kills pedestrian in Arizona, where robots Roam. The New York Times (2018)
51. Marr B (2015) That's data science: airbus puts 10,000 sensors in every single wing!

Asset Management Systems

Integration of Asset Management Standard ISO55000 with a Maintenance Management Model

C. Parra, V. González-Prida, E. Candón, A. De la Fuente, P. Martínez-Galán, and A. Crespo(✉)

Department of Industrial Management School of Engineering, University of Seville, Seville, Spain
adolfo@us.es

Abstract. The family of Standards ISO 55000 on Asset Management aims to support a kind of management oriented to obtain value from assets. Besides these standards, other frameworks have help to define and improve business policies and work procedures for the assets operation and maintenance along their life cycle. With this purpose, the present paper links a Maintenance Management Model well known in this area, with the publication of ISO 55000. As a consequence, justifies that a proper implementation of the Maintenance Management Model fulfils the requirements stated by the Standards ISO 55000, improving consequently the decision making, costs reductions, quality of the operations and increasing business profitability and users satisfaction.

Keywords: Asset management · Operation & maintenance · Reliability engineering · Standards ISO 55000

1 Introduction

While execution of the activities of maintenance and reliability processes are, basically, technical actions, management of these processes are actions associated to administrative decisions. They are oriented to maximize the profitability of the asset, with the purpose of preserving and/or restoring the teams of production to conditions that allow them to meet with a function required during certain periods of time [1, 2].

Approximately 4 decades ago, organizations became aware that in order to manage maintenance and reliability properly it was necessary to include them in the general scheme of the organization and manage them in interaction with the other functions [3]. The challenge then consisted of integrating maintenance within the system of management of the company's assets. The desired picture was that, once reached such integration, the processes of maintenance and reliability received well-deserved importance and would be developed as one function of the Organization: generating 'products' to satisfy internal customers, yielding useful data and information and contributing to the fulfilment of the objectives of the organization. Thus, the concept of "maintenance management system" was born in the 1980s and reliability, whose activities were aimed to benefit from business, rather than focus on them as in the past: as a cost center [4].

A. Crespo Márquez et al. (Eds.): WCEAM 2019, LNME, pp. 189–200, 2021.
https://doi.org/10.1007/978-3-030-64228-0_17

The model of asset management optimization, known as: "Asset Management", is a discipline that emerged at the end of the 1990's and focuses on decision-making of all the life cycle of the physical asset from its creation or acquisition, use, maintenance and renewal or disposal. This in contrast to the approach of traditional management process of maintenance, that has as an object of study the equipment only during their operational life. Therefore, the management of asset links concepts and techniques of different fields, such as finance, engineering, technology, operations, etc.

2 General Aspects and Evolution of Asset Management

Obviously the efforts of organizations to improve the performance of their assets goes beyond the development of systems for the management of maintenance and reliability, it is also about optimizing different aspects that have to do with the life cycle of assets. However, the vision that incorporates the asset management process during its life cycle is extremely beneficial for optimize the maintenance of the assets. The activities of prevention and correction of failures for the improvement of the security of operation of the equipment is greatly influenced by a management linked to the design, construction, mounting, operation, maintenance and replacement of the equipment [2, 5].

The enterprise asset management has been recognized as a discipline since the mid 1990's. It brings together concepts and techniques from different fields, such as finance, engineering, technology, operations, etc.; and focuses on decision making throughout the full life cycle of physical assets, optimizing aspects of different natures on every occasion, but with a integrating vision throughout the life cycle.

The Institute of Asset Management (IAM), an independent and non-for-profit organization has been one of the main promoting bodies of this "crusade". IAM defines asset management as "the art and science of making the right decisions and optimizing the processes of selection, maintenance, inspection and renewal of assets" [6]. It mentions also that a common goal is to minimize the cost of the total life of the asset, and possibly enabling other critical factors too such as risk or the continuity of the business, which must be considered objectively for decision-making.

The proposal of standard PAS 55 was a public specification aimed at optimizing the management of physical assets and infrastructure. The efforts to conceive it started in 1995 when a Committee of Managers, members of the Institute of management of assets and which integrated a variety of representatives of industry, Government and British regulatory bodies, gathered for the first time to define the direction that this standard would take. Their works, revision and publication, conducted through the British Standards Institute (BSI) took 9 years [7]. First published in April 2004, it is to date, the main background of standard ISO 55000 [8]. From 2006, the proposed standard PAS 55 gained recognition and spread its use in the industry when the Regulatory Bureau of Gas and Electric Power in the United Kingdom (UK Office of Gas and Electric Markets) strongly recommended its use in public companies that make up its network of operations. By 2008, most public enterprises of gas and electricity of United Kingdom met the requirements of the proposed standard PAS 55. Subsequently this trend also came to the areas of transport, management of public companies, food, pharmaceuticals, and chemicals, among others. Moreover, of course outside of the United Kingdom several companies took as a reference the proposed standard PAS 55 have increasingly appeared [8]. In terms of its relevance and applicability, it is even possible to make the following analogy: PAS 55 is to asset management what ISO 9001 is to quality management or what ISO 14000 is to environmental management [8].

The proposal of PAS 55 standard defines asset management as " controlled and systematic practices and activities through which an organization optimally manages their assets, their associated performance, risks and expenses through their life cycle, in order to meet the strategic plan of the Organization" [9]. PAS 55 can be applied to any sector of business that manages physical infrastructure and is independent of the function or type of asset. Some examples of companies where it has been applied successfully include roads, airports, trains and petrochemical complexes.

Based on Deming's cycle of planning, doing, checking and acting, this standard can be used also for various purposes: self-evaluations, benchmarking, and improvements in planning, independent audits, certification, selection of contractors, demonstration of competence, etc. Organizations that have adopted the standard PAS 55 proposal reported significant improvements in cost and performance/service issues. PAS 55 provides a clear evidence of a proper management of asset to customers, investors, regulators and other interested parties [5].

Later in 2009, ISO Organization proposes the development of an asset management standard (initially based on the proposed standard PAS 55). It is known today as asset management standards of the series: ISO 55000, 55001 and 55002 (standards adopted from 2014 and whose certifiable standard is ISO 55001), these standards have become the international reference in the area of asset management. The design and implementation of a system of management of asset, in line with the 24 requirements of ISO 55001, is a very broad matter of discussion. This document describes in general the model developed by ISO 55001 and proposes a process of integration between the MMM (Maintenance Management Model, [2], see Fig. 1) with the asset management model proposed by ISO 55000 [5].

3 General Description of the Asset Management Standard ISO 55000

This international standard provides a general vision of the systems of management of asset (i.e., systems of management for the handling of asset). It includes standards ISO 55000, 55001 and 55002. The target audience of this standard is those people needing them as indicated in [10, 11] and [12]:

- Who are considering improving the chain of value of their organizations starting from their bases of asset?
- Who are involved in the establishment, implementation, maintenance and improvement of a system of management of assets?
- Who are involved in the planning, design, implementation and review of the activities of asset management, together with service providers?

The adoption of this set of international standards will enable an organization to achieve its objectives through the efficient and effective management of its assets. The implementation of an asset management system ensures that the achievement of those objectives is consistent and sustainable over time.

Standard ISO 55000 defines asset in the following way: "An asset is an element, thing or entity which has a real or potential value to an organization. The value will vary for different organizations and their shareholders, and may be tangible or intangible, financial or non-financial".

The period that goes from the creation of an asset to the end of its life is called useful life of the asset. Useful life of the asset does not necessarily coincide with the period in which any organization maintains responsibility on it; rather, along its life useful, an asset can provide a real or potential value to one or more organizations, and the value of the asset with regard to the organization can change throughout the useful life of the asset. Asset management allows an organization to recognize the need for and examine the performance of the assets and systems of assets at different levels. Likewise, it allows the application of analytical approaches to the management of an asset throughout the different stages of its cycle of life (which can start with the concept of the need for the asset until its elimination, including the handling of potential responsibilities subsequent to the elimination) [10, 11] and [12].

Key factors in ISO 55000 that make influence on an organization for the achievements of its goals, are cited below:

- Nature and purpose of the organization.
- Its operational context.
- Its financial restrictions and regulatory requirements.
- The needs and expectations of the Organization and interested parties (stakeholders).

The organizations must keep an effective control and efficient policies of asset to generate value through the management of risks and opportunities, to achieve the balance of costs desired, the reduction of risks and the performance. The management of asset translates the objectives of the Organization into activities, plans and decisions related to the asset, using an approach based on risks [10, 11] and [12].

4 Requirements of the Asset Management Model Asset on the Standard ISO 55000

The standard ISO 55000 proposes a model of asset management based on 24 certifiable requirements. Below, the certifiable standard ISO 55000 requirements are quoted:

4. Context of the Organization

    4.1. Understanding the Organization and its context
    4.2. Understanding the needs and expectations of interested parties
    4.3. Determining the extent of the asset management system
    4.4. System of management of assets

5. Leadership

5.1. Leadership and commitment
5.2. Policies
5.3. Roles, organizational responsibilities and authorities

6. Planning

6.1. Actions to address the risks and opportunities in management system asset
6.2. Objectives for the management of assets and planning to achieve them

7. Support

7.1. Resources
7.2. Competencies
7.3. Awareness
7.4. Communication
7.5. Requirements of information
7.6. Documented information

8. Operation

8.1. Planning and operational control
8.2. Change management
8.3. Outsourcing

9. Performance evaluation

9.1. Monitoring, measurement, analysis and evaluation
9.2. Internal audit
9.3. Revision of the management

10 Improvements

10.1. Non-compliance and corrective action
10.2. Preventive action
10.3. Continuous improvement

In the entire version of the standard ISO 55000 proposal, there are 24 requirements, which keep a logical order of elements in accordance with the common framework for quality processes: plan-do-check-act.

5 Integration of the Maintenance Management Model (MMM) with the Asset Management Standard ISO 55000

Although there are no simple formulas for the implementation of an integral model of asset management, nor fixed or immutable rules with validity and applicability for all the assets of production, the 24 requirements needed by the proposal of standard ISO 55000 can be covered by the integral maintenance management model (Fig. 1) proposed at the beginning of this report. In the MMM (which is composed of eight phases), specific actions are described to follow in different steps of the process of management of maintenance that are integrated in a direct form within a process of management of assets [2]. The MMM offers a dynamic, sequential process and in a closed loop that tries to accurately characterize the course of actions to be carried out to ensure the efficiency, effectiveness and continuous improvement of the management of assets from the use and integration of techniques of engineering and maintenance management and reliability. Another version of the MMM, applied to warranty and aftersales management can be find in [13].

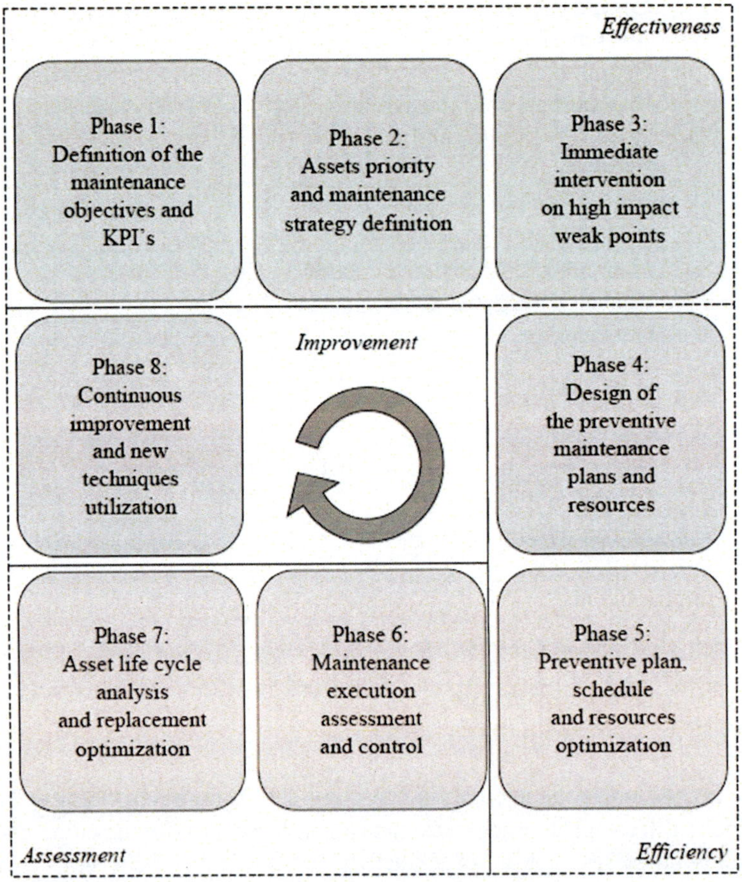

Fig. 1. Model of the process of maintenance management (MMM) integrated into ISO 55000 [2].

In particular, a relationship is made between the phases of the model proposed and the general points of the standard ISO 55000, so that the gradual implementation of the generic model progressively covering the requirements of the standard ISO 55000 may be looked at. According to next paragraphs, the activities to be developed within the 8 stages of the MMM can help organizations, to meet with the 24 requirements demanded by the standard ISO 55000. The following describes in more detail the relationship between the phases of the MMM and the requirements of ISO 55000 (Fig. 2) (Table 1):

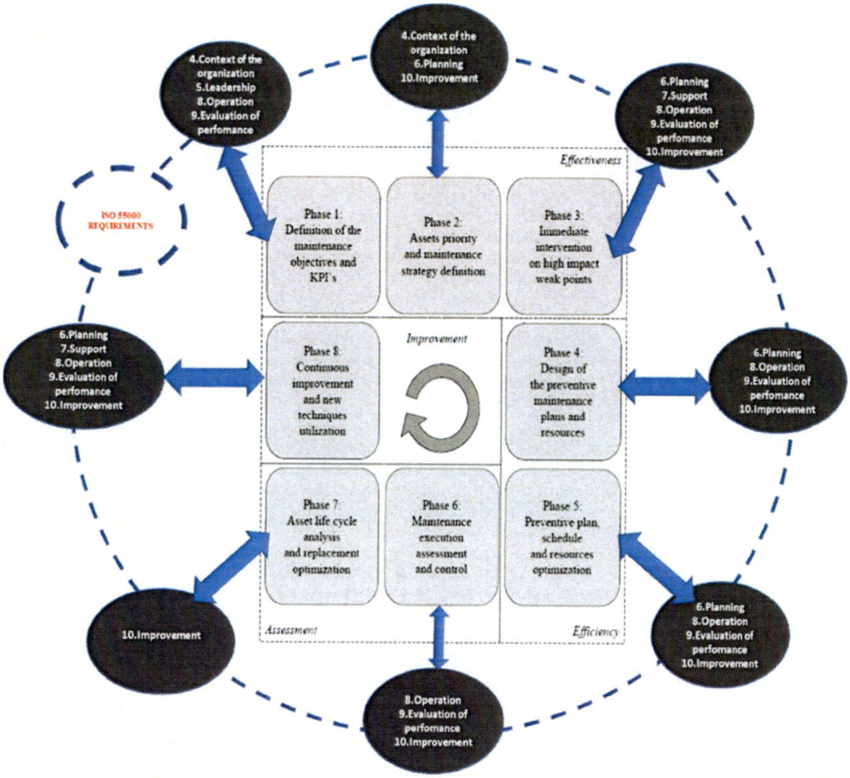

Fig. 2. Relationship between the phases of the maintenance management model (MMM) and the requirements of ISO 55000 [14]

6 Summary of Practical Cases of Integration of the Maintenance Management Model (MMM) Aligned to an Asset Management Process

The model of the process of maintenance management (MMM), integrates other models found in the literature for built and in-use assets, and consists of eight sequential management building blocks [2]. Each block is, in fact, a key decision area for asset maintenance and life cycle management. Within each of these decision areas we can find methods and models that may be used to order and facilitate the decision-making processes. This model has been implemented in various organizations worldwide. Below is a summary of the organizations where the MMM has been implemented integrated into an Asset Management process:

Table 1. Relationship between the phases of the maintenance management model (MMM) proposed and the requirements of ISO 55000 [2] (1/2)

ISO 55000 requirements	Integration of the phases of the MMM proposed with standard ISO 55000
4. context of the Organization 4.1. understanding the Organization and its context 4.2. understanding the needs and expectations of interested parties 4.3. determining the extent of the asset management system 4.4. system of management of assets 5. leadership 5.1. leadership and commitment 5.2. policies 5.3. roles, organizational responsibilities and authorities	Phase 1. Proposes the use of the scorecard (Balanced Scorecard – BSC), proposed by Kaplan and Norton, model that translates the mission of a business unit into its strategy in a set of objectives and quantifiable measures. By implementing the BSC, organizations get to: 1. Formulate policies and strategies for the operation and performance of the maintenance of assets throughout their lifecycle 2. Put into practice the strategies of maintenance and operation, which is translated into objectives at short, medium and long term 3. Develop the plans of action. These are the means to get to the purposes stipulated in the objectives set out in step (2) 4. Establish leadership in the different processes to improve in all areas of the Organization 5. Review and periodically audit the performance of implemented strategies. Monitoring will be made and the casual relations between the measures will be investigated what will be validated at intervals previously established and plans of contingency will be defined Additionally in phase 1, the MMM model proposes that an cohesive organization is designed which supports the process of asset management and is able to implement a holistic process optimization based on the application of techniques of reliability and maintenance, with the assignment of roles, responsibilities, and definition of the leadership of all the activities to be developed during the lifecycle of the asset Phase 2. Proposes the use of models of prioritization, which must comply and align with the expectations of stakeholders (interested parties) and at the same time, cover the legal requirements demanded by the environment of the asset
6. planning 6.1. actions to address the risks and the opportunities in the system of management of assets 6.2. objectives for the management of assets and planning to achieve them 7. support 7.1. resources 7.2. competencies 7.3. awareness 7.4. communication 7.5. requirements of information 7.6. documented information	Phase 2. Proposes at the beginning of a process of improvement, the development and the application of basic models of prioritization of assets based on the analysis of the risk factor (example: qualitative and technical matrix of risks AHP: Analytics Hierarchy, Process, etc.) Phase 3. Proposes the use of the methodology of root cause analysis (RCA) to assess the failures of major impact events, taking as a basis for the definition of solutions, the level of risk caused by failure events to be analyzed Phase 4. Proposes the use of methodology of reliability-centered maintenance (RCM) to optimize maintenance and operation depending on the level of risk plans that generate failures within the context of the operational modes Phase 5. Proposes the use of methods of optimization to be used in the programming and allocation of resources for maintenance and operations. Within the selected methods are the techniques related to processes such as risk analysis: theory of queues, Monte Carlo simulation and probabilistic techniques of point of order from inventory Additionally, at this stage, using continuous improvement methods is proposed in the programming, planning and allocation of resources for maintenance and operations, risk management-based Phase 8. Proposes the use of the systems of information support (ERP, EAM, software of reliability, etc.), to manage and disclose all the documentation and information to be generated by the different assets in their processes of operation and maintenance. The information systems for the management of assets are key tools for their ability to support and facilitate their management, thanks to the transmission and processing of information at high speeds and quantities exceeding the organizations' own borders and strengthening the convergence among sectors. The need for a correct implementation of the support for the management of information systems is the basis for the development of programs to improve reliability, maintenance and operations
8. operation 8.1. operational planning and control 8.2. change management 8.3. Outsourcing 9. evaluation of performance 9.1. monitoring, measurement, analysis and evaluation 9.2. internal audit 9.3. revision of the management	Phase 1. Proposes the use of the Balanced Scorecard-BSC table to measure and review the indicators of economic performance of the Organization and subsequently, integrate them with the technical indicators of operation and maintenance (technical indicators that are developed in phase 6). Additionally, in this phase 1, the use of audits of control and continuous improvement was proposed among which is found: MES (Maintenance Effectiveness Survey), QMEM (Qualitative Matrix of Excellent in Maintenance), etc. Phases 3 and 4. Propose the application of reliability as the RCA and the RCM methods that allow evaluating modes of failure and determine their causes. These methods help to determine the incidents and non-conformities, allow to evaluate the consequences that the failures can cause on safety, the environment and operations and additionally, these techniques propose procedures that help to define actions of improvement and control: corrective, preventive, of redesign and by condition Phases 5. Proposes the application of methods of optimization of maintenance and reliability engineering, which would help to define the processes of planning, programming, outsourcing and the level of training necessary to improve the management of assets in their lifecycle Phase 6. Offers a comprehensive process of measurement, analysis and evaluation of indicators of performance and improvement (indicators of probabilistic assessment: reliability, maintainability, availability, cost and risk) Phase 8. Proposes to establish a process of continuous improvement which should be able to register and to adjust to the constant changes related to techniques and emerging technologies in areas that are considered of high impact as a result of the studies carried out in the previous 8 phases of the proposed maintenance management model

(continued)

Table 1. (*continued*)

ISO 55000 requirements	Integration of the phases of the MMM proposed with standard ISO 55000
10. improvement 10.1. non-conformity and corrective action 10.2. preventive action 10.3. continuous improvement	Phase 2. Proposes at the beginning of a process of improvement, the development and application of basic models of prioritization of assets based on the analysis of the risk factor (example: technical and qualitative risk matrix AHP: Analytics, Hierarchy, Process, etc.) Phase 3. Proposes the use of the methodology of analysis cause root (RCA: Root Cause Analysis) to evaluate them events of failures of greater impact, taking as base for the definition of solutions, the level of risk caused by them events of failures to be analyzed (processes of not conformity and actions corrective) Phase 4. Proposes the use of the reliability-centered (RCM) maintenance methodology, to optimize maintenance and operation depending on the level of risk plans that generate the modes of failures within the operational context (preventive action) Phase 5. Proposes the use of methods of optimization to be used in the programming and allocation of resources for maintenance and operations. Selected methods techniques include related processes such as risk analysis: theory of queues, Monte Carlo simulation and probabilistic techniques of point of order from inventory Phase 6. Proposes a holistic process of probabilistic evaluation of the indicators of: reliability, maintainability, availability, cost and risk Additionally, in this phase a procedure is explained that allows to relate the indicators of reliability and maintainability, with decisions of optimization in the areas of maintenance and operation based on techniques of cost risk benefit analysis (continuous improvement) Phase 7. Proposes a process of cost analysis of life cycle that allows optimizing decision-making associated with the processes of design, selection, development and replacement of assets that make up a production system. The process of life cycle begins with the definition of the different tasks of production for the preliminary design. Then activities are developed such as: plan of production, layout of plant, selection of equipment, definition of processes of manufacturing and other similar activities. Subsequently, prior to the design phase logistics is considered. This phase involves the development of the necessary support for the design and the different stages of production, the possible user support, maintenance plan intended for the use of the asset and the process of divestiture of assets (continuous improvement) Phase 8. Proposes establishing a process of continuous improvement which must be capable of reviewing and evaluating the technical and economic performance of the Organization in a continuous way

- ***Panamá Canal Authority (ACP), since 2012,*** [14]
- ***Colombian Gas Transportation Company, since 2012*** [15]
- ***National Gas Company (ENAGAS), since 2014*** [16]
- ***Electricity Transmission Company of Panamá, since 2019*** [17]

The following paragraphs summarize the most important aspects to take into account, in the alignment processes between a maintenance management model and a comprehensive asset management process [2, 17, 18]:

- View maintenance as a priority and as an internal business opportunity. The process of performing maintenance and managing physical assets must be recognized as a top priority. It must also be viewed as an internal business and not as a necessary evil. It will be viewed as an area that contributes directly to the bottom line when a profit- and customer-centered strategy and continuous maintenance improvement are adopted.
- Develop leadership and technical understanding. Maintenance leaders must understand the challenges of maintenance and provide effective maintenance leadership to operate maintenance as an internal business. Maintenance leadership must continually develop the skills, abilities, and attitudes to lead maintenance into the future.
- Develop pride in maintenance. Maintenance operations will experience fundamental improvements in work ethic, attitude, values, job performance, and customer service to achieve real pride in maintenance excellence. Tangible savings and improvements will occur as a result of continuous maintenance improvement.

- Recognize importance of the maintenance profession. The profession of maintenance will gain greater importance as a key profession for success within all types of organizations as the role of the chief maintenance officer (CMO) becomes well established. Maintenance leaders will be recognized as critical resources that are absolutely necessary for the success of the total operation.
- Maintenance and operations: a partnership for profits. Maintenance and operations will become integrated and function as a supportive team through improved planning, scheduling, and cooperative team-based improvement efforts. Operations will be viewed as an important internal customer. Improved planning and scheduling of maintenance work will provide greater coordination, support, and service to manufacturing-type operations. Maintenance and operations of all types will recognize the benefits of working together as a supportive team to reduce unplanned breakdowns, to increase equipment effectiveness and to reduce overall maintenance costs.
- Improve equipment effectiveness. A leadership-driven, team-based approach will be used by maintenance and manufacturing operations to totally evaluate and subsequently improve all factors related to equipment effectiveness. The goal is maximum availability of the asset for performing its primary function.
- Continuously improve reliability and maintainability. Machines and systems will be specified, designed, retrofitted, and installed with greater reliability and ease of maintainability. Equipment design will focus on maintainability and reliability and not primarily on performance. Design for maintainability will become an accepted philosophy that fully recognizes the high cost of maintenance in the life-cycle cost of equipment. The causes for high life-cycle costs will be reduced through the application of good maintainability and reliability principles during design.
- Maximize use of computerized maintenance management and enterprise asset management. Systems that support the total maintenance operation will improve the quality of maintenance and physical asset management and be integrated with the overall business system of the organization. Computerized maintenance management systems (CMMS) will provide greater levels of manageability to maintenance operations. CMMS will cover the total scope of the maintenance operation providing the means to improve the overall quality of maintenance management. Enterprise asset management (EAM) will provide a broader scope of integrated software to manage physical assets, human resources, and parts inventory in an integrated system for maintenance management, maintenance, procurement, inventory management, human resources, work management, asset performance, and process monitoring.
- Aggressive support compliance to environmental, health, and safety requirements. Maintenance must provide proactive leadership and support to regulatory compliance actions. Maintenance leaders must maintain the technical knowledge and experience to support compliance with all state and federal regulations. The issue of indoor air quality must receive constant attention to eliminate potential problems. Maintenance must work closely with other staff groups in the organization such as quality and safety to provide a totally integrated and mutually supportive approach to regulatory compliance.
- Manage life-cycle cost and obsolescence. The life-cycle costs of physical assets and systems will be closely monitored, evaluated, and managed to reduce total costs. A profit- and customer-centered strategy will achieve significant reductions in total

life-cycle costs through an effective design process prior to purchase and installation. During the equipment's operating life, systems will be developed to continually monitor equipment costs. Information to identify trends will be available to highlight equipment with high maintenance costs.

7 Conclusions

According to above mentioned paragraphs out of the 24 requirements defined by the standard ISO 55000, the maintenance management model (MMM) can help us totally or partially meet the demands of the requirements expected by this standard (the proposal of standard PAS 55 represents the most important background of standard ISO 55000). Finally, the adequate implementation of the maintenance management model (MMM) aligned to a comprehensive asset management process will help organizations to maximize the value of their assets throughout the useful life cycle.

References

1. Crespo Márquez A (2006) The maintenance management framework. Models and methods for complex systems maintenance. Springer, London
2. Parra C, Crespo A (2015) Ingeniería de Mantenimiento y Fiabilidad aplicada en la Gestión de Activos. Desarrollo y aplicación práctica de un Modelo de Gestión del Mantenimiento. Segunda Edición. Editado por INGEMAN, Escuela Superior de Ingenieros Industriales de la Universidad de Sevilla, España
3. Pintelon LM, Gelders LF (1992) Maintenance management decision making. Eur J Oper Res 58:301–317
4. Prasad Mishra R, Anand D, Kodali R (2006) Development of a framework for world-class maintenance systems. J Adv Manuf Syst 5(2):141–165
5. López M, Parra C, Crespo A (2011) Asset management and the PAS 55. In: XIII congress of reliability, 1st edn Spanish Association for the quality, M-45074-2011, vol 1, no 8, Zaragoza, Spain
6. IAM. Institute of Asset Management at. http://www.theiam.org/. Accessed Sept 2010
7. BSI. British Standard Institution in http://www.bsigroup.com/. Accessed Sept 2010
8. Kings-Picknell J (2007) An introduction to PAS 55 optimal management of physical assets. Works Management Tutorial, London
9. PAS 55-1 (2004) Asset Management. Specification for the optimized management of physical infrastructure assets. BSI, United Kingdom
10. ISO 55000:2014 Asset management Overview, principles and terminology
11. ISO 55001:2014 Asset management-Management systems-Requirements
12. ISO 55002:2014 Asset management-Management systems-Guidelines on the application of ISO 55001
13. González-Prida V, Crespo A (2014) After-sales service of engineering industrial assets. A reference framework for warranty management. Springer, 401 p. 1v. ISBN 978-3-319-03709-7
14. Parra C (2015) Review of the basic processes of a maintenance and reliability management model. Project: design and construction of the third set of locks in the ACP (Autoridad del Canal de Panamá). http://dx.doi.org/10.13140/RG.2.2.16830.10566
15. Parra C (2013) Informe final del Proyecto de Confiabilidad. Empresa de Transporte de Gas. Resultados de la Auditoría Integral del Modelo de Gestión de Mantenimiento/ICOGAM: Ingeniería de Confiabilidad Operacional, Gestión de Activos y Mantenimiento. http://dx.doi.org/10.13140/RG.2.2.18429.18409

16. INGEMAN (2019) MGM-Gas. Adaptación de un modelo de Gestión del mantenimiento a la industria del gas natural. Editor: INGEMAN, 2019, 239 p. 1v. ISBN 978-84-09-14105-0
17. Parra C, Crespo A, González-Prida V, Sola A, Kristjanpoller F, Viveros P (2019) Audit models for asset management, maintenance and reliability processes. Case Study Electr Transm Sect. http://dx.doi.org/10.13140/RG.2.2.31929.60006
18. Peters RA (2006) Maintenance benchmarking and best practices, 1st edn. McGraw-Hill, New York, 549 p. 1v. ISBN-13 978-0071463393

Automating Data Driven Decisions for Asset Management – A How to Framework for Integrating OT/IT Operational and Information Technology, Procedures and Staff

Anastasia Kuusk and Jing Gao(✉)

University of South Australia, Mawson Lakes, SA, Australia
{anastasia.kuusk,jing.gao}@unisa.edu.au

Abstract. The paper describes an evidence-based construct hierarchy for engineering asset management based organisations to cost effectively aggregate operational (OT), information technologies (IT) and Industrial Internet of Things (IIoT) to enable automated data driven decision making event triggers between assets, robotics, sensors, business systems and people. The research followed an interpretive epistemology conducted as seven case studies across mining, utility and defence based organisations, a Delphi survey with over fifty Technology and Engineering practitioners to confirm baseline TOP constructs and subsequent application of open, federated data architecture in Australian water and defence industries. The research findings and industry application suggests a cost effective, benefits realization, vendor agnostic way to connect and automate often soiled important data types for reuse and first level automated decision making, improving data quality, integrity, asset reliability and organizational trust in data and automation initiatives. Digital transformation and safety goals can be achieved by people becoming second point decision makers, informed by the automated aggregated data across the organization, providing a valuable extension to the reliance on traditional machine learning and business analytics involving data warehouses and business system data to the exclusion of asset performance. The research extends the application of ISA95/MES to holistic organizational wide emerging frameworks such as RAMI4.0 and

Keywords: OT/IT integration · Data governance · IIoT · Industry 4.0 · Automation · Digital transformation

1 Introduction

Increasingly presentation of asset related data such as resilience and reliability in real time is expected from aggregating and leveraging investment in existing asset, business system, people and sensor generated data to enable automated decision making and achieve organisational Industry 4.0 [2] digital transformation goals. Engineering Asset

A. Crespo Márquez et al. (Eds.): WCEAM 2019, LNME, pp. 201–213, 2021.
https://doi.org/10.1007/978-3-030-64228-0_18

Management (EAM) based organisations are experiencing pressure to leverage financial investments in data governance, technology networks, cloud storage [26] and business systems, move to next generation holistic organisational data governance standards such as RAMI4.0 [22] from well understood asset standards such as ISA95, incorporate edge and sensor hardware advancement and account for cyber security threats. To fulfil such digital transformation organisational goals, organisations need to integrate Operational technology (T), Organisational processes (O) and People (P) domains. Integrating technology requires incorporating asset based operational technologies such as SCADA, PLC's and IoT [1] sensors with information technologies, such as finance, risk, manufacturing execution, product lifecycle management and safety data in business systems.

Information technology (IT) has traditionally been the domain of the office-based technology team, often reporting to a position such as a Chief Information Officer, whilst operational technology (OT) has traditionally been the real time remote asset control domain of engineers. The domains have been implemented in organisations in parallel with separate information and operational technology (T), organisational Processes and standards (O) and People (P) factors [3].

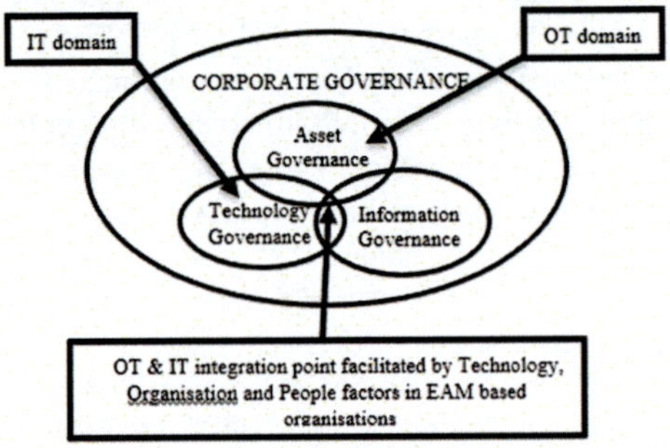

Fig. 1. Multi-disciplinary approach integrated governance research framework

This paper describes a data driven TOP based framework for integrating OT, IT and IoT data driven decision making in engineering asset management (EAM) organisations, reducing the cost and time for such organisations to identify and bridge the gaps of OT/IT integration to achieve digital transformation of manufacturing supply chains, logistics and asset sustainment operations. The benefits of such OT/IT/IoT integrated enterprise data architecture for automated decision making, extending organisations current lense and understanding from business analytics of business system data integrations to incorporating engineering asset management data organisations are increasingly being reported [26, 27]. Example benefits of integrated asset and business system data integrations include

> "generat(ing) reports for management, engineers in the field (see) real time on flows, alarms, gen set hours run etc. via an information historian which in future will load into the Asset Management System which will have the service intervals

(10,000 h), energy consumption, benefits realisation, information to demonstrate growth and dollars for upgrades.... make more productive" (D10)." P. 172 [28]

2 Framework Development Motivation

Since the early 2000's when Gartner coined the term OT/IT convergence, alignment and integration [4] there have been few empirical holistic architectures developed to holistically integrate OT/IT/IoT people, process, data, asset management and technology factors in particularly EAM based organisations. Existing frameworks are available however are disparate by considering only asset management, business office based technology systems or asset based technologies [9].

Disparateness and lack of a holistic view of current asset management people, data and process frameworks is vast across the complex OT/IT holistic system. Asset frameworks include Pas55 [9], ISO5500 [23] Integrated Strategic Asset Management [10] and asset flows in information management [11]. Operational Technology frameworks include ISA95 [9] south to north organisational view of asset based data alongside east to west data views governed by Information Technology NIST CIP cybersecurity [12], COBIT and ITIL ISO38500 related standards [13]. Several data governance frameworks such as DMBOK [14], MIKE2.0 [15, 16] and ISO15489 Records Management are also currently available. The frameworks are often domain specific and applied to and by separate parts of an asset-based organisation.

New frameworks such as RAMI4.0 [22] are emerging that also challenge the north south asset to office view of asset data governance frameworks such as ISA95, driven by technological advancements. Dual Internet Protocol (IP) and ethernet based hardware devices now transverse the OT and IT domains, processing data at the edge of networks instead of centrally [17]. Data from internal and external to the organisation, such as from historians acting as data lake platforms and weather data from internet platforms are 'mashed up' requiring new approaches [26, 27] than organisations architecting networks that keep OT and IT data segregated by internal firewalls (Fig. 2).

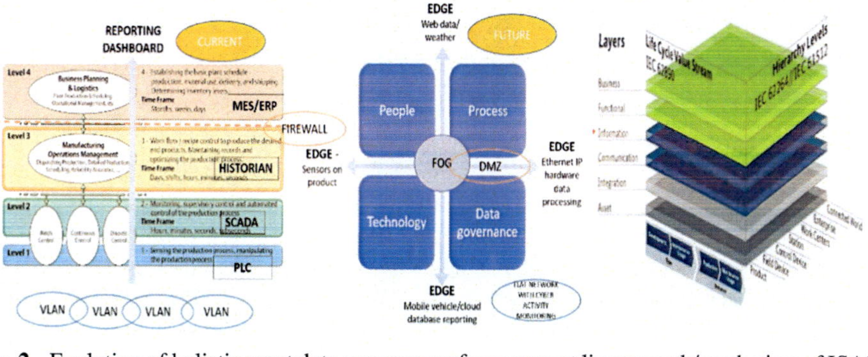

Fig. 2. Evolution of holistic asset data governance from current linear south/north view of ISA95 to TOP and OT/IT/IoT future state RAMI4.0.

Converging the lenses provides organisations with a state of the art comprehensive north, south, east and west organisational wide lense of TOP OT/IT/IoT data [5–8] and therefore a real time, modern, trusted, digital Industry 4.0 view of asset operations supporting functions such as resilience and reliability. Organisations moving to such an integrated data view need a comprehensive, cost effective how to framework to guide the transition from standards based principles of ISA95 to RAMI4.0. With the extent of EAM platform Manufacturing Execution System (MES) and Product Lifecycle Management (PLM) and IoT device, data capture and storage data platforms on offer organisations also need to weigh up the openness of the architecture when aggregating OT/IT and IoT data across organisations for real time data visualisation in the office.

3 Framework Verification and Validation

Providing for how to address automated aggregated data driven decision making gaps of standards such as ISA95, researchers at the University of South Australia School of Computer Science and Asset Institute engaged with technology and engineering practitioners during three Delphi surveys [3] across mining, defence, water and power utility industry and followed up with case studies. This approach provided an iterative building of the integration influencing gap factors framework.

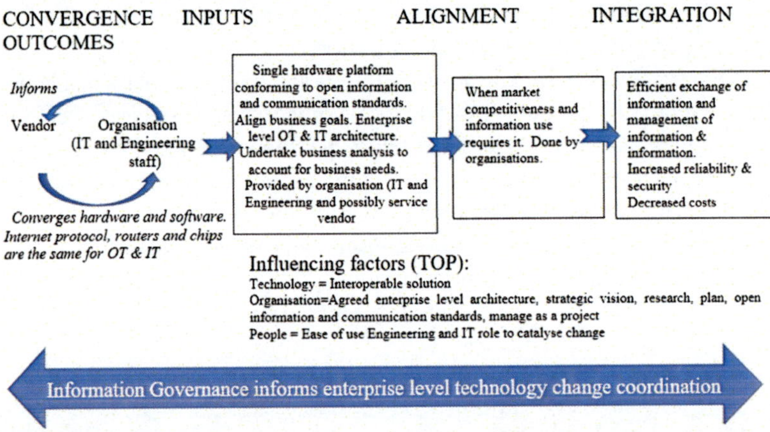

Fig. 3. Iterative building from research results of TOP OT/IT/IoT constructs

As the implementation of IoT sensors for asset activities such as calibration, reliability, resilience and run time has increased [1] since the initial research, application by the researchers of holistic data architecture framework into mining and water utilities has enabled the scalability of the original OT/IT framework to incorporate IoT over time.

In the research phase Over 70 staff from 19 consulting, cybersecurity, mining, energy, utilities, defence and local government authority-based organisations participated in the research to establish an empirical framework of ranked factors that influence successful integration of OT/IT in EAM based organisations. The phase began with a literature review providing definition of integration dimensions and highlighted TOP and OT/IT/IoT gaps that influence successful data governance integration across disciplines (Fig. 4).

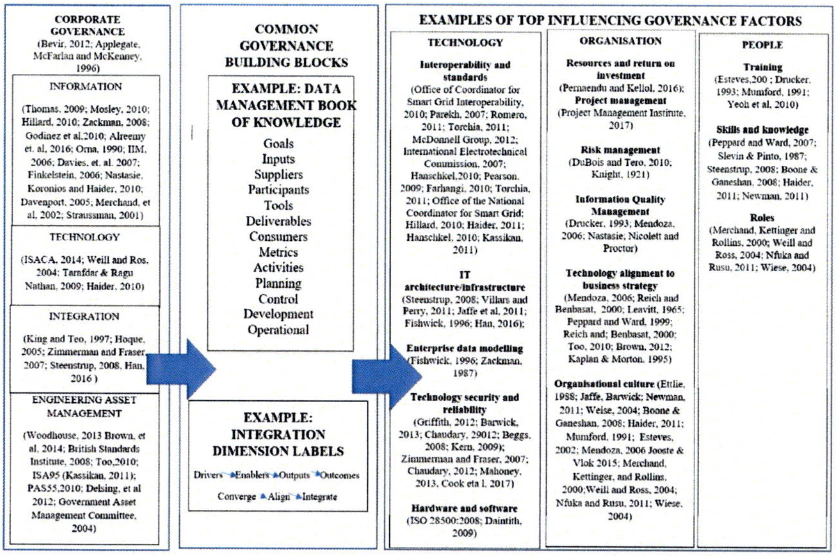

Fig. 4. Overview of TOP and OT/IT data governance asset to office integration gap literature.

Delphi results were received from 30 participants across the three survey rounds within tolerance levels for such research methods [24, 25] analysed using percentage, interval quartile range (IQR), mean, standard deviation statistics [18, 19]. Case studies data was analysed using thematic analysis [20] in accordance with Yin's [21] case study text validation to provide empirically ranked constructs. The Delphi study identified 42 influencing ranked factors, by practitioner consensus, applicable to the EAM context compared to the often studied areas of operational management, such as in manufacturing. The Delphi study also indicated the role of information governance and organisational actors such as management, vendors, engineers and technologists.

The case studies refined the key factors to 38 and provided current, real world applicability. The case studies were undertaken in three iterations. The first was four case studies of water service providers where SCADA (OT) systems run in parallel to IT systems in order to identify a baseline of pathway and influencing factors in the real world. The next two case studies identified if the factors and framework were applicable to other EAM based industries such as defence and power generation. The final case study provided direct application of the framework to an organisation, and also provided further generalisation of the proposed framework for integrating OT and IT to the mining industry.

Table 1. Delphi study findings

Question	Strong	Medium	Low	None
1. Do you agree that convergence is an activity undertaken by vendors, alignment and integration activities by organisations?		A. Vendors converge, organisations align and integrate (66%, (1.58, σ .92)		
3. How should organisations align and integrate?	A. Business analysis (4.36, σ .81, IQR – 1); B. joint business effort (4.27, σ .679 IQR – 1)	C. Standardised platforms (3.64, σ .81, IQR –1)		D. Vendor involvement (3.45, 1.04, –1)
4. What criteria would indicate organisations should transition from converging to aligning?	A. Business needs accounted for (4.36, σ .81, IQR –1); B. Hardware consistent but applications disparate (4.09, σ .7, IQR –.05)	C. One size not fit all; D. costs (43.45, σ .82, IQR –1)		E. Only done by vendors (2, 1, –1.5)
5a. What are the technology influencing factors for aligning and integrating?			A. Interoperable solutions (3.82, σ .87, IQR –1.5)	A. Acceptance of open source solutions (3.09, 1.04, –2)
5b. What are the organisation influencing factors for aligning and integrating?	A. Open information and communication standards (4.18, σ .6, IQR –.5); B. manage as a project (4.18, σ .75, IQR – 1); C. Agreed enterprise level architecture (3.91, σ .54, IQR 0); D. research, plan and execute (4.45, σ .69, IQR –1); E. Strategic vision (4.27, σ .47, IQR – .5)	F. Mutual collaboration (4.09, σ .71, IQR – 1)	F. Robust framework (3.91, σ .7, IQR –1); G. systems thinking analysis (3.91, σ .7, IQR –.5)	
5c. What are the people influencing factors for aligning and integrating?	A. Input from all (4.27, σ .9, IQR –1)	B. Ease of use (3.55, σ .82, IQR –1); C. engineering and IT role to catalyse business change (3.64, σ .81, IQR – 1)	D. Appropriate training (4.18, σ .87, IQR – .51)	E. Acknowledge office of CIO (3.73, 1.01, –1.5)
6. What criteria would indicate organisations should transition from aligning to integrating?	A. When market competitiveness requires it (4.36, σ .67, IQR –1) B. When information use requires it (3.82, σ .75, IQR 0); C. when IT/OT structures aligned (3.82, σ .75, IQR 0)		D. By organisations not vendors (2.36, 1.21, –.5); E. when cost not an issue (3, 1.26, 0); when standards aligned (3.82, .98, – 1.5)	F. When business and IT have consensus (3.91, 1.14, –1.5)
7. How can information governance facilitate integration of OT & IT?	A. Facilitate enterprise level technology change coordination (4.18, σ .75, IQR – 1)		B. Governance informs strategy (4.09, σ .83, IQR –1.5)	

(continued)

Table 1. (*continued*)

Question	Strong	Medium	Low	None
8. Is IT or Engineering best suited to align or integrate OT and IT?	A. Combined (.73, σ .47, IQR −.5)	B. IT (3.5, .97, −1); C. Engineering (2.78, .67, −1)		

In the last two years the baseline framework has been extended to an open, federated data architecture, extending organisations how to tool kit with a workable, cost effective, vendor agnostic, data ownership approach. The framework has been incorporated into digital data strategies in defence and water utilities in Australia providing either an owned data platform framework from which organisations can leverage data to their industries up and down the supply chain or use to compare and contrast vendor claims of interoperability, risks, security and costs of outsourcing aggregated asset data management, analysis and visualisation. The extension of the original framework with application into industry has also facilitated incorporation of IoT into the original OT/IT alignment, convergence and integration model [4].

4 Key Research Findings

The delphi and case study ranked constructs provide a holistic data governance driven framework for integrating OT and IT identified in Fig. 1 below is used by organisations to improve asset resilience and reliability through reducing OT/IT integration financial and time-based technology, people and process risks. The influencing governance factors have been formalised into a ranked checklist that can be applied by EAM based organisations to existing OT/IT people, process and technology architecture. This facilitates OT/IT integration maturity identification; therefore organisations only invest in the OT/IT TOP integration gaps unique to their organisation. Organisations can then achieve resilience and reliability in asset management through data driven decision making.

The research framework, incorporating the literature review, Delphi study and case studies provided an incremental building of theory to contribute to particularly the information systems and asset management bodies of knowledge, by extending existing gaps of such disciplinary frameworks to be easily applied to the real world. The research has taken theoretical frameworks, filled the construct gaps such as whom, when, what and how and validated for the EAM context, lacking in the current focus of the governance literature on other areas of operational management.

The case studies further built on the findings of the Delphi study, testing generalisability, application of the framework in the real world and indicating a dependency and information governance influencing framework. Thus, the difference between the Delphi and case studies was a definition of key influencing constructs on OT/IT integration. Unique insights into construct application in the EAM context were also derived from the research.

The research identified key characteristics of organisations that had achieved OT/IT integration. At the outset of an integration project, key engineering and technology staff should converge as a team to agree on the OT and IT data, asset management and

Table 2. Technology influencing governance perspective

Governance dimension	Variable	Factor
Technology (T)	TOP influencing data governance domains	TOP influencing data governance factors
	T1 Interoperability (to work with other products) and standards	T1.1 Interoperable solutions
	T2 IT architecture/infrastructure	T2.1 Single platform – Unified and integrated data
	T3 Enterprise data frameworkling	T3.1 Acceptance of Open source data and communication standards
	T4 Hardware and software	T4.1 OT controller hardware disparate, applications and other hardware converging – Edge devices
	T5 Security & reliability	T5.1 Increased reliability
		T5.2 Increased security

technology integration standards and communication protocols to be used to cross the OT/IT divide. Engineering and technology investment strategies and projects, including an adherence to open source integration protocols such as OPC UA and hardware, telecommunication and software investment that facilitates data transfer from north, south, east and west of the organisation should be aligned.

Integration is achieved when the organisation has efficient seamless integration of information and data governance from the asset control to the office data reporting environments. Organisations most likely to success at OT/IT integration are driven by the remoteness of their assets – staff cannot for a small amount of time or money access the assets and have an asset engineering manager project manage the integration project because the operations team are more likely to be responsible for ongoing maintenance and use of the integrated environment. Interoperable data and single telecommunication protocols between OT/IT environments facilitate seamless data governance from asset to office. robust integration strategy in an organisation occurs when a risk register is evident (Table 4).

The security versus reliability focus of the two main disciplines, being technologists and engineers, was a constant theme underpinning trust, communication and collaboration. Without collaboration between technologists and engineers, trust, communication and collaboration can impede or facilitate integration of OT and IT. The value of management of information appears to drive the need to collaborate. Systems thinking was not evident in its defined form; however, an organisation having a holistic view of the OT/IT systems - including data flows, OT computer components, engineers, technologists, telecommunication networks and IT office based systems such as asset management systems facilitating maintenance of assets - can facilitate integration.

Table 3. Organisational processes influencing governance perspective

Governance dimension	Variable	Factor
Organisational Process (O)	O1 Resources	01.1 Adequately skilled
	O2 Return on Investment	O2.1 Transition from alignment to integration when cost not an issue
		O2.2 Decreased costs
		O2.3 Costs
	O3 Project Management	03.1 Manage as a project
	O4 Risk Management	O4.1 Robust framework
	O5 Information Quality Management	O5.1 Efficient exchange of data and management of information
	O6 Technology alignment to business strategy	O6.1 Research, plan, execute
	O5 Information Quality Management	O6.2 Strategic vision – IDR performance metrics
	O6 Technology alignment to business strategy	O6.4 Business needs accounted for?
	O7 Organisational culture	O6.5 Agreed enterprise level architecture
		O6.6 Transition from alignment to integration when OT/IT structures aligned
		O6.7 Standardised platforms
		O6.8 Governance informs strategy
		O6.9 Transition from alignment (strategy and planning) to integration (efficient management of info) when standards aligned
		O6.10 Transition from alignment to integration when business and IT have consensus
		O7.1 Information governance facilitates enterprise level technology change coordination
		O7.2 One size does not fit all

Similarly, other factors initially identified by literature and Delphi studies, however not validated by case studies, were evident as well. Training was replaced with ease of use of system solutions, and stakeholder expectations were in the form of technologist and engineering role clarifications of reliability and security. Other aspects of influencing constructs were further defined, such as what "manage as a project" (Project Management office, scheduling and reporting), "open standards" (programming and ISA95) and "competition and interoperability" (single communication networking and data historian) meant in the EAM context. The importance of business and information technology and operational technology strategic alignment, not just the traditional lens of business to technology strategy alignment, was also incrementally identified by the case studies, building on the base constructs defined by the Delphi study.

Roles were also further clarified. The role of engineering manager compared to technologist such as CIO may provide insights into modern views of responsibilities in EAM organisations for successful integration projects and are worthy of further study. The initial Delphi study identification of vendor was complemented in the case studies by defining the place of the consultant after implementation (Fig. 5).

Case studies highlighted that a dependency integration framework is difficult to apply in the real world. As OT and IT are implemented in EAM organisations in parallel, dependency constructs and mapping to terminological groupings such as convergence,

Table 4. People influencing governance perspective

Governance dimension	Variable	Factor
People (P)	P1 Training	P1.1 Ease of use
	P2 Skills and knowledge	P2.1 Acknowledge office of the CIO
	P1 Training	P2.2 Input from all
	P2 Skills and knowledge	P2.3 Mutual collaboration
	P3 Roles	P3.1 Joint business effort
		P3.2 Combined IT & Engineering responsibility to integrate
		P3.3 Technologist responsibility defined
		P3.4 Engineering responsibility defined
	P3 Roles	P3.5 Vendors converge, organisations align and integrate
		P3.6 Organisations transition from alignment to integration not vendors
		P3.7 Vendor involvement in alignment and integration

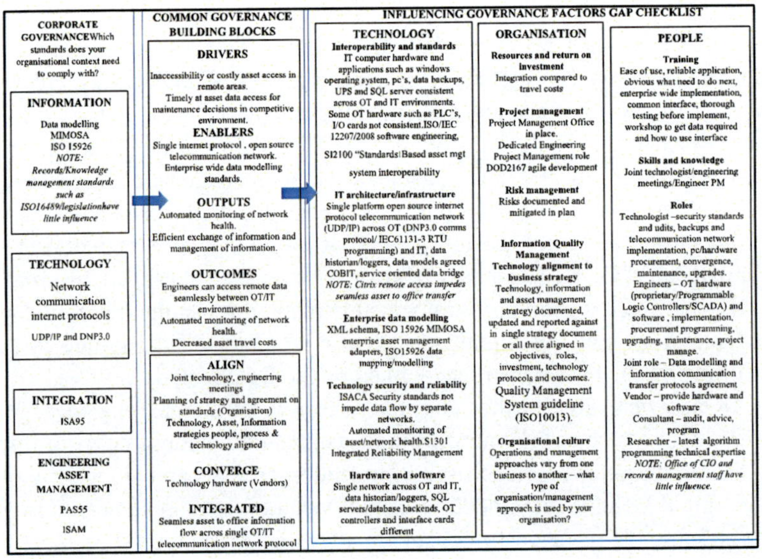

Fig. 5. Baseline post case studies TOP and OT/IT/IoT integrated data framework for EAM.

alignment and integration is very difficult to validate. Therefore, the frameworks identified in the literature review are theoretical only and require granularity as to who is responsible and how/when to apply constructs provided by this research.

5 Research Findings Application

In recent years the research framework of Fig. 3 and Tables 1, 2 and 3 above have been further extended to incorporate the proliferation of IoT devices and subsequent flood of real time data into EAM based organisations OT/IT environments. This was achieved by identifying the organizational priority gaps from the influencing factors in Tables 1, 2 and 3 in water utility and defence organisations OT/IT Industry 4.0 digital transformation journeys and accounting for COBIT. ISA95 and RAMI4.0 Industry 4.0 and cloud data [26] management frameworks (Fig. 6).

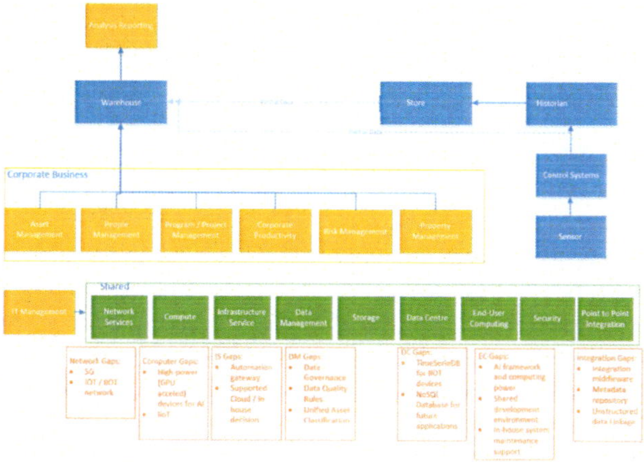

Fig. 6. Current state siloed organizational data architecture

Digital transformation strategies had been defined for organisations at a high level, leaving the need for a 'how to' arrchitecture [27] approach to enable the strategies quickly and cost effectively. By identifying the priority gaps these formed a roadmap of work packages for the organisations to fund and deliver to add integrated, automated, real time data first level decision making and converged network architectures to organisations as benefits and therefore delivering on digital transformation strategies. The recent analysis of the organization TOP gaps has also led to the uptake of future state converged open, federated data architecture adoption that facilitates data acquisition, translation, automated event triggering, decoupling of point to point system integrations and real time data visualisation in the office from the assets (Fig. 7).

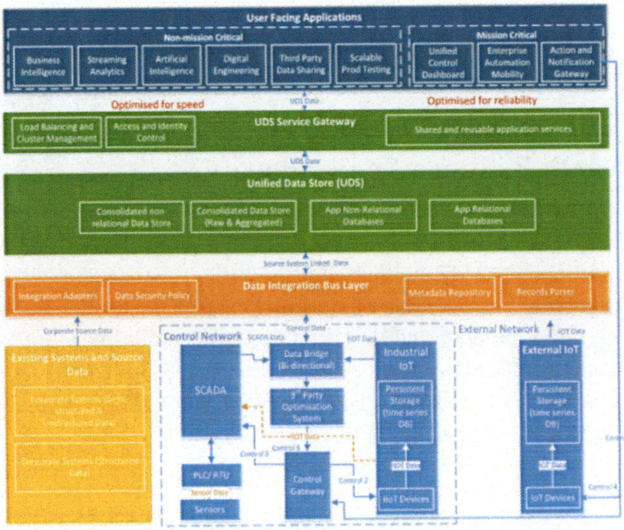

Fig. 7. State of the art OT/IT/IoT integrated, open, federated data architecture

6 Conclusion

Improvements to the automated data driven OT/IT/IoT TOP integration framework for EAM based organisations are ongoing. Incremental framework improvements and further data aggregation insights for asset resilience and reliability will continue to evolve as sensor and edge technology advancements and cyber security threats evolve. application to verify improvement from current organisational disparate OT and IT environments to integrated data driven decision-making including time taken to converge the people, process and technology constructs. Future research will also further improvement the framework, such as measuring outcomes, incorporating the impact of technological advancements such as sensor data, edge and fog computing needing a holistic north/south and east/west data view not provided by the current ISA95 standard application prevailing in EAM based organisations. Such incremental improvement to the emerging OT/IT Integration body of knowledge will continue to inform risk and cost reduction in the areas of asset reliability and resilience whilst also providing an empirical base for IIoT and industry 4.0 developments.

References

1. Winters B (2017) IIoT and process reliability: a real-world approach to the IIoT for process reliability. Chemical Industry Digest
2. Gilchrist A (2016) Industry 4.0: the industrial Internet of Things. Apress, Berkeley
3. Linstone H (1989) Multiple perspectives: concept, applications, and user guidelines. Syst Pract 2:307–331
4. Steenstrup K et al (2008) IT and OT: intersection and collaboration. Gartner, Stamford
5. Kuusk A, Gao J (2015) Factors for successfully integrating operational and information technologies. In: PICMET 2015, Portland, Oregan

6. Kuusk A, Koronios A, Gao J (2015) Consolidating people, process and technology to bridge the great wall of operational and information technologies. In: Engineering asset management: towards certification, Springer, Cham
7. Mendoza L, Marius A, Perez M, Griman A (2006) Critical success factors for managing systems integration. Inf Syst Manage 23:56–75
8. King W, Teo T (1997) Integration between business planning and information systems planning: validating a stage hypothesis. Decis Sci 28:279–308
9. Delsing J, Rosenqvist F, Carlsson O, Colombo AW, Bangemann T (2012) Migration of industrial process control systems into service-oriented architecture. In: IECON 2012-38th annual conference on IEEE industrial electronics society, IEEE, pp 5786-5792
10. Brown K, Laue M, Tafur J, Mahmood MN, Scherrer P, Keast R (2014) An integrated approach to strategic asset management. In: Infranomi, Springer, pp 57–74
11. Haider A (2012) Enterprise architectures for information and operational technologies for asset management. In: Engineering asset management and infrastructure sustainability, Springer, London, pp 315–328
12. NIST (2019) Framework for improving the critical infrastructure cybersecurity. Accessed 2019
13. International Systems Audit and Control Association (2014) COBIT5 framework. International Organisation for Standardisation, Geneva
14. Mosley M, Brackett M (2010) The DAMA guide to the data management body of knowledge (DAMA-DMBOK guide). Technics Publications, Basking Ridge
15. MIKE 2.0 (2013) Method for an integrated knowledge environment. The open source standard for information management. http://mike2.openmethodology.org
16. Hillard R (2010) Information driven business: how to manage data and information for maximum advantage. Wiley, Sydney
17. Shi W, Cao J, Zhang Q, Li Y, Xu L (2016) Edge computing: vision and challenges. IEEE Internet Things J **3**(5)
18. Dalkey N (1969) The Delphi method: an experimental study of group opinion. RAND Corporation, Santa Monica
19. Linstone M, Turoff H (1975) The Delphi method: techniques and applications. Addison-Wesley, Reading
20. Alhojailan MI (2012) Thematic analysis: a critical review of its process and evaluation. West East J Soc Sci 1(1):39–47
21. Yin R (2009) Case study research, design and methods, 4th edn. Sage Publications, Newbury Park
22. Alcácer V, Cruz-Machado V (2019) Scanning the Industry 4.0: a literature review on technologies for manufacturing systems. Eng Sci Technol J
23. International Standards Organisation (2014) ISO55000: asset management: overview, principles and terminology. International Standards Organisation
24. Okoli C, Pawlowski S (2004) The Delphi method as a research tool: an example of design considerations and applications. Information Management 42(1):15–29
25. Pare G, Cameron A, Poba J, Templier M (2013) A systematic assessment of rigor in IS ranking type Delphi studies. Inf Manag 5(5):201–217
26. Cheng B, Zhang J, Hancke G, Karnouskos S, Colombo A (2018) Industrial cyberphysical systems: Realizing cloud-based big data infrastructures. IEEE Ind Electron Mag 12(1):25–35
27. Lara P, Sánchez M, Villalobos J (2018) OT modeling: the enterprise beyond IT. Bus Inf Syst Eng 61(4):399–411. https://doi.org/10.1007/s12599-018-0543-3
28. Govan-Kuusk A (2018) An information governance driven framework for integrating operational and information technology in the engineering asset management context. Thesis, University of South Australia

Framework for Assessing Economic, Environmental and Social Value of Monitoring Systems; Case Water Balance Management in Mining Sector

Minna Räikkönen[1](✉), Teuvo Uusitalo[1], Riitta Molarius[1], Helena Kortelainen[1], and Claudia Di Noi[2]

[1] VTT Technical Research Centre of Finland Ltd., P.O. Box 1300, 33101 Tampere, Finland
{minna.raikkonen,teuvo.uusitalo,riitta.molarius, helena.kortelainen}@vtt.fi

[2] GreenDelta GmbH, Kaiserdamm 13, 14057 Berlin, Germany
dinoi@greendelta.com

Abstract. Mining industry is highly dependent on the state of its assets, and it is also a significant water user and producer of wastewater. Thus, the management of water-related assets and operations is a critical business issue and has an effect both on the growth and profitability in mining companies. Moreover, digitalization influences asset management processes, and implementation of novel IIoT technologies for water balance management are vital for sustainable mining business. However, investments in water-related IIoT technologies must be carefully planned and evaluated to deliver immediate and long-term benefits, to enhance the asset utilization, and to reduce the environmental risk. This research presents an asset management framework and related tool: Framework for assessing Economic, Environmental and Social Value of Monitoring System (FEES). FEES defines the customer value of the water balance system and supports the economic, environmental and social impact and risk assessment in mining sector, and demonstrates the costs and risks as well as the benefits achieved by a potential customer. The paper contributes to the asset management research field by providing a multidimensional approach to assess water-related investments in mining industry. From a long-term perspective, applying a multidimensional approach to the assessment generates several benefits, including advancing the ability to communicate the value of water-related investments to business ecosystem, local societies, people, investors and other stakeholders as well as to support the goal of sustainable growth. The proposed model provides practitioners with the opportunity to adopt a proactive approach to assess water balance system's value.

Keywords: Impact · Assessment · Value · Risk · Water balance management system · Mining · Asset management

1 Introduction

Mining industry is highly dependent on the state of its assets, and it is also a significant water user and producer of wastewater. Furthermore, the connection of mining assets with

A. Crespo Márquez et al. (Eds.): WCEAM 2019, LNME, pp. 214–226, 2021.
https://doi.org/10.1007/978-3-030-64228-0_19

the surrounding environment can cause unpredictable and seasonal impact on the site's water volumes and qualities (Fig. 1). Thus, the management of water-related assets and operations in mines is a critical business issue and has an effect both on the growth and profitability. Moreover, digitalization influences asset management processes in mining companies. On the user side, operators of mining assets continuously seek ways to optimise useful life of composite equipment, machinery, and systems that incorporate fourth industrial revolution technologies in the form of artificial intelligence, IIoT (Industrial Internet of Things), industry 4.0 and 5G networks [1]. However, yet the mine operators' ability to monitor and predict the status and behaviour of the vast water volumes and related assets is often suboptimal and adequate precautions are not usually taken. Therefore, asset investments towards improved water balance management are vital for sustainable mining.

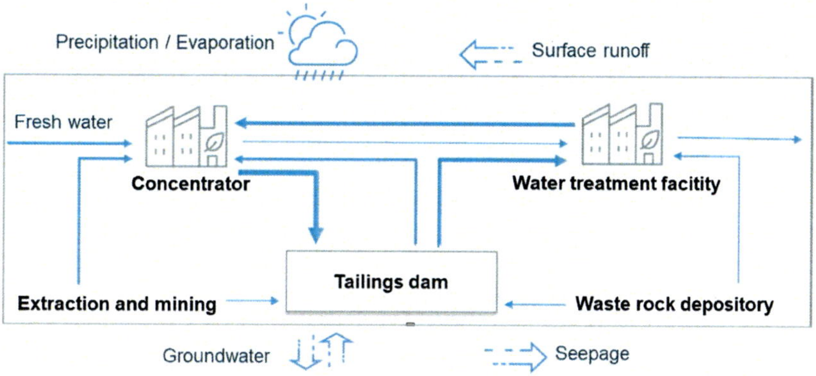

Fig. 1. Mine site's water balance in relation to mining assets and surrounding environment.

In general, predictive water balance management system (WBMS) can be considered as an IIoT-based investment where the cost-effectiveness is of central concern and where the benefit is largely the enhanced management of water and related assets as well as the avoidance or reduction of negative environmental consequences. Thus, the ability to communicate the various impacts of water-related investments to employees, authorities, residents, shareholders and other stakeholders can give a competitive advantage for a company. Furthermore, demonstration and communication of the value of the water balance management can bridge the divides between technology providers, mining companies, monitoring authorities and citizens living in the neighbourhood.

In the context of mining, achievable benefits of a potential customer or end user vary according to the site's location, business environment and economic, social and environmental drivers. However, it is evident that assessment of economic, environmental and social impacts as well as risk reduction is often very challenging and complex task. Typically, the assessments focus on the demonstration of the short-term effects rather than emphasize the long-term-impacts [2, 3]. Moreover, the investments should be evaluated and prioritized not only in monetary terms, but also with regard to dependability, safety, sustainability, and social acceptability, for example [4–6].

2 Aim, Research Question and Methodology

This paper contributes to the field of asset management research by providing a multi-dimensional approach to assess water-related investments in the mining industry. The objective of the paper is to establish a framework to define the customer value of the WBMS and to support the economic, environmental and social impact and risk assessment of water-related investments in mining sector as a whole. The underlying research question is: How could a scientifically grounded framework for conducting an integrated economic and risk assessment be constructed? Our leading hypothesis is that business benefits of WBMS can be significant, but many of the benefits, especially the cost savings generated by reduced water-related risks, are either unknown or insufficiently assessed.

The research was inspired by 'Dynamic predictive solution for sustainable water balance management in mining (SERENE)' project, funded by the European Institute of Innovation and Technology (EIT). The project responds to the growing need to improve water balance management in the mining sector. The SERENE project developed a new IIoT-based water management system that helps the mining industry monitor the entire mine site's water balance in real-time. The system also provides the industry with the possibility to create short-term forecasts for the site's water volumes and water quality in different production and environmental conditions.

The paper adopts a top-down approach where the requirements for the value assessment are derived from the business decision-makers' needs and stakeholder requirements. To strengthen the emergence of practical solutions, the research was conducted in close co-operation with a mining company in Northern Europe and a company developing the water balance management technology.

In our research, content analysis and interviews were used to study and to increase knowledge of various aspects of water balance management in mining and to construct a conceptual framework for integrating different assessment methods. The empirical work verifies and further develops the conceptual ideas into methods to solve practical problems concerning assessment of economic, environmental and social impacts of water-related assets and operations. Case studies are well-suited to studies where the research context and environment are complex and the aim is to provide a thorough understanding of the phenomenon in question [7]. The case study is an appropriate choice especially in the research projects where the existing theories are not capable of explaining the research questions [8].

Our case study includes the economic assessment of a WBMS that is upscaled and piloted within the SERENE project. The system utilizes remote sensing, advanced data analytics and IIoT framework to provide a solution that enables monitoring of an entire mine site water balance in real time. Existing technologies are combined with these new technologies in order to achieve a holistic system covering the mine, with water quality, quantity and the hydrological cycle incorporated into the system [9]. The second part consists of the water risk assessment conducted at the ore enrichment site of the case study area to support the asset management processes in mines.

The input data needed for the assessment methodology is based on the primary data from various IT systems and experts of the case mine and the technology provider. Experiences and learnings from the case study were combined with the results of the literature study, and thereafter the methodology was finalized. Consequently, the key steps in developing the practical method were problem solving and building a solution for the technology provider.

3 Water-Related Risks in Mining Areas

Both water scarcity and the management of vast water volumes in mines can cause significant risks for mining assets, operations and eventually for mining companies' entire business. Water balance takes into consideration water sources, the consumption of water and the water sinks. In mining, water sources consist of external water sources, moisture in the run of mine ore, precipitation, and mine dewatering. Major water sinks include water lost in the tailings, concentrates, and discharged water [10]. Water balance accuracy affects on the decision making in water-related investments as it can influence economic and environmental performance and management of risks [11].

The water balance in mining operations can be either water positive (water accumulates on the site) or water negative (net loss of water) depending on the available water resources at a mine site. Water negative sites require continually to supply water from external water sources into the water system [12]. In Nordic countries with high precipitation [13], the prevailing state is a surplus of water in the mining area. Extra water needs to be collected, cleaned and returned to the environment. The discharge limits (volume and concentrations of various substances) are given in the environmental permit in terms of both annual averages and peak values.

Water-related risks are linked with either water quality or quantity. They may threaten the functionality of mines or related enrichment plants, humans, or the ambient environment. The consequences of risks may vary from erosion of soil and formation of sinkholes to loss of biodiversity and contamination or pollution of the environment. In worst cases, the consequences may have severe health effects and loss of lives of mine staff or people living around the area due to poisonous spills [14]. Due to contaminated water, dam failure is one of the most important risks [15, 16].

Smaller spills from tailing areas into the environment may also have serious effects due to their impact on company reputation and, companies' social licence to operate (SLO). SLO is seen as a signal from communities to grant different levels of acceptance of mining activities in a certain area [17]. Local communities, NGOs and civil society as a whole have numerous expectations that might go beyond being compliant to minimum legal standards and environmental requirements, such as respect for indigenous water rights, minimization of environmental harm, and engagement of the community in mining companies' routines [18]. Identification and management of environmental and social impacts and risks are fundamental aspects in the creation of a positive working environment. It benefits mining companies by, cost minimization, reduced risk of litigation, avoidance of operation disruption and creating better reputation [19]. Acceptance and trust in the mining industry are affected by communication, economic benefit distribution, and balance of benefits over impacts [20, 21].

With this background, there is a clear need to manage the water-related assets, related operations and to secure water management throughout the whole life cycle of mine operations, which may last many decades after closure of the mine [22]. Thus, decision-makers within companies need to understand both current and future changes in water availability and management in order to be prepared for future risks due to growing water demand, climate variability, policy developments, and many other factors. Implementation of a water management system as well as continuous improvements and preparedness against long-term stress factors in mines could be effective measures for tackling the challenge of fluctuating and excess water quantities [23]. There is a place for on-line monitoring and control of water flows. This kind of water management needs to be an integral part of a mine´s asset and control management systems in the future. A water management system can utilize new tools for environmental water management such as water footprint and handprint, eco-efficiency indicators, risk, financial and economic assessment, as well as robust and practical tools to identify, understand and manage water-related risks and opportunities in the future across a range of plausible scenarios [22–25].

In our project, we conducted a water-related risk assessment in our case company. The risk assessment method applied was fault tree analysis (FTA). FTA translates a physical system into a structured logic diagram (fault tree), in which certain specified causes lead to one specified 'top event' of interest [26]. Event and logic symbols are used to construct the logic diagram. A fault tree typically consists of four steps [27]: 1) System definition, 2) Fault-tree construction 3) Qualitative evaluation, 4) Quantitative evaluation. The purpose of this case study was to identify the different events that influence water-related risks and to determine whether a WBMS is able to manage them.

4 Value of the Water Balance Management System

The question of value can be examined at many different levels, taking into account aspects such as economic, social and environmental value. Value can be considered to consist of two components, perceived use value and exchange value. Perceived use value is based on customer's perception of the products or service's use. Exchange value is the amount paid for the product or service by the customer to the producer [28]. Value needs to be considered from the perspective of different stakeholders and in a manner that involves relationships, exchanges and interactions. Rana et al. have defined value in the following manner: "Value is the set of benefits derived by a stakeholder from an exchange" [29]. This definition is also adapted in our research as it reflects the idea that a value network generates economic, environmental and social value through complex dynamic exchanges between one or more enterprises, customers, suppliers, strategic partners and the community.

In Fig. 2, the main value elements of the WBMS in question is presented. The value elements were identified and the structure developed in the SERENE project in co-operation with researchers, the mine company, and the technology provider for the WBMS. Thus, it covers both practical and academic needs.

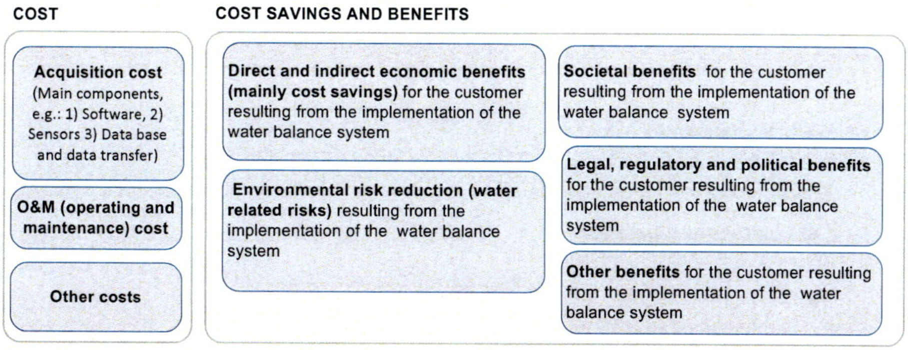

Fig. 2. Costs, cost savings and benefits.

The main value elements refer to acquisition and commissioning, operation, maintenance and other costs as well as direct Operation and Maintenance (O&M) cost savings and environmental risk reduction. It should also be considered that, for example, societal and environmental benefits generate value for all parties in the value ecosystem – not only for the customer. The acquisition cost includes from the mining and processing sites' (customer) point of view the purchase cost of the WBMS and from technology provider's point of view, the total investment cost i.e. the cost for developing and testing the solution and required return on investment. The O&M cost of a solution is the cost associated with operating and maintaining the WBMSs during its entire lifetime.

From the cost savings point of view, mining companies gain the capability to ensure water availability for production, to manage large volumes of excess water, to access to right quality process water, to identify and predict water-related risks, and to manage water-related operations and assets, for example tailings dams. Risk is usually expressed in terms of risk sources, potential events, their consequences and their likelihood [30]. In our research, value element "risk" is focused on water-related risks i.e. risk reduction resulting from the implementation of the WBMS. Five main water-related risks regarding water positive areas that were considered in the study are: dam failure, exceeding the yearly permission limits, exceeding the momentary permits, ground water pollution and the use of raw material. Moreover, the WBMS improves water recycling, facilitates water handling, and provides possibilities for mines to identify water quality or quantity related production losses. It also helps to minimize raw materials usage and water discharge and to optimize chemical usage. Most of all, mining companies gain capabilities to fulfil current and future environmental permit regulations to maintain their license to operate. (see Table 1).

Table 1. Main cost and cost saving categories of the WBMS.

Cost
WBMS acquisition cost
WBMS service fee/licence cost
WBMS operation and maintenance costs (costs that are not included in service fee)
Cost savings and benefits
Water-related risk reduction
Lower electricity consumption due to lower pumping need (raw mater, water discharge)
Lower chemicals consumption due to lower water usage
Lower need for laboratory testing
Lower need for laboratory investment
Higher production time (incl. breakdowns, unplanned & planned shutdowns) due to fewer process disturbances
Higher safety

5 Assessing and Demonstrating Value

Financial methods are most commonly used for value assessment. Slightly less popular but still common is to compare a financial measure against a hurdle rate (or acceptance level) to support the decision-making. Traditional financial methods incorporate payback analysis, discounted cash flow (e.g., internal rate of return (IRR) and net present value (NPV)) and return on investment [2]. Additionally, a variety of life cycle cost analysis methodologies for assessing the life cycle cost (LCC) of physical assets. Life cycle cost (LCC) evaluation considers the costs that incur over the whole product lifecycle from the concept and design phases to the operation and maintenance, and to the disposal/recycling phase. The methodology can also be applied merely to a part of the lifecycle [31]. There are also other alternative methods for valuing direct tangible and intangible impacts, for example, travel cost willingness to pay methods which usage, however, can be very time consuming. They all vary in how precise they describe the value [32].

Value assessment in our research can be seen as decision-support method that supports customers' asset management process especially in investment planning phase and the technology providers' marketing purposes (Table 2).

Table 2. Assessing and demonstrating the value of the WBMS.

What - To support technology provider's marketing and investment decisions (from the customer, i.e. mining and mineral processing site, viewpoint) on WBMS
Why - To demonstrate and communicate WBMS costs, cost savings and risk reduction to the potential customer
Who - Technology provider's marketing and sales representatives. Customer's managers responsible for investment planning and decision-making for water balance management
When - Mainly the early phases of investment decision-making

As there was no specific value assessment method available that integrates risk and economic assessment, the methodology was developed during our research. The assessment model, is a practical framework (Framework for assessing Economic, Environmental and Social Value of Monitoring System, FEES) for the integrated life cycle cost analysis (LCC) [31] and risk assessment [30] of the WBMS. This is because the essential purpose of the WBMS is in achieving cost savings and water-related risk reduction by improving water balance management (Fig. 3). The proposed FEES framework provides practitioners with the opportunity to adopt a proactive approach to assess WBMS's value and to support investment and asset management decision-making.

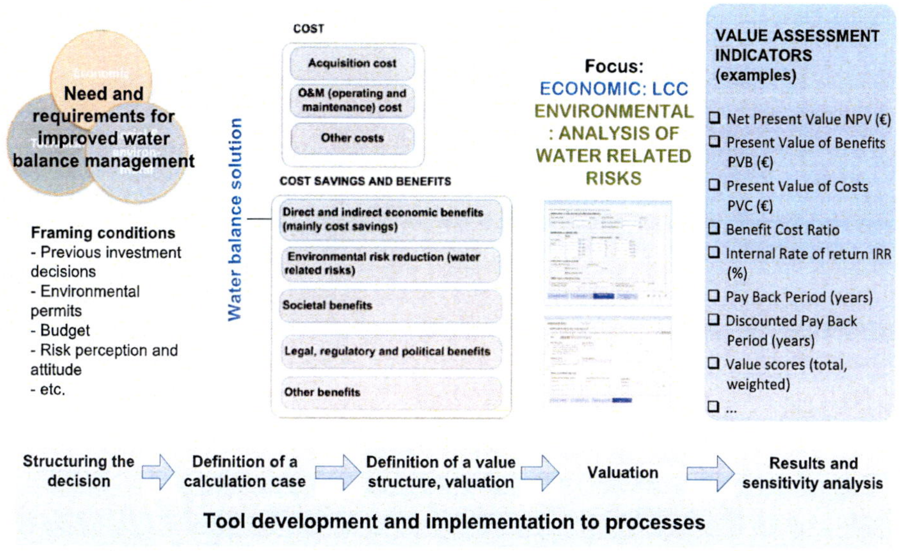

Fig. 3. Asset management framework for assessing the economic, environmental and social impact and risks of WBMS.

The FEES framework consists of the following four steps: (1) Basic information (calculation parameters, customer O&M information), (2) Cost and cost savings related data (cost and structure, valuation of costs and cost savings), (3) Results (key financials and graphs) and (4) Sensitivity analysis (sensitivity parameters, numerical results, graphs). The essential part of the FEES is the definition of the value structure and assessment of risks, costs, and benefits. More importantly, the FEES demonstrates a linkage between economic and environmental viewpoints by incorporating the results of a risk analysis into economic analysis (Fig. 3, Fig. 4 and Fig. 5). Based on the methodology, a FEES prototype tool that enables value assessment of WBMS for various customers was developed (Fig. 4).

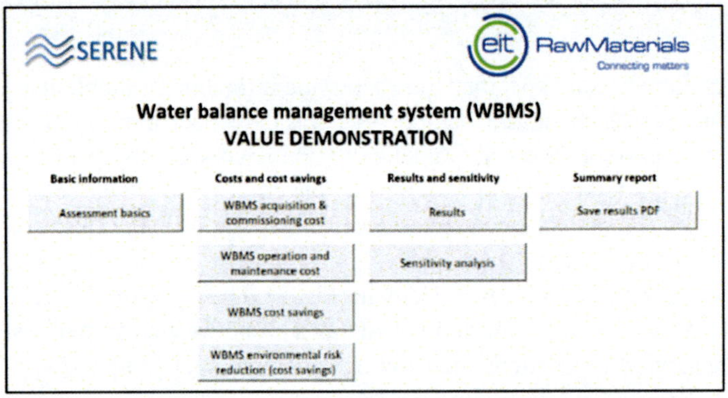

Fig. 4. Water balance management value assessment – FEES prototype tool.

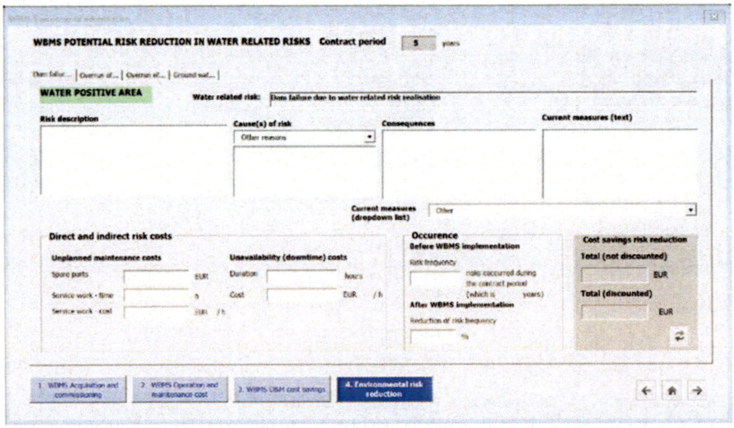

Fig. 5. FEES value assessment – assessment of water-related risk reduction.

Figure 5 presents an example of one of the FEES tool's module – a quantification of water-related risk – and related user interface. Categorization and assessment of risks and related costs could be helpful for developing an understanding of the potential concerns facing the stakeholders and for calculating the benefits of water-related risk reduction. The costs include both unplanned maintenance costs (spare parts and work) and unavailability costs. Occurrence is calculated based on the risk frequency (risks realized during the contract period) and the reduction of risk frequency due to implementation of the solution. Annual cost savings are calculated by multiplying the occurrence and the cost and dividing this by the contract period (years). The water-related risks that were identified and assessed were incorporated in the FEES tool. In our study, the decision situation concerns acquisition of the WBMS. The result indicators that are calculated by the FEES tool are: the total, annual and cumulative life cycle cost and cost savings (both non-discounted and discounted values) are calculated as result indicators. In addition, more detailed costs are calculated: both non-discounted and discounted values of reduced pumping cost, reduced cost of wastewater treatment, reduced chemical cost, reduced laboratory testing cost and investment needs for laboratory assets as well as cost savings regarding the reduction of water-related risks and process disturbances. The most important indicator is the cumulative cost savings generated by WBMS, and this indicator can also be selected to be used as a basis for establishing the managerial decision.

Value assessment is typically done before the costs and cost savings have occurred. Therefore, the results of calculations are partly dependent on the estimates of future values, which always incorporate uncertainty. To reduce the uncertainty related to decision in question, sensitivity analysis can be used. The sensitivity analysis module of the FEES tool determines what will happen if parameters are changed (what-if analysis). The what-if analysis can be made for one or more calculation parameters at a time. By making sensitivity analysis, the decision maker has an opportunity to validate whether the available results and data are accurate enough to provide information for decision-making. It is evident that there are a couple of alternative solutions that fulfil the customer's technical requirements for the better water balance management. In order to approve the purchase, its life cycle costs cannot exceed the life cycle savings and other benefits. Thus, the customer of the technology provider, typically a mining company, would be interested in gaining information on the cost of ownership and in what timeframe, if at all, would the cost be paid back through decreased operation and maintenance costs and reduced water-related risk.

6 Conclusions

The developed FEES framework and related tool form a practical model for value demonstration, communication and assessment of a WBMS. The framework combines economic and risk assessment approaches and investigates the potential impacts, risks, and values of the introduction of a WBMS on a mine site. The main benefits of developed framework are related to its' features on how environmental, social, and cost aspects can be incorporated in the asset management decision-making, and how the water-related risk reduction potential of WBMS can be quantified.

The SERENE project gave us a detailed view and an industrial setting on the implementation of IIoT investment in the mining sector. However, the depth of the research was limited, and accordingly, all the important viewpoints on IIoT cost savings and risk reduction potential might not have yet come up in the research. Thus, the further research and co-operation with companies will enable both the gathering of the necessary feedback for upgrading the framework and also improve the exploitation potential of the approach in the mining industry. In all, analysing the life cycle costs, cost savings and benefits of a WBMS is not a simple task. For example, the availability, accessibility and usability of data needed can be restricted and the conducting of the analysis can be very time consuming. Furthermore, the establishment of a cost structure that comprises all relevant parameters for, and the development of WBMS cost functions might be challenging. This is especially if the assessment is conducted in the very early phases of a life cycle when the uncertainty is high.

The effective and good communication and demonstration of the WBMS's value to customer and other key stakeholders represents a critical area in business negotiations. Communication of the value contributes to reaching the selling objectives and to building the brand that customers and other stakeholders perceive. Furthermore, demonstration and communication of cost savings and benefits of a solution can bridge the divides between technology providers, customers and other stakeholders. However, it is important to take into consideration that customers can diverge in their responses to the same information. Therefore, it is important to get a good understanding of how value is perceived by each of customer. Moreover, achievable benefits of a potential customer or end user may vary according to the site's location, business environment and economic, social and environmental drivers.

To sum up, the FEES value assessment framework and related tool provide an easy-to-apply evaluation method to assess the customer value of WBMS and thus to support the acquisition decision of the customer. Furthermore, the application of the FEES increases the transparency of value calculations and decision-making.

Acknowledgements. This activity has received funding from European Institute of Innovation and Technology (EIT), a body of the European Union, under the Horizon 2020, the EU Framework Programme for Research and Innovation. The authors want to give their sincere thanks especially to Susanna Horn, Sonja Kuokkanen and Eija Saari from Outotec. The authors also wish to express their gratitude to all people who contributed to the research in the EIT SERENE project.

References

1. Amadi-Echendu J, Dakada M, Ramlal R, Englebrecht F (2019) Asset replacement in the context of servitization. In: 2019 IEEE echnology & engineering management conference, TEMSCON 2019, pp 136–142
2. Götze U, Northcott D, Schuster P (2015) Investment appraisal: methods and models, 2nd edn. Springer, Heidelberg
3. Cooper R, Edgett SJ, Kleinschmidt EJ (2001) Portfolio management for new product development: results of and industry practices study. R&D Manage 31(4):361–380
4. Martinsuo M, Killen CP (2014) Value management in project portfolios: identifying and assessing strategic value. Proj Manage J 45:56–70
5. Keeney L, Raiffa H (1993) Decisions with multiple objectives: preferences and value trade-offs. Cambridge University Press, Cambridge
6. Räikkönen M, Välisalo T, Shylina D, Tilabi S (2016) Supporting asset management decision-making – new value creation perspective. In: Proceedings of the 10th world congress on engineering asset management (WCEAM), lecture notes in mechanical engineering, Springer, pp 479–486

7. Yin R (1994) Case study research: design and methods, 2nd edn. Sage Publishing, Beverly Hills
8. Eisenhardt KM, Graebner ME (2018) Theory building from cases: opportunities and challenges. Acad. Manage. J. 50(1):25–32 Linked references are available on JSTOR for this article
9. Keenan J, Kemp D, Owen J (2019) Corporate responsibility and the social risk of new mining technologies. Corp Soc Responsib Environ Manag 26(4):752–760
10. Gunson AJ, Klein B, Veiga M, Dunbar S (2010) Reducing mine water network energy requirements. J Clean Prod 18(13):1328–1338
11. Kunz NC, Moran CJ (2016) The utility of a systems approach for managing strategic water risks at a mine site level. Water Resour Ind 13:1–6
12. Northey SA et al (2017) The exposure of global base metal resources to water criticality, scarcity and climate change. Glob Environ Chang 44:109–124
13. Neitola R, Pasanen A, Taskinen A, Krista P, Turunen K, Raaikainen A, Hämäläinen M (2016) Current and new water management solutions at northern mines
14. Zhao H, Xia B, Fan C, Zhao P, Shen S (2012) Human health risk from soil heavy metal contamination under different land uses near Dabaoshan Mine, Southern China. Sci Total Environ 417–418:45–54
15. Viseu T, Betâmio de Almeida A (2009) Dam-break risk management and hazard mitigation. WIT Trans State Art Sci Eng 36:211–239
16. Burritt RL, Christ KL (2018) Water risk in mining: analysis of the Samarco dam failure. J Clean Prod 178:196–205
17. Litmanen T, Jartti T, Rantaala E (2014) The social license to operate. Exploring and testing its assumptions, preconditions and limitations in the case of mining in Finland. In: Conference: green mining annual seminar & mining in Finland and sustainable extractive industries action plan implementation, House of Estates, Helsinki
18. Bice S (2014) What gives you a social licence? An exploration of the social licence to operate in the Australian mining industry. Resources 3:62–80
19. Vanclay F, Hanna P (2019) Conceptualizing company response to community protest: principles to achieve a social license to operate. Land 8:101
20. Jartti T, Litmanen T, Lacey J, Moffat K (2017) Finnish attitudes toward mining. Citizen survey - 2016 Results. University of Jyväskylä and CSIRO
21. Mancini L, Sala S (2018) Social impact assessment in the mining sector: review and comparison of indicators frameworks. Resour. Policy 57:98–111
22. Hamilton R (2019) From water management to water stewardship-a policy maker's opinion on the progress of the mining sector. Water (Switzerland) 11(3):438
23. Wessman H, Salmi O, Kohl J, Kinnunen P, Saarivuori E, Mroueh UM (2014) Water and society: mutual challenges for eco-efficient and socially acceptable mining in Finland. J Clean Prod 84(1):289–298
24. Northey SA, Mudd GM, Saarivuori E, Wessman-Jääskeläinen H, Haque N (2016) Water footprinting and mining: where are the limitations and opportunities? J Clean Prod 135:1098–1116
25. Hoogmartens R, Van Passel S, Van Acker K, Dubois M (2014) Bridging the gap between LCA, LCC and CBA as sustainability assessment tools. Environ Impact Assess Rev 48:27–33
26. Haimes YY (2008) Risk Modeling, Assessment, and Management, 3rd edn
27. Lee WS, Grosh DL, Tillman FA, Lie CH (1985) Fault tree analysis, methods, and applications - a review. IEEE Trans Reliab 34:194–203
28. Bowman C, Ambrosini V (2000) Value creation versus value capture: towards a coherent definition of value in strategy. Br J Manag 11(1):1–15

29. Rana P, Short S, Evans S, Granados M (2017) Business models and business modelling: state of the art. In: Liyanage JP, Uusitalo T (eds) Value networks in manufacturing: sustainability and performance excellence, Springer, Cham, pp 75–93
30. SFS-ISO (2018) 31000 Risk Management – Principles and Guidelines
31. IEC (2017) IEC 60300-3-3: Life cycle costing. IEC, Geneva, Switzerland
32. Frezatti F, de Souza Bido D, da Cruz APC, Barroso MFG, de Camargo Machado MJ (2013) Investment decisions on long-term assets: integrating strategic and financial perspectives. Eur Account Rev 22(2):297–336

Social Complexity and Systems Intelligence in Asset Management Systems

Ricardo Peculis(✉), Tieling Zhang, and Richard Dwight

SMART Infrastructure Facility, School of MMMB, Faculty of Engineering and Information Sciences, University of Wollongong, Wollongong, NSW 2522, Australia
{peculis,tieling,radwight}@uow.edu.au

Abstract. It is well understood that the success in managing assets depends on well-established engineering principles, techniques and processes. Nevertheless, socio-organisational factors also play important roles in asset management systems and asset management activities. Their influence expands well beyond the engineering aspects. This chapter focuses on the discussions on the potential to take into account the social complexity and Systems Intelligence in asset management systems development. It suggests that the momentum to bring engineering and governance together to create the conditions for achieving a successful asset management system is Systems Intelligence, as our ability to behave intelligently in the context of complex systems involving interactions, dynamics and feedbacks is insufficient. This chapter presents eight dimensions of Systems Intelligence within the context of the asset management system, discusses the asset management system as a socio-technical system and as a 'system of systems' and applies basic principles from systems theory as a way to improve our Systems Intelligence to be better prepared to deal with social complexity. It shows how Systems Intelligence can be applied to increasing the chances of success in managing assets. On the basis of these discussions, the future research directions are identified.

Keywords: Asset management system · Social complexity · Systems Intelligence

1 Introduction

1.1 The Need for Systems Intelligence

This chapter postulates that Systems Intelligence, as defined by Hämäläimen et al. [1], can complement the framework and guidelines presented in ISO 55000 series [2], the international standard for asset management. Asset management exists in the context of socio-technical systems, where people in teams, groups and organisations exert their preferences and apply their knowledge in ways that influence the way asset management is performed. Socio-organisational factors, such as motivation and behaviour, influence the performance of organisations by promoting, or not, collaboration, cooperation and learning [3]. Understanding how systems work and how motivation and behaviour interact dynamically is important to manage socio-technical systems. Systems Intelligence connects all these aspects.

A. Crespo Márquez et al. (Eds.): WCEAM 2019, LNME, pp. 227–238, 2021.
https://doi.org/10.1007/978-3-030-64228-0_20

ISO 55000 [2] establishes the Asset Management System as the coordinated elements that define asset management policies, objectives and processes to achieve these objectives. Together with business, financial and regulatory constraints, ISO 55000 acknowledges the influence of the expectation of the organisation and its stakeholders in how to extract value from their assets. Kriege and Vlok [4] confirm that ISO 55000 recognises that leadership, culture, motivation and behaviour can assist asset management in achieving intended goals and ISO 55002 [2] recommends that Human Resources (HR) should support asset management needs. The authors in [4, p. 437] also report that ‘it remains unclear to which extent and in which areas exactly HR is affecting the asset management system and identify five critical areas within HR with significant influence in asset management and the asset management system, without offering an explanation of how these critical areas work dynamically together and influence each other. These areas are: (a) Organisational Culture; (b) Motivation and Leadership; (c) Learning and Development; (d) Knowledge Management; and (e) Change Management. To answer the open question posed by Kriege and Vlok [4], this chapter complements the framework in ISO 55000 by considering asset management as part of a larger complex socio-technical system activities, where conflicting motivations often abound, undesirable behaviour may exist and lack of knowledge is the norm rather than exception.

The structure of this chapter is as follows: Sect. 2 describes asset management socio-technical system; Sect. 3 presents important concepts extracted from systems theory and introduces Systems Intelligence; Sect. 4 expands ISO 55000 with aspects that conform with Systems Intelligence; and Sect. 5 concludes the chapter.

2 Understanding the Asset Management Socio-technical System

2.1 The Asset Management Socio-technical System

Asset management provides value for the organisation through the asset. Public assets need to meet strict budget, quality of service and safety goals. Not meeting intended objectives often results in undesirable consequences, whether it is commercial, social or political in nature. According to Hastings [5] asset management aims to answer three basic questions: (1) Does it work? (2) Is it safe? (3) How does it support the business aim? While the nature of engineering focuses on the first two questions addressing the intended effectiveness for the asset, engineering must support answering also question (3) which is associated with the business objectives such as efficiency, profit, customer satisfaction and market share. Differences between engineering and governance create conflicts between engineers and senior management [6]. While engineering tasks focus on technical aspects, governance defines ways to ensure organisations run in the interest of owners, often prioritise short-term goals to maximise immediate productivity and value [6]. Conflicting motivation can cause behaviour that drives decisions that steer the system away from meeting declared intended objectives [3]. The success of asset management depends on how well the asset is managed in the complete life cycle of the asset: from translating the need into specification, acquiring, operating, maintaining, retiring and disposing of the asset. Engineering and business competencies are paramount but they are not sufficient. Furthermore, understanding the system as a whole is also necessary.

Figure 1 shows a graphic representation of the management of physical assets from a social system concept where five classes of actors apply their own specific knowledge to influence the asset management activities in accordance with their own interest [5]. Engineers are concerned with the technical aspects, and often lack of business awareness. Finance specialists see assets as items in a balance sheet and engineering as activities to be outsourced. Senior managers focus on marketing, finances and political issues, and prioritise short-term imperatives. The public always 'want more'; they care about social and environmental issues, but lack of appreciation and understanding about planning, finances and what it takes to acquire and operate assets. Finally, the lobbyists favour specific solutions to achieve their interests.

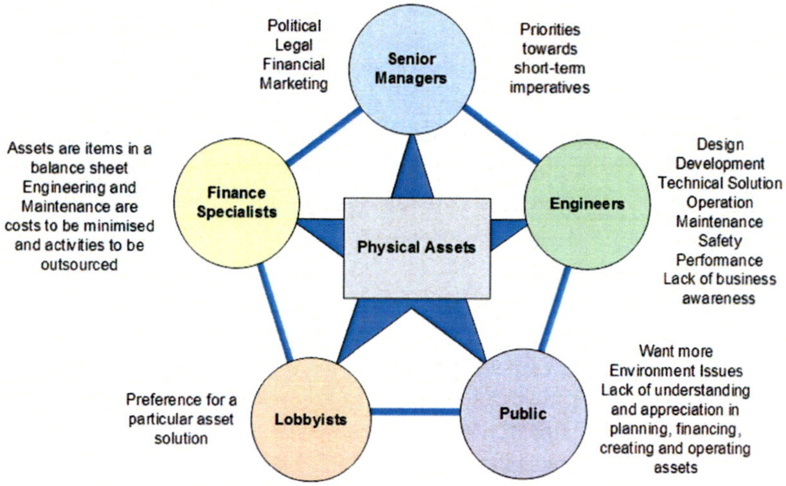

Fig. 1. Asset management social system (adapted from Hastings [5], Sect. 1.1).

Commercial and public assets are influenced by internal and external stakeholders dealing with technical, financial, economic, social and political pressures [3, 5–7]. Managing public assets, in particular, suffers from pressures during the planning phase that may result in 'great planning disasters', as described by Hall [7]. If the asset survives the troubled planning phase, asset management has to deal with adverse consequences throughout the remaining life cycle of the asset.

2.2 The Dynamics of Motivation, Behaviour and Action

Asset Management involves a series of activities applying knowledge, skills and experience to find solutions to satisfy engineering and business needs. To understand the causes of success and failure of knowledge-based activities we need to look into the nature of tasks and the social system that executes them to understand the dynamic between motivations, behaviour and action. The process comprises a series of transformations that transform artefacts, e.g. description of the needs and specifications, from one domain into artefacts in another domain. Lack of knowledge, skills and experience causes distortions that propagate from the definitions of the expressed need to the implemented solution and in the end the solution is likely not to satisfy the need [3]. To correct such distortions it takes longer time and costs more than what was expected. To execute effectively transformations in knowledge-based activities it requires knowledge, skills and experience in three areas: Domain, Technology and Teamwork. 'Domain knowledge' is about understanding the customer's need and the field where the need exists, e.g. public transport, mining, water, energy, etc. 'Technology knowledge' is about application of technology, e.g. engineering, software, business transformations, to find a solution that will satisfy the customer's need. 'Teamwork knowledge' is about knowing how to work together. Knowledge-based activities may fail due to lack of knowledge to engineer the solution, and lack of management knowledge to recognise and plan for this deficiency.

The asset management social system presents a dynamic balance between motivation and behaviour of its actors. The various actors in the system are motivated in accordance with their own preferences and goals, as shown in Fig. 1. Motivation is a private characteristic of each actor and cannot be observed until it is reflected on specific behaviour. Behaviour is any noticeable change or response of a person or a system. Motivation drives behaviour and this drives action [3]. Action is what really matters reflecting into the application of engineering and management knowledge, skills and experience that support the asset management activities or not. Conflicting motivations explain the difficulty of reaching decisions of consensus. Undesired behaviour is harmful to promoting learning, knowledge sharing, cooperation and actions as they do not favour effective and successful asset management.

The theory of behaviour aims to provide ways to predict the likely behaviour of a person in accordance with a classification of personality. Social systems will behave as the result of the collective behaviour of individuals. The behaviour of the system is not expected to respond linearly to individual behaviour, as individuals may influence each other, even in feedback and changing the behaviour of the individual that has started the process of change. Several theories exist to classify people in accordance with personality and behaviour styles which help to describe personal characteristics and predict how a person may behave and perform under certain circumstances [3]. Among those the Life Styles Inventory (LSI) [8] offers a classification of behaviour styles, that helps to understand the dynamics of motivation, behaviour and action. The Life Styles Inventory assesses twelve life styles attributes that form a continuum correlated with the four areas of concern and three characteristics of behaviour defined as 'Constructive', 'Passive' and 'Aggressive'. The life styles are also classified in accordance with their interaction style as Aggressive, Passive or Constructive.

The asset management social system behaves as a problem-solving group searching for engineering solutions to manage complex physical assets within also complex and often conflicting constraints. The effectiveness of problem-solving groups depends on the group interaction style [9] which defines the style of the group as a whole unit, as a system, resulting from the dynamic interaction of its members in accordance with their roles and individual personality styles. The group interaction style reflects the atmosphere of the group, the group's own personality that will promote cooperation or persuasion in finding the solution for the problem. Group members with more power, e.g. in management or leadership roles, political influence or with stronger personality styles are likely to influence other members in the group. Across problem-solving groups, solution quality increases when the group shows a constructive interactive style and decreases with a passive interaction style; and the acceptance of the solution increases with a constructive interaction style and decreases with both passive and aggressive interaction styles. The dynamics of group interaction styles, illustrated in Fig. 2, tells us that constructive behaviour promotes constructive behaviour, and aggressive behaviour suppresses the constructive and promotes passive behaviour, while passive behaviour promotes passive behaviour.

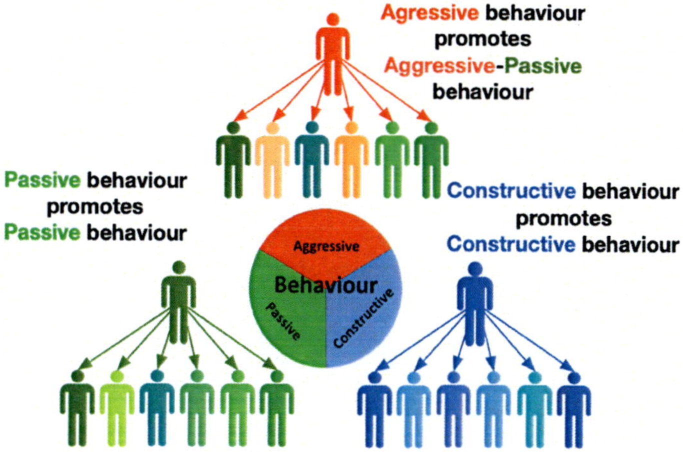

Fig. 2. The dynamics of group interaction styles.

Colours are used to represent the three primary interaction styles: Aggressive (red), Constructive (blue) and Passive (green). Aggressive actors present the characteristic of "just do it as you were told". These actors do not ask for help or help others, and they are not interested in cooperation and how the problem is solved, "just do it". Constructive actors, however, have an attitude of cooperation, helping other actors and do not hesitate to ask help if needed. These actors aim to reach better solutions for the problem, even if it needs more effort and takes longer to achieve. Passive actors have a "don't care" attitude and are satisfied to do what they were told without questioning, even when they believe there are better options. The hue of colours in Fig. 2 represents the degree or intensity of the interaction style of each actor as a combination of the three primary styles. The dynamics of group interaction styles explain that when knowledge-based projects are not performing as expected and interests are at risk, decisions are made within constraints, e.g. unrealistic cost and schedule, which are unlikely to address the lack of knowledge; and often produce undesirable behaviour that decreases motivation and increases passivism, discourages learning and cooperation and worsens the situation.

3 Systems and Systems Intelligence

3.1 Asset Management System as a System of Systems

The asset management socio-technical system has many interconnected components influencing each other. The technical components, or the asset, behave as they were engineered and constructed. Various components comprised in the asset operate as a system, performing a function that was not possible when they were isolated. The asset may change behaviour due to usage and aging, but they are not driven by their own motivation. Therefore, changes in behaviour of the asset can be predicted and should be managed in accordance with their lifecycle. The social system influences the asset and changes itself as people change in accordance with their own motivations, experience and knowledge. Understanding how the socio-technical system works is important to successfully manage the asset.

To complicate even further, management of public infrastructure assets may exist in the context of 'system of systems' that are systems comprised of constituent systems (CS) being managed and operated independently [10]. Constituent systems in the 'system of systems' aim to achieve their own interests and collaborate with other CS motivated by self-interests. A public transport system may comprise several transportation modes such as heavy and light rail, busses and ferries, managed and operated by separate and independent organisations, often with conflicting interests and motivations. Finding the conditions that facilitate the transformation of self-interests in common interests adds complexity to manage assets in 'system of systems', and embracing Systems Intelligence comes to aid this task.

3.2 Systems Intelligence

Hämäläimen et al. [1] define Systems Intelligence as 'our ability to behave intelligently in the context of complex systems involving interaction, dynamics and feedback' [1, p. 15]. The concept of Systems Intelligence is complemented by defining eight dimensions of Systems Intelligence [1, p. 19], shown in Fig. 3. **(1) Systems Perception** is our ability to see the systems around us and understand how various parts in the system are interconnected and influence each other; **(2) Attunement** is the capability we have to feel and tune into systems, modifying ourselves in ways that would change the system for achieving the intended goals; **(3) Reflection** is our capacity to reflect on our thoughts and think about our thinking, aiming to find ways to improve the system and make the system behave towards its intended purpose; **(4) Positive Engagement** is the character of our communicative interactions that will influence and change other parts of the system; **(5) Spirited Discovery** is about passionate engagement with new ideas that will bring solutions to the challenges that always permeate systems; **(6) Effective Responsiveness** is our talent at taking timely, appropriate actions that transform and make the system behave at its best; **(7) Wise Action** is our ability to behave with understanding and a long time horizon, finding and implementing strategies that bring benefits in the long run; **(8) Positive Attitude** encompass our overall approach to life in systems. Systems Intelligence is an emerging competency for engineering, it can be taught and learned and should be included in the engineering curriculum [11].

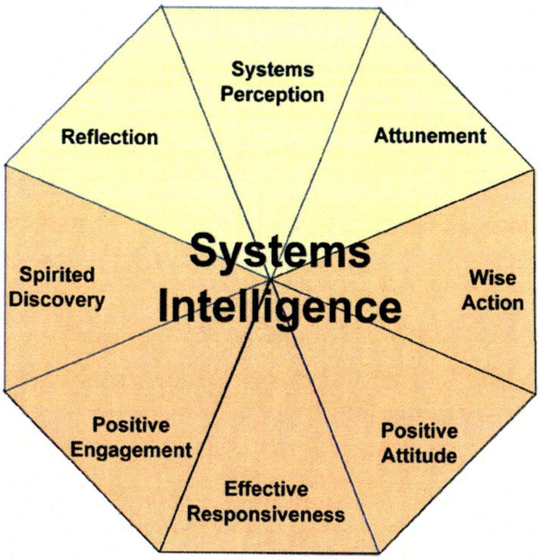

Fig. 3. The eight dimensions of Systems Intelligence [1, p.19].

The importance of systems understanding for managing socio-technical systems is not new and has been expressed by the works of Beer [12], Deming [13] and Senge [14] to name a few. Systems Intelligence helps us to deal with a powerful truth about systems which states that *'the purpose of a system is what it does'* [12, p. 218], as the emergent property of the interconnected components, and not present in any components in isolation. Systems always do what they are capable of doing, not more or nor less. The implication this insight brings is that if what the system is capable of doing does not reflect the intended purpose for the system, then the system will have to be changed. Changing socio-technical systems is about creating systemic structures that establish conditions that drive beliefs and motivations causing people to behave in desirable ways. The way the people behave determines the behaviour of social systems. What people in the system often do not realise is that they can influence the system structure by changing their own behaviour and therefore the behaviour of the system.

4 Expanding Asset Management Framework with Systems Intelligence

4.1 Framework

Framework is a structure of principles, theories, assumptions, concepts, values, and practices that constitutes a way of viewing reality. The framework proposed by ISO 55000 [2] is based on Planning, Operation, Performance Evaluation and Improvement, which is based on PDSA (Plan, Do, Study, Act) model [13]. ISO 55000 acknowledges the importance of the 'Context of the Organisation' (needs and expectations), Leadership and Support (resources, competence, awareness, communication and information) in asset management. ISO 55000 framework is practical in nature, and it does not offer principles or theories to explain how the asset management socio-technical system works. Deming [13] in his 'System of Profound Knowledge' emphasises the importance of appreciating the system as a whole, and in developing theories about how the system works. According to Deming [13], management is about prediction and theory/model is the key to make prediction possible.

The effectiveness and ultimate success of asset management depend on more than engineering and other technical competencies, with dependencies from what could be a very complex social system. Systems Intelligence offers a frame of reference that aids dealing with such complexity. Systems Perception, the first of the eight dimensions of Systems Intelligence, is a central concept, while Attunement and Reflection support the first. Together these three dimensions of Systems Intelligence guide us to develop a better understanding about the asset management socio-technical system. The other five dimensions of Systems Intelligence agree with 'constructive behaviour' addressed by 'Framework for Steering Infrastructure Projects to Success' [3] and the 'Theory of Collaborative Rationality' [15], the latter suggests that a consensus solution should be the goal and often the best for the system. This chapter proposes to expand the framework offered by ISO 55000 with concepts extracted from Systems Intelligence as shown in Fig. 4.

The proposed 'Framework for Steering Asset Management to Success' (FSAMS) extends the application of the 'Framework for Steering Infrastructure Projects to Success' [3] into asset management. FSAMS incorporates the principle that reflects 'the aim of the system', i.e. 'the system does what it is capable of doing' [12]. The first three dimensions of Systems Intelligence, shown in yellow, support the application of the 'system principle'. FSAMS includes three integrated management processes (Knowledge, Motivation and Behaviour) for creating the conditions necessary for 'steering' the socio-technical system towards the 'intended aim of the system' in asset management.

The framework offers a theory that explains how the asset management socio-technical system works. The theory involves the dynamic of motivation, behaviour and action [3]. Action comprises activities that acquire and apply knowledge. To perform a task effectively and efficiently requires the application of specific knowledge, skills and experience, named collectively 'Knowledge'.

4.2 Process of Knowledge Management

The Process of Knowledge Management (PKM) identifies the knowledge required, the knowledge available, the gap of knowledge and how to acquire it. PKM addresses the three areas of knowledge presented in Sect. 2.2: Domain, Technology and Teamwork. Domain and Technology (including engineering and science) knowledge is what is needed to engineer and implement the solution effectively. Teamwork knowledge is what is needed "to make it happen" and includes understanding how the asset management socio-technical system works. The Process of Knowledge Management embraces Systems Intelligence as a competency for effective engineering and asset management, and plan for Systems Intelligence to be taught, learned and shared, as suggested in [11].

The Process of Knowledge Management deals with knowledge acquisition and ignorance reduction, explained by "The Five Orders of Ignorance" [16]. At Zero Order of Ignorance (Zero-OI) all is known and there is nothing to be discovered or learned. Zero-OI exists in very simple tasks, which is not the case in asset management. At First-OI

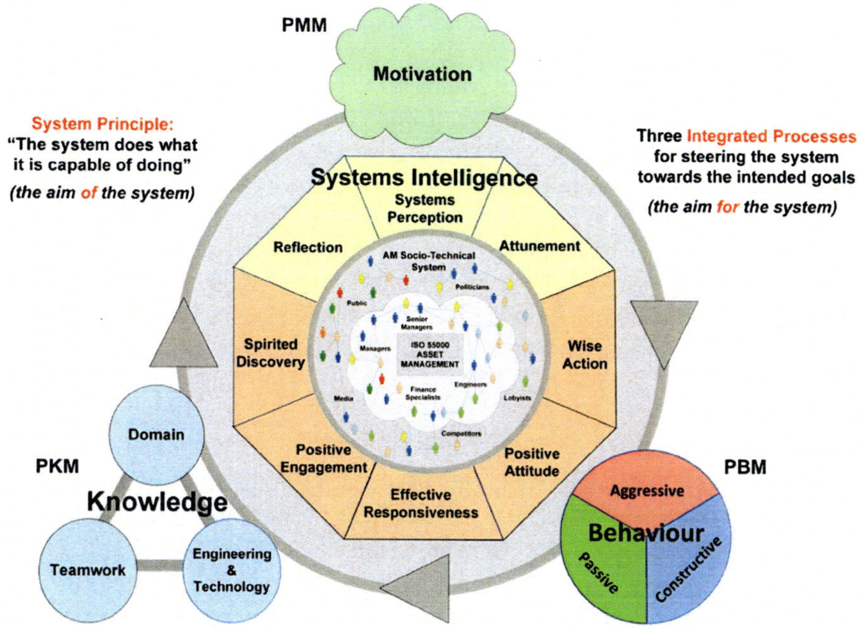

Fig. 4. Framework for Steering Asset Management to Success (FSAMS).

there is a known gap of knowledge that can be acquired through learning and investigation, but nothing is truly 'unknown'. Asset management operates at Second-OI or Third-OI. At Second-OI, lack of knowledge exists but the gap of knowledge can be acquired, and there are 'unknowns' which can be revealed through available processes of discovery. Third-OI is like Second-OI without suitable processes of discovery. PKM identifies what the Order of Ignorance the asset management socio-technical system is in and creates processes of learning and discovery to reduce the Order of Ignorance. PKM takes into account the time and costs needed to acquire the knowledge required and to resolve the "unknowns".

4.3 Process of Motivation Management

The objective of the Process of Motivation Management (PMM) is to create conditions to motivate actors in the asset management socio-technical system to produce constructive behaviour that promotes learning, cooperation, knowledge sharing and ignorance reduction. The Process of Motivation Management identifies key stakeholders that could influence, for better or for worse, the behaviour of the asset management socio-technical system and their motivations. These are likely the actors in the five classes (engineers, finance specialists, senior managers, the public and lobbyists) described in Sect. 2.1 and shown in Fig. 1. PMM also identifies conflicting interests often presented in asset management 'system of systems' as described in Sect. 3.1, and creates conditions to align self-interests with the common goals.

The Process of Motivation Management identifies the ways in which actors in the asset management socio-technical system could be motivated. Most distinctive types of motivation are *intrinsic motivation,* which is associated with performing an activity because it brings satisfaction rather than a reward or consequence, and *extrinsic motivation,* which is the driver for performing an activity in order to attain a separate outcome [17, pp. 55–56 & p. 60]. *Intrinsic motivation* is concerned with self-determination, competence, task involvement, curiosity, enjoyment and interest; and extrinsic motivation is concerned with recognition, competition, money and other tangible incentives [18]. Understanding how the actors are motivated, the Process of Motivation Management develops strategies based on intrinsic and extrinsic motivation factors. The Process of Motivation Management also applies the five behavioural dimensions of Systems Intelligence represented in orange in Fig. 4 (Positive Engagement, Spirited Discovery, Effective Responsiveness, Wise Action and Positive Attitude) to develop motivation strategies.

4.4 Process of Behaviour Management

The Process of Behaviour Management (PBM) identifies behaviour that is favourable to achieve the intended goals and avoid those that do not. Constructive behaviour facilitates learning and collaboration, while aggressive behaviour constrains constructive behaviour and promote passiveness [3]. Aggressive behaviour is present when there are conflicting motivations and the system displays results that diverge from expectations in the form of low profitability, cost overruns, schedule delays, poor quality and loss of market share. PBM may also include processes for assessing and monitoring individual and group interaction styles [8, 9].

The arrows in Fig. 4 show that motivation drives behaviour resulting in the action of acquiring and sharing knowledge. When the adequate knowledge is available, including system understanding, motivation also changes for the benefit of the whole system. FSAMS offers principles, theories and processes that allow us to understand the reality of the asset management socio-technical system. The processes in FSAMS provide guidance to create the conditions to steer the asset management system in the ways that increase the chances of success.

5 Conclusions

This chapter showed that asset management takes place in a complex socio-technical system that cannot be controlled, but it can be steered towards the intended goals by creating the appropriate conditions [3]. The insights offered by Deming's 'System of Profound Knowledge' [13], the 'Framework for Steering Infrastructure Projects to Success' [3], and the 'Theory of Collaborative Rationality' [15] were adapted and applied into the 'Framework for Steering Asset Management to Success' in order to manage physical assets more effectively.

The open questions posed by Kriege and Vlok [4] presented in Sect. 1.1 have been answered by explaining the dynamics that exists between motivation, behaviour and actions that reflect, or not, into learning and knowledge sharing by the adoption of Systems Intelligence. Organisation Culture influences Motivation & Leadership and these are drivers for creating the conditions to foster constructive behaviour that promotes Learning & Development. The processes of motivation and behaviour management, respectively represented by PMM and PBM, proposed in FSAMS, address the dynamics of motivation and behaviour steering Organisation Culture to improve the asset management socio-technical system. The process of knowledge management deals with Knowledge Management and Learning & Development as specific and effective actions to acquire knowledge. Change is the norm in complex systems and embracing Systems Intelligence facilitates Change Management.

FSAMS embraces Systems Intelligence concepts that can be learned by asset managers, engineers and stakeholders in the asset management socio-technical system. The ideas here presented aim to instigate new discussion and consideration. Further work includes assessing the impact of each of the eight dimensions of Systems Intelligence on the development of successful asset management systems, the practical application of the proposed FSAMS and ultimately to report results with case studies.

References

1. Hämäläimen RP, Jones R, Saarinen E (2014) Being better better: living with systems intelligence. Aalto University Publications, Nord Print, Helsinki
2. International Standard Organisation ISO (2014–2018) International Standard 55000 Series: ISO 55000:2014 Asset management - Overview, principles and terminology; ISO 55001:2014 Asset management - Management Systems - Requirements; ISO 55002:2018 Guidelines for the application of ISO 55001. Geneva, Switzerland
3. Peculis R (2017) Coping with social complexity of infrastructure projects. In: ISNGI conference proceedings 2017, international symposium for next generation infrastructure, London, UK, pp 283–293
4. Kriege L, Vlok PJ (2015) Human resources within ISO 55000 – the hidden backbone to the asset management system. In: 9th WCEAM research papers, Springer, Cham, pp 435–445
5. Hastings NAJ (2010) Physical Asset Management. Springer, Dordrecht
6. Mardiasmo D, Brown KA, Burgess K (2008) Asset management and governance: an analysis of fleet management process issues in an asset-intensive organization. In: International conference on infrastructure systems: building networks for a brighter future, Rotterdam, Netherlands, 10–12 November 2008
7. Hall P (1980) Great planning disasters. Penguin Books, New York
8. Cooke RA, Rousseau DM (1983) The factor structure of Level I: life styles inventory. Educ Psychol Measur 43(2):449–457
9. Cooke RA, Szumal JL (1994) The impact of group interaction styles on problem-solving effectiveness. J Appl Behav Sci 30(4):415–437
10. Peculis R, Shirvani F (2017) Infrastructure system of systems integrity. In: ISNGI conference proceedings 2017, international symposium for next generation infrastructure, London, UK, pp 294–303
11. Lappalainen P (2017) Stirring up engineers' systems intelligence: a case study of life-philosophical pedagogy. Int J Eng Pedagogy (iJEP) 7(3):61–72
12. Beer S (2002) What is cybernetics? Kybernetes Int J Syst Cybern 31(2):209–219
13. Deming WE (2000) The new economics: for industry, government, education, 2nd edn. MIT Press, Cambridge
14. Senge PM (2006) The fifth discipline: the art & practice of the learning organisation, 2nd edn. Random House Australia, Sydney
15. Innes JE, Booher DE (2009) Planning with complexity: an introduction to collaborative rationality for public policy. Routledge Taylor & Francis Group, New York

16. Armour P (2000) The five orders of ignorance. Commun ACM 43(10):17–20
17. Ryan RM, Deci EL (2000) Intrinsic and extrinsic motivations: classic definition and new directions. Contemp Educ Psychol 25(1):54–67
18. Bi XB, Zhou N (2010) Research on inspiration mechanism of knowledge sharing based motivations. In: 2010 international conference on e-business and e-government (ICEE), Guangzhou, China, pp 1883–1886

System Safety Engineering Approach for Autonomous Mobile Machinery

Risto Tiusanen(✉), Eetu Heikkilä, and Timo Malm

VTT Technical Research Centre of Finland Ltd. (VTT), P.O. Box 1300, 33101 Tampere, Finland
risto.tiusanen@vtt.fi

Abstract. Autonomous machines are complex systems that are able to perform independent decision-making and to operate without operator's continuous activities. Increasing autonomy and system complexity create new challenges to safety engineering. In this article, we introduce and discuss about a safety engineering approach that is being developed in VTT for autonomous work-machine applications in close collaboration with machine manufacturers and system suppliers. The main motivation has been to support the early development phases of novel automation solutions and system operating concepts. The system approach focuses on system-level safety issues arising from the shift from manual mobile machines to autonomous machinery systems. It utilizes elements from system safety engineering methods and the latest safety standards for autonomous machinery, as well as the goal-based safety case approach to support the management of safety goals and requirements. The approach and methods are being applied to identify and analyze autonomy related safety risks in several industrial cases.

Keywords: Autonomy · Work-machine · System safety · STPA · Safety case

<u>Summary of the Modifications</u>

- *The importance of the complexity and the lack of a large amount of data for a meaningful analysis are discussed in the end of chapter 1. The two proposed articles (Dekker et al. 2011 and Paté-Cornell et al. 2012) have been studied and cited in the text.*
- *Discussion upon the impact of the organizational and human performance in the context of autonomous mobile machinery as well as linkage to asset management is added to the chapter 1.*
- *This article focuses on the system safety engineering approach for autonomous mobile machinery including the new system theoretic approach STAMP and its associated STPA method. The research work continues and in coming articles, we will present examples of applications of the major elements of the approach. We hope that those examples will then help to understand its application.*

1 Introduction

The development towards a higher automation level in logistic systems and material handling is a clear global trend. Needs to increase productivity, efficiency and

A. Crespo Márquez et al. (Eds.): WCEAM 2019, LNME, pp. 239–251, 2021.
https://doi.org/10.1007/978-3-030-64228-0_21

safety are driving forces for autonomous machinery in industrial applications using mobile work-machines. This means that machine manufacturers' business environment is changing from machine sales towards system business. Instead of optimizing single machine's operational capability and ensuring its safety, the system suppliers have to take into account the overall goals of the logistic processes, system requirements and constraints and the operational environment. Machine manufacturers are looking for business growth in the maintenance of these automated or autonomous systems and, in some applications in its operation. Changes in business concepts bring in new challenges and responsibilities to design and maintain safe operating concepts and operating environments.

It has been strongly addressed that fully automated cars and trucks that drive us, instead of us driving them, will become a reality in coming years. The development of autonomous cars is progressing through several levels of driver assistance and automation technologies in the coming years. This includes everything from no automation (where a fully engaged driver is required at all times), to full autonomy (where an automated vehicle operates independently, without a human driver) [1]. The current level of autonomy in cars is somewhere between so called 'partial autonomy' and 'conditional autonomy'. It means that the driver does not need to control steering or acceleration all the time, but the driver is expected to be aware of the traffic situation and capable of taking back control whenever needed.

In the industry utilizing mobile work machines, the goal in most cases is not the full autonomy. The level of autonomy sought depends on the objectives and opportunities of the business in the industry in question and the requirements of the work processes and operating environment. The increase of machine autonomy will have a great importance to people's work tasks and roles on sites when operating and controlling the mobile machines in the future. Instead of manual drivers, there will be remote operators and system operators in control rooms and experts in maintaining automation and information systems. One interesting and challenging vision of advanced operating concepts in logistic systems is to enable autonomous machinery, manual machines and manual workers to operate and collaborate in the same open work area. This challenges organizations jointly develop work processes and human-machine interactions to achieve desired production goals and ensure safety in all situations.

Increasing machine autonomy poses also new challenges for asset management. Instead of management of individual machines and equipment, in the future, companies operating machines must be able to maintain their extensive automation systems involving several information systems and communication systems together with system suppliers and subcontractors. It is important to create the conditions to be able to manage the entire automated logistics chain over its lifetime and to avoid sub-optimization of individual functions. On the other hand, machine automation together with digitalization bring new possibilities for data collection, diagnostics, monitoring and data analysis to enable optimal control of the work processes and implementation of preventive maintenance concepts for the entire automated machinery systems.

In one of the biggest mobile work-machine sector, earth moving machines and mining machines, the term autonomous operation has been defined as follows: Autonomous mode is 'the mode of operation in which a mobile machine performs all machine safety-critical and earth-moving or mining functions related to its defined operations without operator interaction' [2]. Machines operated in a lower level of autonomy are called semi-autonomous machines. Those mobile machines are intended to operate in autonomous mode during part of their operating cycle and which requires active control by an operator to complete some of the tasks assigned to the machines e.g. filling a bucket, gripping a container or position a drill pit.

Another big industrial sector utilizing mobile work-machines is cargo handling in ports and terminals. Currently automated cargo handling systems are designed for isolated areas but the demand is towards open operating environments and areas where there can be automated and manual operations at the same time [3]. Container terminals were among the first users of automated work machines and the systems have been in production use already for over two decades. In container terminals, the operational area is limited to a fully controlled environment in which it is relatively easy to install the specialized infrastructure required by automated container handling equipment. Unlike automated cars, the decision-making can be centralized through software layers from the terminal operating level all the way to the control of the machines [3].

Safety critical systems enabling the automated or even autonomous operations are based on advanced and intelligent software solutions and safety functions use information from several interacting systems [4]. Different operating environments and work processes require different solutions to ensure safe operation of the autonomous mobile machinery systems. The problem is that current safety engineering methods developed for automated machinery do not cover or consider real autonomy aspects [5, 6]. The key element in machine autonomy is adaptability to dynamically changing environment based on the available information. In other words, autonomous behaviors cannot be fully predetermined [7].

Autonomous mobile machinery is a novel concept with very limited experiences and data available from previous applications. This adds to the challenge of performing safety evaluations, and effectively, bringing new technologies to the market. The lack of data that is explicitly related to autonomous mobile machines, however, does not mean that any meaningful analysis cannot be performed. Various insights from other domains and relevant previous technologies can be used to reduce the level of uncertainty [8].

The characteristics typical to autonomous systems are seen to increase the system complexity. Dekker et al. [9] and Leveson [10] argue that to manage complexity, traditional Newtonian methods are not sufficient for performing safety analyses. This needs to be considered also in the system safety engineering approach. In practice, the concept of complexity can be addressed from several viewpoints. Leveson [10], points out some perspectives on complexity that are relevant in autonomous systems design. For example, interactive complexity, which is related to the interaction among system components, needs to be considered as new components and software are introduced to enable autonomous functionality. Another important factor is dynamic complexity, which relates to changes in the system or its operating environment over time.

In this article, we introduce and discuss about a safety engineering approach that is being developed in VTT in Finland for autonomous work-machine applications in close collaboration with world leading mobile machine manufacturers and system suppliers.

2 Assessment of Machine Autonomy Related Safety Risks

There are many challenging safety aspects on the way towards more automated and even autonomous machinery systems. Each step towards full autonomy introduces new safety risks compared to lower level autonomy and manual operation. New approaches and methods are needed to identify, assess and manage autonomy related safety risks from system concepts to specific procedures and functions, and to design feasible and acceptable safety solutions. Product orientation should be changed towards system thinking also in safety engineering. According to the general systems engineering approach, safety engineering should be a continuous top-down process in all the system development phases from concept evaluation to detailed design of the safety critical operations and functions [11]. The safety engineering methods should be selected not only to support the objectives of each system development phase but also to support the overall risk assessment process, traceability, and reuse of the analysis and assessment results [12].

The safety engineering approach applied in Finland for highly automated machinery systems has been utilizing well-known traditional risk analysis methods such as Preliminary Hazard Analysis (PHA) [12], Operating Hazard Analysis (OHA) [12] and Hazard and operability study (HAZOP) [13]. This approach has been successfully applied to automated machinery systems in underground mining applications, large-scale material handling applications and in container terminals in ports [14]. Our experience on increasing autonomy in machinery applications is in line with the findings in literature [3–5]. New methods are needed to identify and assess machine autonomy related safety risks and specific aspects raising from the increased autonomy e.g. in safety critical decision-making, machine learning and independent perception and control actions. The focus should be shifted from analyzing chains of events to identifying problems arising from unsafe interactions between different elements in socio-technical systems [15]. To address the above issues, systems-theoretic approaches, focusing on safety control instead of risk, have been proposed as a potential new basis for performing more comprehensive safety analyses.

2.1 Update of System Safety Approach for Autonomy Issues

Alongside the development of traditional methods, VTT has studied the applicability of Systems-Theoretic Accident Model and Processes (STAMP) approach and the Systems-Theoretic Process Analysis (STPA) method to the safety design of autonomous machine systems at different stages of the system life cycle. The overall 'Top-Down' approach and PHA, OHA and HAZOP methods have been supplemented by the application of STAMP and STPA methodologies in order to take better account of the system complexity and machine autonomy aspects (see Fig. 1).

STAMP approach, which is developed in the Massachusetts Institute of Technology (MIT) in US, describes complex sosio-technical systems as a hierarchical control structures. STAMP is said to be an alternative to the chain-of-failure-events that underlies the traditional safety analysis techniques (such as FTA, ETA, HAZOP, FMECA, and HFACS). STAMP addresses safety as a system control problem, instead on focusing on system failures and focuses on enforcing constraints on component behavior and interactions. STAMP model describes systems using control loops, which form a so called "safety control structure" and constitute a hierarchical system model including human operators and technical control elements [10].

The STAMP approach is accompanied with the STPA hazard analysis method, which is a systematic procedure to identify flaws within the safety related control actions. STPA is based on systems theory and it is a relatively new analysis method. It aims to identify and analyze unsafe control actions (UCA) that, in a particular context and worst-case environment, will lead to a hazard. STPA has been proposed as an alternative or complementing method for the traditional analysis approaches [16].

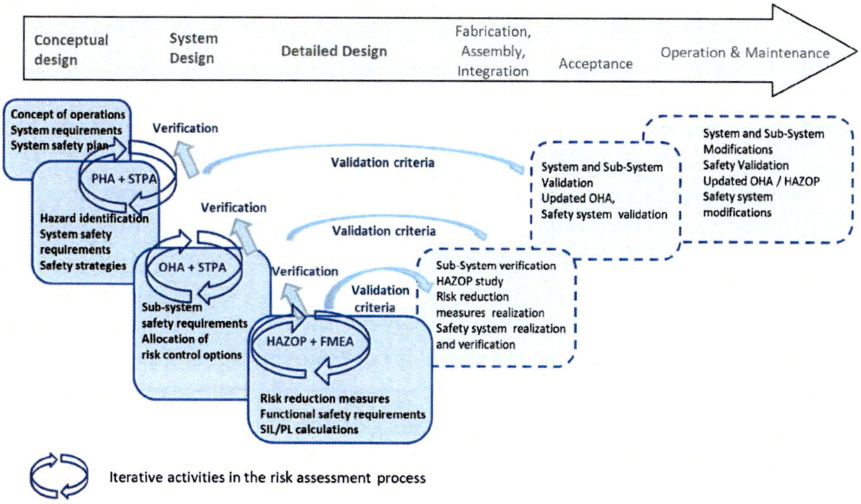

Fig. 1. An outline of the iterative safety engineering activities linked to the development phases of an autonomous machinery system.

2.2 Conceptual Design - Identification of Risks Related to Autonomy

Conceptual design of autonomous machinery system is done in the overall system level. The aim is to cover all work-process and operations, operation environment issues, interfaces to other systems, and all foreseeable activities in the operation environment in the coming system life cycle phases. Concept level evaluation includes among others feasibility studies of different automation concepts, safety and reliability risk analyses and rough estimation of system life cycle costs in the overall system level. The purpose of safety engineering in the overall system level is identification of safety critical uncertainties, threats and hazards [14]. The PHA method has been further developed for the identification and assessment of the machine autonomy related risks affecting the overall machinery application. In addition to analyzing the potential hazards of the physical systems and general safety risks in the operating environment, other types of losses and unwanted events are taken into account by applying STPA methodology.

The first task in STPA procedure is to define system-level losses and system-level hazards. A loss in STPA methodology involves something of value to stakeholders.

Losses may include a loss of human life or human injury, property damage, environmental pollution, loss of mission, loss of reputation, loss or leak of sensitive information, or any other loss that is unacceptable to the stakeholders [15].

2.3 System Design - Identification of Operational Risks and Unsafe Control Actions

System design in upper system level aims setting requirements for detailed design and allocating design requirements to all subsystems according to the selected operating concept [14]. At this stage in autonomous machinery development there might be two or three alternative concepts that will be developed and evaluated in parallel. This gives the possibility to compare them from techno-economical point of views. The concept evaluation supports also the decision making regarding the level of autonomy. As part of the concept evaluation, the OHA method has been developed to support the identification of potential risks in operating concepts in different autonomy levels. The aim has been to identify hazardous events caused by human errors, technical failures or external effects, and to estimate the risks and to specify risk control options. New operational safety risks have been identified especially in operations where automated or fully autonomous machines and manual operations are mixed in the same operating area [3]. It has been noted that the traditional OHA method does not cover the interactions between the system elements and actors because it is based on the system functional description and operating procedures. It does not include views that control action could be unsafe in certain conditions or the lack of certain control action could affect safety in planned operation.

Recently, in close collaboration with system suppliers and control system designers, the suitability of the STPA method to supplement OHA method in system design phase has been studied. The hierarchical control structure created according to the STPA methodology has proven to be a clear improvement in analyzing the operational risks and possible control problems. The system model in STPA has proven to be a very useful way to describe an automation system at a level that is understood by different design parties. In this context, it is good to remember that the STPA control structure is not a physical model. The connections in the model show information that can be sent, such as commands and feedback. They do not necessarily correspond to physical connections [15].

The increasing autonomy in forms of machine learning capability and independent functionalities for perception, adaptive decision-making and dynamically changing system controls sets challenge for safety engineering in this system level. STPA methodology providing support for the identification of unsafe control actions and creation of related scenarios and ending up to the constraints and requirements for safe control actions and safety functions looks promising and worth exploring further.

2.4 Detailed Design - Safety Related Functions and Interfaces

Lower system level safety engineering in autonomous machinery applications deals with functional safety issues of safety related functions, specific external safety systems and the design of human–technology interfaces (HTI). Safety engineering in this level focuses on systematically analyzing deviations, failures and possible problems in defined system operations, functions and technical designs. For this stage, the HAZOP method [13] has been applied to identify and analyze possible human errors and technical problems (both named as deviations). Their causes and consequences are systematically analyzed by means of the 'guide words' and the standard procedure.

The HAZOP studies for the safety related control functions and operator interfaces have mostly been conducted at functional level. Our experience is that HAZOP method can be applied to operational procedures just as well as technical system designs. It involves a structured team-work-based technique that is effective for exploring interactions between parts of a system. HAZOP method has also proven its applicability in the analysis of new control systems and novel technologies and for exploring interactions between subsystems. It also provides input for functions that require more detailed failure analyzes like FTA or FMECA [14].

It is good to note that STPA can also be extended and continued to the detailed design level. The hierarchical control structure (system model) can be created down to the detailed level and the STPA can be done on signal and software module level. Ongoing research in VTT on these methodologies will give new information on the applicability of STPA also in the detailed design level. The output of safety analysis in this stage should give direct feedback of the designs, improvement proposals and supplement the system safety evidence with the accepted safety measures and solutions selected for the implementation.

3 Management of Safety Goals and Requirements – Safety Case

According to the systems engineering approach, risk management decisions in the system-development phase are made systematically as the system development proceeds. In practice, the decisions to reduce the safety and availability risks are based on comparison of alternative solutions at different layers of protection and prediction. One promising approach for the management and documentation of safety requirement in autonomous machinery systems is the Goal-based design approach that has successfully been applied and demonstrated in safety qualification e.g. in marine sector [16, 17].

VTT have studied methodology for building and visualizing such demonstrations by using the goal-based safety case approach. In this approach, the safety requirements (represented as goals) and the related safety evidence are linked together in a visual manner, essentially forming a structured safety case database. This provides the means for demonstrating the actions that have been taken to fulfil the safety requirements, and indicating the safety evidence documentation related to each goal or sub goal.

An overview of a proposed safety qualification procedure utilizing goal-based safety case approach is presented in Fig. 2. The safety case can be represented with various visualization languages, such as the Goal Structuring Notation (GSN) [18].

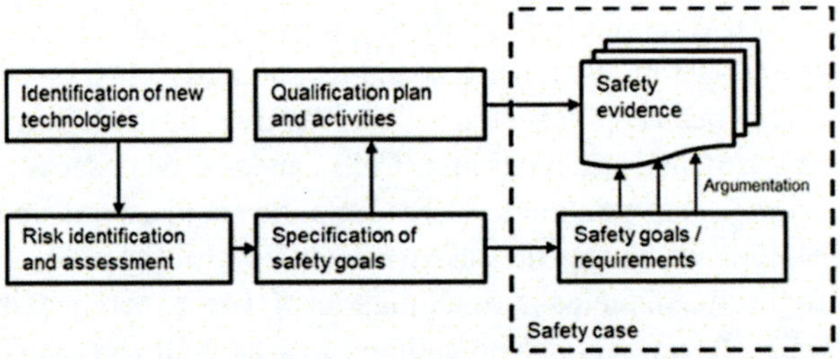

Fig. 2. An overview of a safety qualification process based on the safety case approach [19]

The goal-based approach has a strong communicative importance through visual representation of system's safety requirements, making the link between the safety requirements and evidence easily comprehensible [20]. When developing autonomous machinery systems, the goal-based model can be used as a means of communication between the system designers, end users and approving authorities to confirm that sufficient confidence for safe operation of the system in all foreseeable operating conditions has been achieved.

4 Support from Safety Standards

The approaches for standardizing autonomous systems vary significantly between the different fields of industry [6]. In many fields, the paradigm of standardization is seen to be shifting from traditional prescriptive standards to the direction of more general performance guidelines [21]. While this may give increasing freedom for technology developers in some aspects of design, it is also likely to increase their responsibility for ensuring and especially demonstrating the safety and reliability of autonomous technologies.

Many of the standards related to safety requirements for autonomous machine systems are published quite recently and there is not yet much experience with their application. In the areas of public transport and vehicle traffic there are safety standards providing essential guidance for designing safe vehicles. However, for example ISO 26262:2018 [22] is developed for vehicles that ultimately have a human driver responsible for safe operation of the vehicle. In autonomous vehicle or machinery systems operating without a driver there are many essential issues that still need commonly accepted and standardized guidelines. Figure 3 gives an overview of safety standards supporting the development of autonomous machinery systems.

Coopman (2019) [23] has expressed needs for standardization among others for autonomous vehicle technologies, including machine learning and sensor fusion, that exhibit complex, non-deterministic, and potentially unpredictable behaviors. Design guidelines are also needed for handling changes to the operating environment that will require continual updates to achieve adequate safety.

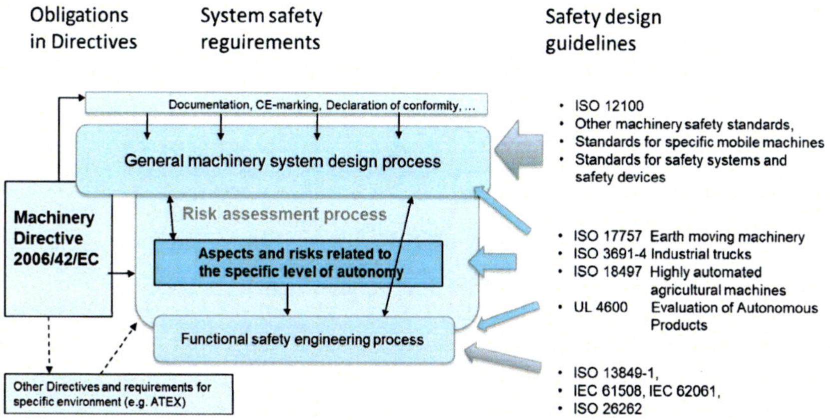

Fig. 3. Safety standards supporting the design of autonomous machinery

There are different objectives and boundary conditions for the development of machine autonomy in different industrial sectors. This guides standardization for safety engineering and sets needs and requirements for the necessary methods. Autonomous transport systems operating indoors are based on onboard intelligence and reliable environmental perception. Safety planning focuses on analyzing and validating their technical safety functions. Higher autonomy levels of machine combinations used outdoors in agriculture are based on the increasing ability of the automation system to detect, alert and, if necessary, respond to hazardous situations. At present, safety in these machines is a matter of user perception and professionalism and safety engineering must be focused strongly on human factors and driver assisting systems.

In the context of autonomous machinery systems used outdoors, two recently published standards are highlighted and discussed in following chapters: ISO 17757:2019 [2] and UL 4600:2020 Standard for Evaluation of Autonomous Products [24].

4.1 ISO 17757 - System Approach for Risk Assessment

ISO 17757:2019 [2] provides safety requirements for Autonomous machines and semi-Autonomous machines used in earth-moving and mining operations, and their Autonomous or semi-Autonomous machine systems (ASAMS). ISO 17757 presents the objectives for the safety design of an autonomous machine system, at the system level, taking as a starting point the requirements, risks and boundary conditions set by the work process and the operating environment. It states that the integration of ASAMS into the site planning process is important. ASAMS are complex systems, because of the complexity of the logistic processes themselves, their relation to people, manned operations and the layers of safety that need to be built into them [2]. Supporting infrastructure and operating area requirements should be identified early in the project, as automation systems can have specific needs (e.g. fueling facilities, control rooms, communications network). The standard introduces the concept of an autonomous operating zone (AOZ) (See Fig. 4), controlled by the access control system, were monitored manned machines and monitored persons could work at the same time with autonomous machines.

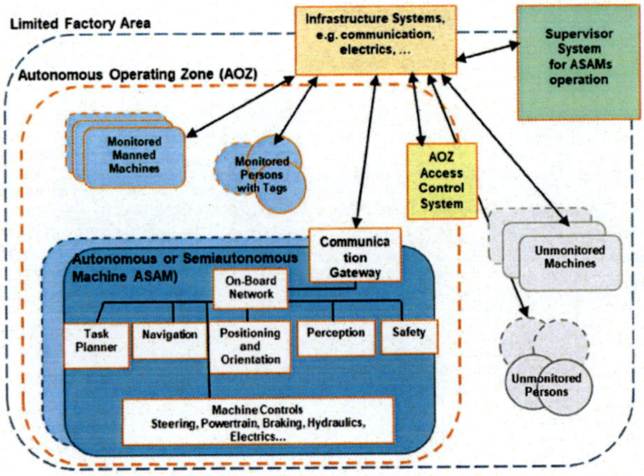

Fig. 4. Main elements in autonomous or semi-autonomous machine systems (ASAMS). Modified from ISO 17757:2019 [2]

ISO 17757 specifies safety criteria for both the autonomous machines and their associated systems and infrastructure, including hardware and software, and provides guidance on safe use in their defined operational environments during the overall system life cycle. The standard also emphasizes the importance of human-technology interaction and consideration of human factors to ensure overall safety, which fits well with the general system safety principles that have been applied for years in complex safety critical applications in aviation, transportation and energy sectors.

4.2 UL4600 - Safety Case Approach

UL 4600:2020 has been developed to ensure that autonomous are products and especially self-driving cars are safe. It concentrates on ensuring that a valid safety case is created [22]. In UL 4600:2020 approach, the information is collected from design and validation processes and the results are standardized and organized into a coherent safety case. That safety case is meant to provide credible evidence that a highly autonomous vehicle is indeed appropriately safe for deployment (See Fig. 5).

A safety case includes three main parts: goals, argumentation, and evidence. Goals describe what it means to be safe in a specific context, such as generic system-level safety goals e.g., ‘a car shall not hit pedestrians’ and element safety requirements e.g., ‘ensure correct computational results despite potential transient hardware faults’. Arguments are a written explanation as to why a goal is achieved e.g., ‘the system can detect and avoid pedestrians, including ones that are unusual or appear in the roadway from behind obstacles’. Evidence supports that the arguments are valid, typically based on analysis, simulations, and test results [24].

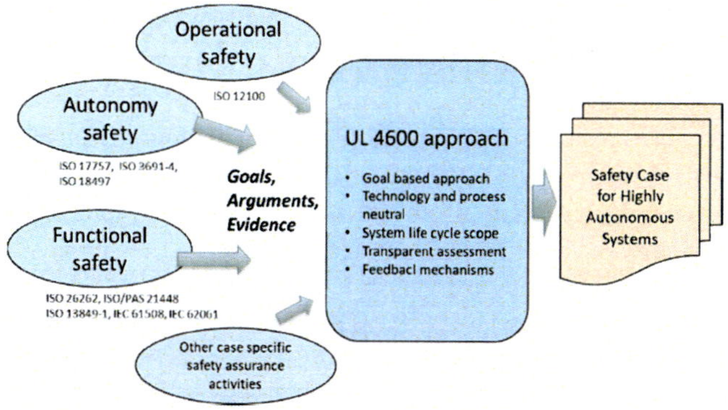

Fig. 5. Simplified illustration of the safety case approach according to UL 4600:2020

5 Discussion

The system safety approach utilizing PHA, OHA and HAZOP methods has been successfully applied earlier in several automated work-machinery applications in mining and cargo handling sectors [14]. Currently, the approach and methods are being developed in Finland by introducing a systemic analysis method STPA, which is a relatively new hazard analysis technique based on an extended model of accident causation called STAMP [4, 10, 15]. Preliminary results of applying STPA has shown that both PHA and STPA aim to identify and define system-level characteristics, operating conditions or system behaviors that are needed to prevent system hazards and losses or control risks. The output of the concept level assessment of risks related to autonomy aspects forms the basis for system safety goals for the next life cycle phases and more detailed safety engineering activities. STPA complements the perspectives of traditional PHA and provides a formal presentation to connect losses - system level hazards - risk control options.

The system level analysis should give information of possible causes of operational risks and descriptions of possible unsafe control actions associated with autonomous operation. Our experience is that STPA supplements OHA by allocation of safety goals and risk control options to sub systems and supports the reasoning of the requirements for safety related functions and control actions. From the overall risk management point of view the system level STPA results can give valuable information for the asset management planning and allocation of maintenance activities and resources. As our studies continue in the detailed design level, an interesting research question is, does STPA bring benefit for the analysis of autonomy related functionalities compared to HAZOP studies.

Our approach is in line with the principles and safety-engineering guidelines introduced in ISO 17757:2019 standard [2]. The standard defines a process for risk assessment and safety requirement specification in autonomous earth moving machine and mining machine systems, but it does not name any specific methods for safety engineering activities. It would be useful if the standard could give more guidance how the intelligence required by autonomous operation should be divided between the 'On-board' system and the site level 'Supervisor' system. In addition, more guidance for the design of the interactions between the main system elements could be helpful.

Safety case approach described in UL4600:2020 [24] aims to gather all the needed safety evidence and to produce a sufficient safety demonstration of the system. In our current research, we study how the goal-based safety case approach can support the safety requirements management in an entire business ecosystem developing autonomous machinery applications. The model and visualization of the allocation of safety goals according to [18] seems promising. It could be used as a means of communication between all stakeholders in the design teams, not only safety engineers and system designers. On the other hand, the goal-based approach, in the current form, does not offer any tools for prioritizing the different levels of safety goals. The safety engineers and designers need to have an understanding of the significance of the various goals when allocating them to system hierarchy levels and subsystems.

6 Conclusions

The systematic and clearly phased risk assessment process supports the development of machine autonomy when selecting operating concepts, selecting technological solutions and when designing interactions between an artificial intelligence and human operators in safety critical decision-making situations. Robust safety engineering procedures need to be in place to not only identify and control safety risks, but also to document and communicate safety-related aspects between all relevant stakeholders.

STPA method, including the modelling of the hierarchical control structure and bringing in a new perspective for the analysis of autonomy aspects by the identification of unsafe control actions is a welcome addition to system safety approach when dealing with higher levels of autonomy.

The safety case approach is heavy and laborious to implement since it is developed for large-scale safety critical systems like in aviation or space industry. It should be further developed to fit for the needs in machinery automation sector. A lighter version of safety case procedure to manage goals, arguments and evidence is certainly necessary in the future when the level of autonomy increases.

Standardization for autonomous machinery is currently developing strongly in many industrial branches, vehicles and transport in the forefront, and it already now gives support and guidelines for the identification of new autonomy related safety aspects, risk assessment processes and evaluation of risk mitigation alternatives.

Acknowledgement. Research on safety engineering methods and safety requirement management in autonomous machinery systems has been done and in ongoing in Finland in a national co-innovation project financed by Business Finland, VTT and participating companies.

References

1. NHTSA. https://www.nhtsa.gov/technology-innovation/automated-vehicles-safety. Accessed 24 June 2020

2. ISO 17757:2019 Earth-moving machinery and mining – Autonomous and semi-autonomous machine system safety
3. Hämäläinen J, Yli-Paunu P, Peussa P (2018) Automated container terminals and self-driving cars: Industry outlook. https://www.kalmarglobal.com/48eaed/globalassets/media/215590/215590_White-paper-Autonomous-cars-web.pdf
4. Leveson G (2018) Safety analysis in early concept development and requirements generation. In: 28th annual INCOSE international symposium, Washington, DC, USA, 7–12 July 2018
5. Heath T (2018) Autonomous industrial machines and the effect of autonomy on machine safety. Master of Science Thesis, Tampere University of Technology, 62 p, March 2018
6. Tiusanen R, Malm T, Ronkainen A (2020) An overview of current safety requirements for autonomous machines - review of standards. In: Automaatiopäivät 23 conference, Oulu, Finland, 15–16 May 2019. In Automation in Finland 2019 Special Issue. Open Engineering
7. Magnusson G, et al (2018) Modelling requirements of autonomous system. Fraunhofer. https://ntrs.nasa.gov/archive/nasa/casi.ntrs.nasa.gov/20190001490.pdf
8. Paté-Cornell ME (2012) On "black swans" and "perfect storms", risk analysis and management when statistics are not enough. Risk Anal 32(11):1823–1833
9. Dekker S, Cilliers P, Hofmeyr J-H (2011) The complexity of failure: implications of complexity theory for safety investigations. Saf Sci 2011(49):939–945
10. Leveson N (2012) Engineering a safer world: Systems thinking applied to safety. MIT Press, Cambridge
11. INCOSE (2015) Systems engineering handbook: a guide for system life cycle processes and activities. Wiley, Hoboken ISBN 978-1-118-99940-0
12. Vincoli JW (2006) Basic guide to system safety. Wiley, Hoboken
13. IEC 61882:2016 Hazard and operability studies (HAZOP studies) - Application guide
14. Tiusanen R (2014) An approach for the assessment of safety risks in automated mobile work machine systems. Dissertation, VTT, Espoo
15. Leveson N, Thomas J (2018) STPA handbook. https://psas.scripts.mit.edu/home/get_file.php?name=STPA_handbook.pdf
16. DNV 2011 (2011) Qualification of New Technology. Recommended Practice. DNV-RP-A203. Det Norske Veritas AS
17. Heikkilä E (2016) Safety qualification process for autonomous ship concept demonstration. Master's thesis. Aalto University
18. Kelly T, Weaver R (2004) The Goal Structuring Notation. A Safety Argument Notation. University of York
19. Montewka J, Wróbel K, Heikkilä E, Valdez-Banda O, Goerlandt F, Haugen S (2018) Challenges, solution proposals and research directions in safety and risk assessment of autonomous shipping. In: Proceedings of PSAM 14, Los Angeles, CA, September 2018
20. Heikkilä E, Tuominen R, Tiusanen R, Montewka J, Kujala P (2017) Safety qualification process for an autonomous ship prototype: a goal-based safety case approach. In: Weintrit A (ed) Marine navigation - proceedings of the international conference on marine navigation and safety of sea transportation, TRANSNAV 2017, CRC Press, pp 365–370. https://doi.org/10.1201/9781315099132-63
21. Danks D, London A (2017) Regulating autonomous systems: beyond standards. IEEE Intell Syst 32(1):88–91
22. ISO 26262:2018 Road vehicles - Functional safety - Parts 1-10
23. Koopman P (2019) An Overview of Draft UL 4600. Standard for Safety for the Evaluation of Autonomous Products. Edge Case Research
24. UL 4600:2020 Standard for Evaluation of Autonomous Products. UL Standards

Covid-19 Perspectives

Check for updates

Engineering Asset Management at Times of Major, Large-Scale Instabilities and Disruptions

Dragan Komljenovic(✉)

Institut de recherche d'Hydro-Québec (IREQ), 1800, boul. Lionel-Boulet, Varennes, QC J3X 1S1, Canada
komljenovic.dragan@ireq.ca

Abstract. Contemporary organizations function in a complex business and operational environment composed of closely interdependent systems. They are also complex by their internal structure, management and deployed modern technologies. This complexity is not always well understood, and cannot be efficiently controlled. As the complexity and interdependencies increase, man-made systems become more unstable creating conditions for cascading, system-level failures causing serious threats to both themselves and society in general. Such breakdowns may consist of a) serious physical damages and destruction of their physical assets (caused by natural disasters, extreme weather phenomena and climate change, malicious human actions, etc.), and/or b) large functional disruptions with no physical damages of assets (caused by major organization's internal disturbances, market crashes, pandemics, disruptions of supply chains, etc.). Those sources of risks are basically external to organizations. They are unable to control them, but are deeply affected by those risks.

The latest case of the COVID-19 pandemic demonstrates the above. It is affecting both all sectors of life and businesses worldwide. It convincingly shows that we need to think, plan and act globally in order to deal with such situations that will also take place in the future. Thus, organizations have to find ways of coping with this reality to remain economically viable. We are of opinion that the concepts of structured Asset Management (AM) and resilience put together may provide an efficient framework in this regard.

Two case studies in a major North American electrical utility demonstrate the applicability of this approach: i) during an exceptional ice storm with significant damages of its physical assets, and ii) coping with challenges of COVID-19 with no destruction of its physical assets.

Keywords: Complexity · Major disruptions · Asset management · Resilience · Pandemic

1 Introduction

In the modern and deeply connected world, various man-made systems have created highly interdependent networks that are not fully understood, and cannot be adequately

A. Crespo Márquez et al. (Eds.): WCEAM 2019, LNME, pp. 255–270, 2021.
https://doi.org/10.1007/978-3-030-64228-0_22

controlled [12, 24, 39, 42]. The type of connections between networks and entities may be physical, informational, energy, material, geospatial, policy/procedural/logistic, societal, financial/monetary, etc. [20, 24, 33]. Such a context particularly applies to critical infrastructures: transportation systems, power, energy and communication grids, as well as water and gas distribution systems [56]. Meanwhile, other man-made systems such as finance, health care, education, supply chains, cyber and social networks, etc. also grow in complexity and interconnectedness. They are closely linked to above mentioned critical infrastructures [20, 33].

Consequently, contemporary organizations operate in an increasingly complex operating and business environment that is characterized by deep intrinsic uncertainties related to the often rapid evolution of markets/customer expectations, changing regulatory framework, new technologies including the advent of the Industry 4.0 concept, rapidly changing natural environment, change in the political environment, malicious human actions, climate change, changes in demand, advent of new competitors, pandemics, etc.

Such a complex world offers extraordinary opportunities for advances and prosperity in all spheres of life and activities. However, this complex world has become highly fragile and vulnerable to failure at all scales, posing serious threats to society, even when external shocks/disruptions are absent [12]. With the growing complexity as well as density and strength of interactions, man-made systems can become unstable, creating potentially uncontrollable situations, and cascading failures even when decision-makers are highly competent, possess data and technological means at their disposal, and do their best.

The challenges related to the complexity, cascading and system-level breakdowns of contemporary man-made systems led to research works aiming at better understanding and managing them that resulted in numerous publications. Certain research works discuss the global impact of the complexity on overall human activities including growing importance of extreme events as well as emergent and systemic risks [12, 33, 35, 39, 41, 42, 46, 54]. Some other authors analyze the impact of complexity to technological and engineered systems highlighting the need to further study them in order to enable a more efficient management [20, 21, 23, 24, 33, 38, 56]. Complex technical and non-technical mechanisms that trigger the causal relationships of catastrophic system-level failures are still not fully understood and can create new types of emerging risks through unexpected behaviors and combinations of various influence factors (technological, natural, human). Those new, previously unknown or not considered risks could pose utmost challenges to resilience, safety and business continuity of modern organizations, and the society as a whole. This topic has already been discussed in research works [5–7, 17, 18, 26, 39, 47]. However, it should be emphasized that certain risks may be positive (opportunities), and should be managed as such [17, 18, 49].

Systemic failures within organizations profoundly affect their operational and business environment. The may consist of a) serious physical damages and destruction of their physical assets (caused by natural disasters, extreme weather phenomena and climate change, malicious human actions, etc.), and/or b) large functional failures/disruptions with no physical damages of assets (caused by major organization's internal disturbances for various reasons, market crashes, pandemics, political instabilities, significant disruptions of supply chains, etc.). Majority of above mentioned sources of risks are basically external to organizations which do not control them, but are deeply exposed to and affected by those risks (Fig. 1).

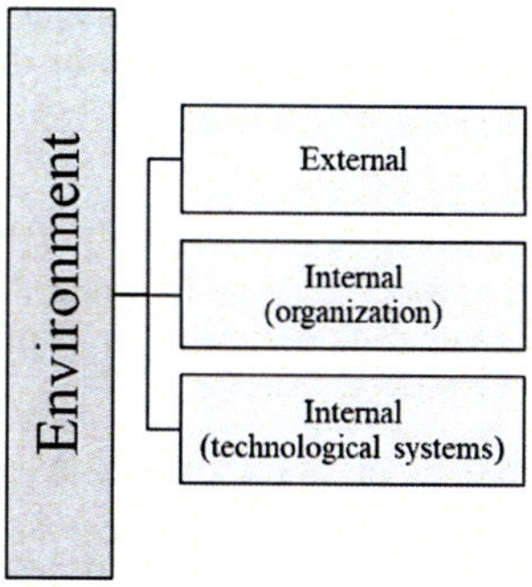

Fig. 1. Types of operating and business environment regarding an organization

To make these systems convenient, fundamental changes are required in ways they are studied, modeled, exploited and managed. Shifting the attention from a component-oriented view to an interaction-oriented, holistic view will allow better understanding of complex man-made systems, and the emergent phenomena characterizing them. This paradigm shift will enable new solutions to both long-standing and emergent problems.

The latest case of the COVID-19 pandemic demonstrates the above. It has affected life, businesses and growth in virtually every sector around the world although there was no destruction of physical assets. New research on this topic is emerging, but one still does not understand it true scale [3, 26, 28, 30, 31, 40, 43, 44, 51, 55] (Fig. 2).

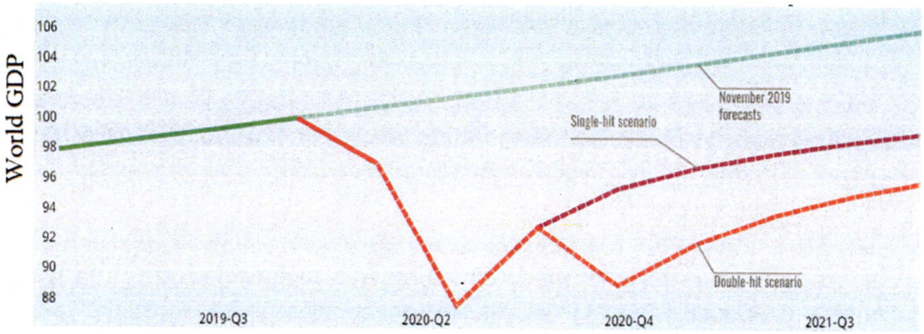

Fig. 2. Projections of COVID-19 impact on the world GDP [30]

Supply chains are also profoundly affected or disrupted worsening an already serious worldwide situation [10, 27, 52]. Moreover, both the confinement and high, sudden unemployment create additional tensions within and between societies/countries that further amplify both negative impacts of the pandemic as well as existing conflicts and tensions potentially creating new ones. Mankind finds itself in an unchartered territory.

It is worth highlighting that since 2007 the World Economic Forum (WEF) has never identified the pandemic risks as a serious threat in its *Global Risk Report* showing the limits of predictions under complexity and deep uncertainties [54].

Meanwhile, global shocks and large-scale disruptions cannot be excluded in the future. The COVID-19 case convincingly demonstrates that we need to think, plan and act globally in order to deal with similar situations, even under extreme uncertainties embedded in our lack of complete understanding of the epidemiologic, biological and mutational aspects of this menace.

In such a context, organizations face significant challenges in designing an adequate asset management system (AMS). The paper aims at proposing a novel resilience-based engineering asset management (AM) relevant at times of major, large-scale disruptions and instabilities.

The remaining of the paper includes the proposed resilience-based methodology related to engineering asset management. It is followed by two case studies that demonstrate interrelations between AM and resilience in a major North American electrical utility: i) during an exceptional ice storm with significant damages of its physical assets (consequence of type a) above), and ii) coping with the challenges of COVID-19 where no destruction of physical assets happened. The paper ends with a discussion regarding needs for future research and anticipated benefits of the proposed methodology.

2 Methodology

This Section presents the proposed resilience-based approach in engineering asset management (AM) at times of major, large-scale disruptions and instabilities. Organizations have to find ways of coping with the new reality ("new normal") to remain economically viable and assure their sustainable development. It seems that the concept of resilience has gained popularity and application among contemporary organizations in the context of ever-growing complexity of operational and business environment exposed to deep uncertainties as well as new emerging and systemic risks. It is still a field of research under strong development, but there are several publications which already provide significant insights [1, 4, 5, 8–11, 16, 27, 29, 32, 34, 37, 40, 47, 48]. There are numerous definitions of resilience across the literature, but the ISO definition is retained in this paper. It defines the resilience as an adaptive capacity of an organization in a complex and changing environment [19].

The resilience has four properties: 1. robustness, 2. redundancy, 3. resourcefulness, 4. rapidity, and four interrelated dimensions: 1. technical, 2. organizational, 3. social and 4. economic [4, 5, 53, 56]. It covers four phases regarding the occurrence of a disruptive (adverse) event: 1. planning (preparation, anticipation) – before an adverse event; 2. absorption – loss of performance while an event occurs; 3. recovery – recover the performance after event; 4. adaptation – lessons learnt and continuous improvement after an adverse event (Fig. 3).

As presented in Fig. 3, the resilience also includes the robustness that is defined as the ability of a system to withstand damage (ability to absorb a shock) and continue operating [1]. The level of loss of the performance following an adverse event will be lesser if a system design is more robust (absorption phase). The ability and time to

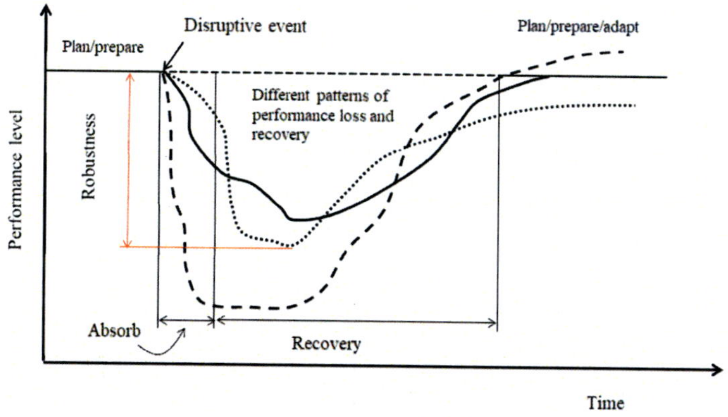

Fig. 3. General concept of resilience

recover from damages is a function of both the magnitude and nature of the performance loss, overall preparedness and available resources. The recovery time is longer if fewer resources are available and/or the level of performance loss is greater. Thus, there are various possible recovery patterns (Fig. 3). The achieved level of performance after recovery may also differ. Systems and organizations that successfully adapt and learn from internal and external return of experience may achieve better levels of performance through adaptation and continuous improvement. The resilience is a constant process that itself must be adaptive and improving.

As depicted in Fig. 1, there are elements in the overall business and operational environment belonging to the external environment of an enterprise (e.g. natural, business, legal, regulatory and political environment, market factors, pandemics, etc.). This environment cannot be accurately predicted, controlled, nor strongly influenced. Its complexity is high. Deep uncertainties and opacity are prevailing here. Nevertheless, it usually exercises a major impact on the performance of an organization and is prone to emerging and extreme risks. Resilience management is a recommended approach in managing risks and uncertainties of this environment. The concept of the Complex System Governance (CSG) may also be quite useful in this regard [6, 21, 38].

Effective and efficient organizations use a structured approach to their AM in order to balance competing priorities, manage various influence factors (external and internal), and ensure an equilibrium between long-term benefits and immediate needs [15]. An Asset Management System based on the ISO 55000 family of standards helps an organization to establish a coherent approach and coordinated allocation of appropriate resources and activities [14]. An efficient AM is also able to take into account the risks of extreme and rare events in the overall strategy and asset management decision-making, and there are already discussions and research on how AM may help in dealing with COVID-19 [2, 14, 15, 24, 28, 40, 45, 49].

Figure 4 shows the proposed, high-level holistic resilience-based asset management approach that integrates relevant functions and activities of an organization, and relates them to the main four phases of the resilience concept.

The proposed approach integrates and harmonizes the following functions and activities of an organization [22] (Fig. 4):

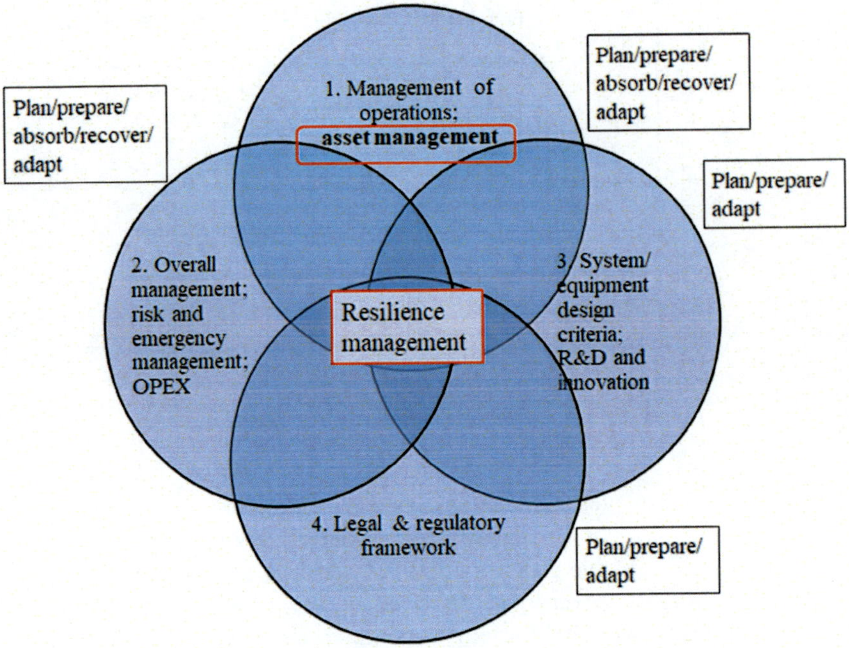

Fig. 4. Proposed resilience-based engineering asset management

1. *Operation management; maintenance, monitoring and inspection management, asset management:* An adequate performance in those enterprise's functions brings an adequate operating environment supporting a sustainable development even in cases of major disruptions. They positively contribute to all four resilience phases (plan, absorb, recover, adapt).
2. *General (overall) management, risk and emergency management, return of experience and lessons learnt* are essential in the overall resilience management of an organization. The excellence in those functions increases the overall resilience in all four phases. As far as risk assessment and management are concerned, it is recommended to follow general prescriptions of the Standard ISO 31000 [17], but other approaches such as business continuity management and emerging risks management may also complement it [16, 18, 48]. The excellence in those functions and activities brings an overall improved performance of an enterprise even in the case of major disruptions because of its ability to control and manage its main functions and processes. In this regard, attention should be paid to achieve and improve human and organizational performance since it may also be a major contributor to accidents and disasters, but also positively contribute in cases of an organization's hardship.
3. *Establishing an adequate equipment and system design criteria; R&D and innovation*; given complexity of operating and business environment, deep uncertainties related to occurrence, magnitude and nature of disruptive events or large-scale instabilities, it is crucial to foresee future operating conditions as accurately as possible and integrate them into design criteria for events that may threat physical integrity of assets. Disruptive events with no physical damages of assets (e.g. pandemics, market

crashes, political instabilities, significant disruptions of supply chains, etc.) require the knowledge of nature of such events as well as vulnerabilities of an organization exposed to them. It seems that traditional methods of analyzing them are not always good enough [5, 12, 24, 33, 43, 45, 48, 56], and new ones should be elaborated. Since uncertainties regarding future operating conditions of an organization are deep, sufficient margins should be provided, but also optimized in order to assure its economic viability and sustainable development. For that purpose, R&D and innovation are vital factors in both improving and gathering knowledge and develop new methodologies that can contribute to a better understanding and modeling of less known phenomena. Those new approaches would provide new insights, support a sound decision making at all management levels and enable assessing both the efficiency of asset management system and the adequacy of preventive and mitigating measures with regard to the overall performance and resilience of the organization.

4. *The fourth element relates to the legal and regulatory framework* which is not under the control of an organization. However, an adequate legal and regulatory framework is of a key importance for safe and sustainable operation acting through laws, regulatory directives, prescriptions and requirements.

The relationships between these four functions and activities are multiple and of different types such as physical, spatial, informational, financial, resource exchanges etc. Consequently, changes in their nature or characteristics may cause unexpected impacts that are often non-linear due to the complexity of their connections.

Thus, the overall asset management and resilience strategy in cases of major disruptions and instabilities is composed of an array of interacting and interdependent activities and functions within a multilevel structure of an organization, and outside. The cumulative effect of various actions produces a result that is superior to a simple sum of individual effects [24]. It is worth emphasizing that is important to have an overall good performance in all the activities presented in Fig. 4. The excellence or a good performance in just one or two domains cannot usually make up for serious weaknesses in other activities [22]. The proposed methodology favors an overall culture of preventive actions within an organization. However, it should be stressed that organizations ought to consider balancing preventive and mitigation strategies to ensure sufficient safety margins, adequate performance, economic viability and long-term sustainability. It is also worth highlighting that the methodology is technologically neutral and may be easily adapted to any type and size of the organization with its specific management, business and operational environment.

3 Case Studies

Two case studies demonstrate interrelations between AM and resilience in Hydro-Quebec (HQ): i) during an exceptional ice storm with significant damages of its physical assets, and ii) coping with the challenges of COVID-19 where no destruction of physical assets happened, but the overall functioning of the organization is deeply affected at all levels [13, 23–25, 50]. Both case studies relate to external operating and business environment that the enterprise does not control, but is exposed to the consequences (Fig. 1).

HQ is a major North-American electrical utility owned by the Government of Quebec [23]. It manages 63 hydroelectric power plants with approximately 38 GW installed generation power, and 353 hydroelectric generators. It operates the most extended and complex transmission line network in North-America, with over 34,000 km high voltage (49 kV to 735/765 kV) lines (more than 11 000 km of 735-kV), and 533 transmission substations. Approximately 85% of concentrated load is located in the south within the larger Montréal (Metropolitan) loop. HQ has 15 strategic inter-connections with neighbour grids for exports/imports (Ontario, New Brunswick, NE USA). Its distribution grid comprises 118,000 + km MT and 107,000 + km LT lines, 680,000 + pole mounted transformers and 3000 + distribution substations (Fig. 5).

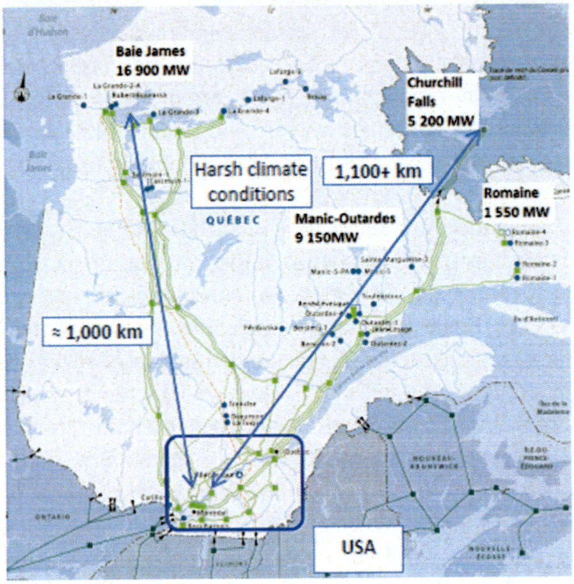

Fig. 5. Hydro-Quebec's generation and transmission assets [23, 50]

3.1 Disruptions Due to an Exceptional Ice Storm 1998

The Occurrence of a Disruptive Event (Fig. 3): Between January 5 and 10, 1998, Québec experienced exceptionally harsh weather as three successive storms left up to 110 mm of ice over the south of the province. The 1998 ice storm was exceptional because of two unusual situations that occurred hundreds of kilometres away from Québec. First, El Niño caused a large mass of warm air to form over the Gulf of Mexico. Due to prevailing winds, this warm air mass moved to Québec followed by another one. Second, the usual pattern of west-to-east prevailing winds stopped for a few days as a result of a major high-pressure system that stationed over Newfoundland and Labrador. The 1998 ice storm consisted of three successive freezing rainfall events within a very brief span of time, covering a space of about 400 000 km^2, of which 130 000 km^2 were in Québec.

According to Environment Canada, during these three episodes of freezing rain, the average ice accretion was between 50 and 70 mm.

The Absorption Phase (Fig. 3): Storms damaged 24,000 poles, 900 steel towers and 3,000 km of lines and left 1,393,000 customers without power [23, 24]. The ice storm has had an overall cost to the public finances of $1.656 billion with two-thirds or $1.028 billion, was borne by the Québec government, directly or through HQ.

The Recovery Phase (Fig. 3): The whole society was mobilized in recovery activities. Every afternoon at 5, the Quebec's Premier Lucien Bouchard, HQ's President and CEO, André Caillé, and a representative of the Organisation de la sécurité civile (ORSC) held a press briefing to inform the public about the condition of the grid. The aim was to provide accurate information and keep the public informed at every stage of the response. On January 10, the ORSC set 'Operation Ice Storm' into motion, which comprised several work units, each charged with a priority mission to assist disaster victims. Some 750 volunteers from government departments and agencies took care of administration, food, financial assistance, firewood, generators, accommodation and information. Close to 9,000 soldiers were called in to help pick up branches, dispose of broken parts of transmission and distribution lines, transport new components for rebuilding lines and ensuring safety. The media played an important role in the ice storm. From the very outset, HQ set up 30 missions to be deployed in the affected areas. Each mission consisted of some 120 people, including a mission chief, a building supply procurement manager, about 50 soldiers, tree trimmers, line crews and a community relations officer. On February 6th 1998, the power restored to last customer highlighting the end of the recovery phase.

The adaption/improvement phase (Fig. 3) and the general resilience-based AM approach (Fig. 4): Following the 1998 ice storm, HQ has undertaken a comprehensive programme to reinforce its grid. This activity includes R&D efforts to better understand icing events and mitigation measures to strengthen facilities and assets. Test lines were built at HQ's research institute, IREQ, in order to replicate icing conditions, and to test and validate specific designs and parameters. New maps were produced for extreme winds, ice accumulation, frosts and their combinations. Results were incorporated into construction standards and methods, while various research projects (IREQ, universities) helped to make the power system more robust. Innovations include the new generation of insulators to better protect facilities and interphase spacers that curb the effects of galloping and high-amplitude oscillations along overhead conductors. There are also works to develop and integrate new technologies or improve existing ones which would help increase efficiency of preventive and mitigating measures (e.g. smart grid features). Several projects to secure power supply to customers were implemented such as diversifying generation sources (e.g. interconnections) and supply corridors (e.g. Montreal loop). More than 900 km of lines were rehabilitated to more robust design criteria (greater than the CSA Standard) and with the installation of anti-cascading pylons. New posts and 295 km of new lines were added while existing 552 km of lines were also reinforced. A new 1,250 MW interconnection to Ontario (Outaouais substation) was built. Other preventive and mitigation measures include: i) Vegetation control to prevent power failures by maintaining clearance around power lines; ii) Sygivre, a real-time ice storm management system that detects ice storms, tracks their development and keeps potential users informed; iii) De-Icing means: Remotely Operated De-Icing All Weather Vehicle, Lévis Substation De-Icing, De-Icer actuated by Catridge, On-Load Network De-Icer, and Joule Effect De-Icing; iv) Towers in reserve at the Lines' Emergency Bank; v) Emergency measures plan updated [23].

HQ transmission grid conception and exploitation philosophy have also been revised and improved at the system level during years 2000s [50]. Four basic principles and successive defence barriers have been reviewed: *Principle No. 1:* Service continuity must be assured following events most likely to occur on the power system. *Principle No. 2:* Hydro-Québec's power system must include ways of avoiding system-wide power failures under extreme contingencies. *Principle No. 3:* Strategic equipment must not sustain any damage in the event of a general outage to ensure that system restoration is always an option. *Principle No. 4:* Hydro-Québec's transmission system must be designed so as to allow the system to be restored within a reasonable period after a catastrophic event (Fig. 6).

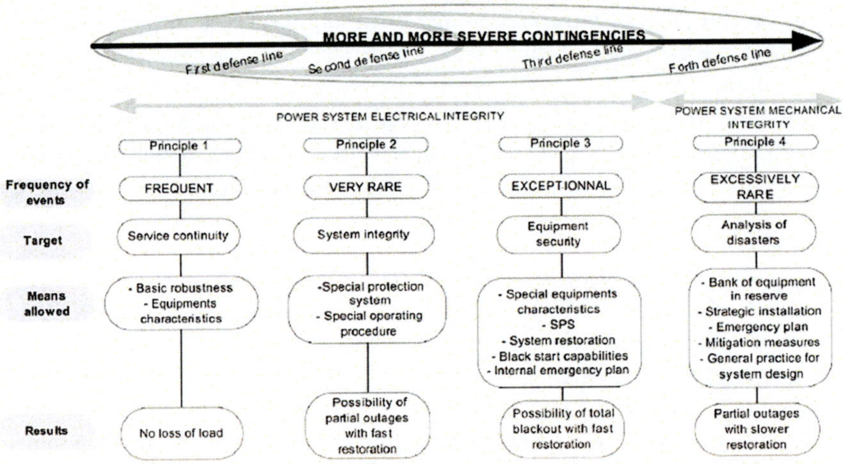

Fig. 6. Hydro-Québec's concept of successive lines of defense [50]

Structured Asset Management programs have been elaborated and introduced helping an adequate fitness for service of HQ critical assets. It also supports building the overall resilience of both the physical asset systems and organization. They are continuously improved based on the return of field experience and R&D efforts.

It is worth emphasizing that the Province of Quebec has been exposed to several important ice storms since mid-2000s, but there were no important service interruptions due to the augmented resilience of the whole system.

3.2 Management of COVID-19 Impacts at Hydro-Quebec

Although there is no destruction of physical assets, the ongoing COVID-19 pandemic and crisis brings another set of challenges and issues. They are related to the organization's ability to carry out normal operation, business and maintenance activities in a context where all the sectors of life and business are severely hit [3, 13, 28, 30, 40]. The organizational and economic resilience of the organization are rather tested in this context. The magnitude of the ongoing crisis is still unknown since it is still unfolding and continuously brings new surprises and unexpected scenarios worldwide. The current situation is a mix of the absorption and recovery phases, with some elements of adaptation (Fig. 3). It seems that existing models are of a limited use [43]. There are already some initial publications on how AM may help organizations in handling the COVID-19 crisis [28, 40], but the discussions and research continue.

As far as Hydro-Quebec is concerned, the magnitude of losses is still unknown. The initial estimates provided by its CEO anticipate up to 1 billion $ reduction in revenues in 2020 only [13]. Hydro–Quebec is on the list of essential services established by the Government of Québec. However, the enterprise is adapting to the new context, and has introduced several measures to mitigate the consequences of the crisis [13]: a) workforce: online (remote) work is generally introduced for employees who are able to accomplish their tasks in that way. Other workers (operating and maintenance crews) indispensable to maintain the operation have to respect the new security/safety rules aimed at protecting their health. Daily and short meetings are held to keep in touch and ensure both their wellbeing and the quality of work. There are undergoing discussions upon new ways of performing work (more importance will likely be given to work from home). The foreign business travel is banned. Despite more difficult working and operating conditions, basic maintenance tasks are carried out, and the level of the system performance meets defined requirements; b) customers: a majority of big industrial customers are generally hardly hit due to market instabilities and disruptions. There are discussions with HQ in order to define support measures. Small customers and population in general are provided with support measures regarding modalities of payment including the abandon of the service cut off for nonpayment; c) no visitors are allowed to Hydro-Quebec's offices and installations; d) Supply chain and suppliers (providing goods and services): Even in light of the events related to COVID–19, HQ is planning a number of tender calls to carry out its projects. This message holds for the entire supply chain for projects, including construction activities and the procurement of goods and professional services. The payment delay is also reduced in order to help HQ's suppliers. This orientation also plays an important role in supporting the economic recovery in Quebec; e) R&D and innovation project as well as the collaboration with the pairs of industry continue since this aspect is important not only in assisting the enterprise in the COVID-19 crisis, but also in preparing and adapting it to other future disruptive events that will likely occur.

3.3 Discussion Regarding Integration of the Resilience-Based AM Approach

Two case studies clearly demonstrate needs for further developing the knowledge and understanding of extreme and disruptive events in order to adequately cope with them and ensure the resilience, economic viability and sustainability of Hydro-Quebec. In this regard, the development and integration of the resilience-based AM approach (Fig. 4) becomes relevant. On top of the improved operational and management practices as well as R&D and innovation at HQ, further recommendations and initiatives were made in an ongoing process to cope with future extreme risks (weather and/or others [23, 24]), climate change and deep uncertainties: a) Pursue the aim of excellence in general, risk and operation management as well as asset management at all organizational levels through a continuous improvement; b) Continue participating in various pertinent industry working groups to ensure leadership among electrical utilities and define future works; c) Continue participating in relevant standardization bodies in order to influence orientations in the standardization and regulatory framework; d) Continue participating at pertinent conferences and other forums enabling further contacts, feedbacks and balisage among pairs; e) Continue R&D and innovation efforts, collaboration with IREQ, universities, other research institutions. R&D and innovation are important factors in increasing the overall resilience; f) Continue developing and integrating new technologies or improve existing ones which would help increase efficiency of preventive and mitigating measures (e.g. smart grid features, AI); g) Develop new analysis methods and models able to quantify impacts of extreme events in a complex operational and business environment and its deep uncertainties. Those new approaches should provide new insights and enable assessing the adequacy of preventive and mitigating measures with regard to the overall performance and resilience of the organization and support a sound decision making at all management levels. These activities may involve i) understanding short- and long-term impacts of extreme events (weather and others) and climate change on assets and their systems such as required performance, reliability and availability, physical and structural integrity, duration of their useful life; ii) improving or developing new risk analysis and aggregation methods in the context of the complexity and deep uncertainties, including the characterization of systemic and emergent risks, and iii) enhancing multi-criteria decision-making methods in order to better integrate the impact of overall complexity and interdependences to their outcomes [23].

It is worth highlighting that the first case study covers an event occurred more than 20 years ago. The enterprise has had sufficiently time to analyze both the context and relevant factors/lessons in order to prepare and adapt itself to similar situations. The COVID-19 crisis is ongoing. The knowledge is still insufficient for a complete understanding of the phenomenon. Only further analyses and research will enable a better comprehension of this crisis helping the enterprise to improve its resilience to similar future events. The proposed approach may help in this regard.

4 Conclusions

Enterprises worldwide are constantly forced to produce more at lower costs. This way, we created highly dependent and complex economic, financial, technological and cyber systems that offer tremendous opportunities and possibilities to carry on business efficiently. Meanwhile, those man-made systems have shown that they are also fairly fragile at times of major disruptions and instabilities – they are prone to system-level breakdowns and cascading failures. The latest COVID-19 crisis clearly demonstrates this feature. It seems that classical methods of analyze are quite limited in capturing the whole range of this phenomenon. New ways of thinking and new methods of analyzing complex systems become necessary. The resilience-based approaches are gaining the importance in this context. The paper proposes a resilience-based asset management methodology that has potential of holistically approaching the problem of how to handle extreme and disruptive events in organizations. Two case studies in a major North-American electrical utility (Hydro-Quebec) demonstrate the applicability of the proposed methodology. Suggestions for future research are also provided.

References

1. Albino DK, Friedman K, Bar-Yam Y, Glenney IV WG (2016) Military strategy in a complex world, V2, Cornell University Library and New England Complex Systems Institute, (New England Complex Systems Institute Report 02 February 2016), pp 1–22, https://arxiv.org/abs/1602.05670. Accessed 24 June 2020
2. Amadi-Echendu JEL (2004) Managing physical assets is a paradigm shift from maintenance. In: IEEE international engineering management conference, pp 1156–1160
3. Aven T, Bouder F (2020) The COVID-19 pandemic: how can risk science help? J Risk Res 1–6. https://doi.org/10.1080/13669877.2020.1756383
4. Aven T (2019) The call for a shift from risk to resilience: what does it mean? Risk Anal 39(6):1196–1203
5. Aven T (2016) Risk assessment and risk management: Review of recent advances on their foundation. Eur J Oper Res 253(1):1–13
6. Brocal F, González-Gaya C, Komljenovic D, Katina PD, Sebastián MA (2019) Emerging risk management in industry 4.0: an approach to improve organizational and human performance in the complex systems. Complexity 2019. https://doi.org/10.1155/2019/2089763. Article ID 2089763
7. Brocal F, González C, Sebastián MA (2018) Technique to identify and characterize new and emerging risks: a new tool for application in manufacturing processes. Saf Sci 109:144–156
8. Ciapessoni E, Cirio D, Pitto A, Panteli M, Van Harte M (2019) Defining power system resilience, reference paper. Electra 306, 32–34
9. Cotton-Barratt O, Daniel M, Sandberg A (2020) Defence in depth against human extinction: prevention, response, resilience, and why all matter. Glob Policy 1–12. https://doi.org/10.1111/1758-5899.12786
10. Golan MS, Jernegan LH, Linkov I (2020) Trends and applications of resilience analytics in supply chain modeling: systematic literature review in the context of the COVID-19 pandemic. Environ Syst Decis 40(2):222–243. https://doi.org/10.1007/s10669-020-09777-w
11. Hosseini S, Barker K, Ramirez-Marques JE (2016) A review of definitions and measures of system resilience. Reliab Eng Syst Saf 145: 47–61
12. Helbing D (2013) Globally networked risks and how to respond. Nature 497:51–59
13. Hydro-Québec: COVID-19 Handling/impact. https://www.hydroquebec.com/covid-19-en.html?fromslider=true&slide=1, https://www.hydroquebec.com/suppliers/covid-19.html, https://montrealgazette.com/business/local-business/hydro-quebec-cuts-2020-outlook-as-covid-19-weighs-on-demand, https://www.lapresse.ca/affaires/entreprises/2020-06-15/covid-19-un-impact-de-plusieurs-centaines-de-millions-chez-hydro-quebec. Accessed 25 June 2020
14. ISO 55000: 2014 Asset Management – Overview, Principles and Terminology, ISO Standard, January 2014
15. ISO/TC 251, Achieving the UN Sustainable Development Goals, 1–4, 2018, https://committee.iso.org/home/tc251. Accessed 21 June 2020

16. ISO 22301:2019: Security and resilience—Business continuity management systems—Requirements, International Standard
17. International Organization for Standardization (ISO): Risk management–guidelines. ISO 31000:2018
18. International Organization for Standardization (ISO): ISO 31050 Guidance for Managing Emerging Risks to Enhance Resilience: Thriving in a World Growing in Uncertainty ISO 31050-Leaflet_v09aj14082018, 2018, https://www.eu-vri.eu/filehandler.ashx?file=16526. Accessed 24 June 2020
19. ISO Guide 73 (2009) Risk management - vocabulary; management du risque – Vocabulaire. International Standard 2009
20. Katina PF, Pinto CA, Bradley JM, Hester PT (2014) Interdependency-induced risk with applications to healthcare. Int J Crit Infrastruct Prot 7(1):12–26
21. Keating CB, Katina PF, Bradley JM (2014) Complex system governance: concept, challenges, and emerging research. Int J Syst Syst Eng 5(3): 263–288
22. Komljenovic D, Stojanovic L, Malbasic V, Lukic A (2020) A resilience-based approach in managing the closure and abandonment of large mine tailing ponds. Int J Min Sci Technol (in press)
23. Komljenovic D, Delourme B, Lavoie M (1998) Ice storm Canada, January 1998, case study series of extreme weather, dynamic resilience to extreme weather, learning from best practice examples, world energy council, Abu Dhabi, September 2019. https://www.worldenergy.org/assets/downloads/Ice_storm_Canada_-Extreme_weather_conditions_SEP2019.pdf. Accessed 26 June 2020
24. Komljenovic D, Gaha M, Abdul-Nour G, Langheit C, Bourgeois M (2016) Risks of extreme and rare events in asset management. Saf Sci 88:129–145
25. Le S (2020) l'année la plus difficile de l'histoire d'Hydro?, April 2020. https://www.lesoleil.com/opinions/point-de-vue/2020-lannee-la-plus-difficile-de-lhistoire-dhydro-4596f5bfc11b9223bb15d9884ee7262b?utm_campaign=lesoleil&utm_medium=article_share&utm_source=facebook&fbclid=IwAR1OgnZM6m-85suqTkpRnvkrDEi5G8p0L4CEK-AF9pDx-6Se08_FJzHtiBM. Accessed 24 June 2020
26. Lewiner C, Vié P (2020) Energy & utilities, operating in the "new normal". Key considerations for mitigating risks and seizing opportunity in a post-crisis world, Capgemini. https://resources.industrydive.com/Energy-and-utilities-operating-in-the-new-normal. Accessed 22 June 2020
27. Linkov I, Carluccio S, Pritchard O, Ní Bhreasail Á, Galaitsi S, Sarkis J, Keisler JM (2020) The case for value chain resilience. Manage Res Rev. https://doi.org/10.1108/MRR-08-2019-0353
28. Marquez AC (2020) Strategies for COVID-19 pandemic recovery. Applying engineering asset management principles. https://idus.us.es/handle/11441/94965
29. Moreno R, Panteli M, Mancarella P, Rudnick H, Lagos T, Navarro A, Ordoñez F, Araneda JC (2020) From reliability to resilience, planning the grid against the extremes. IEEE Power Energy Mag 41–53. https://doi.org/10.1109/mpe.2020.2985439
30. OECD (2020) Tackling coronavirus (COVID–19) contributing to a global effort. http://www.oecd.org/coronavirus/en/, https://read.oecd-ilibrary.org/view/?ref=134_134029-llcg91quqm&title=A-collapse-in-output-followed-by-a-slow-recovery-World-GDP. Accessed 21 June 2020
31. OECD (2020) A systemic resilience approach to dealing with Covid-19 and future shocks. http://www.oecd.org/coronavirus/policy-responses/a-systemic-resilience-approach-to-dealing-with-covid-19-and-future-shocks-36a5bdfb/. Accessed 21 June 2020
32. OECD/IEA (2015) Making the energy sector more resilient to climate change, Paris. https://webstore.iea.org/making-the-energy-sector-more-resilient-to-climate-change. Accessed 15 June 2020

33. OECD (2011) Reviews of risk management policies: future global shocks, improving risk governance. https://www.oecd.org/governance/48329024.pdf. Accessed 22 June 2020
34. Park J, Seager TP, Rao PSC, Convertion M, Linkov I (2013) Integrating risk and resilience approaches to catastrophe management in engineering systems. Risk Anal 33(3):356–367
35. Paté-Cornell ME (2012) On "black swans" and "perfect storms", risk analysis and management when statistics are not enough. Risk Anal 32(11):1823–1833
36. Petchrompo S, Parlikad AJ (2016) A review of asset management literature on multi-asset systems. Reliab Eng Syst Saf 181:181–201
37. Patriarca R, Berström J, Di Gravio G, Costantion F (2018) Resilience engineering: current status of the research and future challenges. Saf Sci 102:79–100
38. Pyne JC, Keating CB, Katina PF, Bradley JM (2018) Systemic intervention methods supporting complex system governance initiatives. Int J Syst Syst Eng 8(3):285–309
39. Renn O, Lucas C, Haas A, Jaeger J (2017) Things are different today: the challenge of global systemic risks. J Risk Res 22:1–16
40. Roberts, C (2019) Developing organization resilience through asset management to respond to COVID-19, IAM. https://theiam.org/news/developing-organization-resilience-through-asset-management-to-respond-to-covid-19/. Accessed 19 June 2020
41. Roege PE, Zachary AC, Mancillas J, McDonagh JA, Linkov I (2014) Metrics for energy resilience. Energy Policy 72:249–256
42. Rzevski G, Skobelev P (2014) Managing complexity. WIT Press, Billerica, p 01821
43. Siegenfeld AF, Taleb NN, Bar-Yam Y (2020) Opinion: what models can and cannot tell us about COVID-19. In: Proceedings of the national academy of sciences of the United States of America (PNAS), first published 24 June 2020. https://doi.org/10.1073/pnas.2011542117
44. Sittamparam R (2020) Cambridge risk expert: Covid-19 is a 'black elephant'. https://www.spearswms.com/cambridge-risk-expert-covid-19-is-a-black-elephant/. Accessed 22 June 2020
45. Syed Z, Lawryshyn Y (2020) Multi-criteria decision-making considering risk and uncertainty in physical asset management. J Loss Prev Process Ind 65(101064):1–16
46. Taleb NN (2010) The black swan: the impact of the highly improbable, 2nd edn. Random House Trade Paperbacks, New York
47. The National Academies Press (2012) Disaster resilience: a national imperative. https://www.nap.edu/catalog/13457/disaster-resilience-a-national-imperative. Accessed 22 June 2020
48. Torabi SA, Rezaei Soufi H, Sahebjamnia N (2014) A new framework for business impact analysis in business continuity management (with a case study). Saf Sci 68:309–323
49. Trindade M, Almeida N, Finger M, Ferreira D (2019) Design and development of a value-based decision making process for asset intensive organizations. In: Mathew J, Lim C, Ma L, Sands D, Cholette M, Borghesani P (eds) Asset intelligence through integration and interoperability and contemporary vibration engineering technologies. Lecture notes in mechanical engineering. Springer, Cham, pp 605–623
50. Trudel G, Gingras JP, Pierre JR (2005) Designing a reliabile power system: Hydro-Québec's integrated approach. Proc IEEE 93(5):907–917
51. Trump BD, Linkov I (2020) Risk and resilience in the time of the COVID-19 crisis. Environ Syst Decis 40(2):171–173. https://doi.org/10.1007/s10669-020-09781-0
52. Wong CWY, Lirn TC, Yang CC, Shang KC (2019) Supply chain and external conditions under which supply chain resilience pays: an organizational information processing theorization. Int J Prod Econ. (in press)
53. Woods DD (2015) Four concepts for resilience and the implications for the future resilience engineering. Reliab Eng Sys Saf 141:5–9
54. World economic forum (2020) The global risks report. https://www.weforum.org/reports/the-global-risks-report-2020. Accessed 22 June 2020

55. World energy council: COVID-19: post-crisis futures short and medium-term scenarios. https://www.worldenergy.org/news-views/entry/ceo-view-planning-in-the-face-of-critical-uncertainties. Accessed 22 June 2020
56. Zio E (2016) Challenges in the vulnerability and risk analysis of critical infrastructures. Reliab Eng Syst Saf 152:137–150

The End of Megaproject Certainty: Post COVID-19 National Infrastructure Management

S. G. Pudney(✉), D. E. Mills, and S. Mudunuri

Southern Cross University, Gold Coast Airport, Terminal Dr, Bilinga, QLD 4225, Australia
{steve.pudney,david.mills}@scu.edu.au

Abstract. Uncertainty and its profound impact on the management of infrastructure project portfolios is introduced. The advent of COVID-19 is depicted as a defining event in the management of infrastructure. This study applies the Black Scholes option valuation model and real options analysis to determine value for money in front-end engineering for Australian megaprojects. The findings indicate that despite the fact all current Australian infrastructure projects examined had a positive net present value and a benefit to cost ratio greater than one, almost sixty percent of the planned expenditure was for projects for which real options analysis indicated the engineering as being poor value for money. The paper concludes with recommendations to manage the national portfolio of infrastructure projects as a pipeline of carefully chosen pre-engineered options, some of which are constructible projects and others being non-asset, demand management solutions.

Keywords: Black Scholes model · Decisions under uncertainty · Megaproject management · Project Portfolio management · Real options analysis · Real options theory

1 Introduction

More than two thousand years ago, Heraclitus postulated that change is the only constant. Think back to your childhood. For many of us, the only ways to communicate over distance were telegram, landline telephone or written letter. How different is communication now? The evidence for change is strong yet its likelihood is often neglected as a factor in major infrastructure decisions. Whilst change in demand for various services is a given, the degree and nature of that change is uncertain. Should the level of uncertainty in demand for services be a significant, or even the dominant factor, in relation to our design of infrastructure solutions? If so, how should decisions regarding major infrastructure to deliver services take uncertainty into account?

Over the last decade, the Queensland government made several attempts at defining a solution to public transport bottlenecks in the Brisbane rail network. Construction of the Cross-River Rail (CRR) project eventually started in September 2019 at an estimated cost of AUD 5.4 billion. Billed as a key to unlocking the public transport gridlock for Brisbane commuters, the project now appears misplaced against the backdrop of COVID-19. Many

A. Crespo Márquez et al. (Eds.): WCEAM 2019, LNME, pp. 271–287, 2021.
https://doi.org/10.1007/978-3-030-64228-0_23

Brisbane commuters, in consultation with their employers, have chosen to work from home. Some employers are seeing measurable improvements in workforce productivity of those working from home and the opportunity to reduce the cost of office rent. The result is a significant decline in the demand for public transport services in Brisbane. Nevertheless, the CRR construction is underway and scheduled for completion in 2024. Would another option be more appropriate at this juncture? Should the CRR project be delayed or even abandoned?

Decisions to invest in large scale infrastructure assets are always made with the expectation of future returns, be they pecuniary, social, environmental or a combination. These decisions are complex and difficult to get right because they typically involve a partially or fully irreversible commitment of substantial capital sums in the face of significant uncertainty. This uncertainty is mainly with respect to future returns on the investment which is tied closely to the demand for services emanating from that infrastructure. Demand can be heavily influenced by the evolution of customer needs, the impact of disruptive technologies and even government policy changes.

Given the relentlessness of uncertainty and the continuing need for services to be prudently delivered through infrastructure assets, the authors see the need for a new paradigm of research into megaproject portfolio management at national level. To this point in time, infrastructure megaproject research has been focused mainly on aspects of singular megaprojects, such as stakeholder management, risk management, sustainability, complexity management or governance [1]. This study represents an exception to that general pattern by analyzing the management of a portfolio of nationally significant infrastructure megaprojects within an environment of uncertainty.

The COVID-19 pandemic has been a watershed event. The last comparable global pandemic was the Spanish influenza over 100 years ago. Infrastructure management, indeed the nature of infrastructure itself has changed so much in the intervening years. The Brisbane CRR project is but one infrastructure project. The advent of COVID-19 has called into question the soundness of a number of proposed projects and the process for prioritizing new infrastructure projects itself. Should the processes for infrastructure management be reassessed in the light of our knowledge of the effects of COVID-19?

This manuscript is organized into sections in order to convey the logical development of a framework for infrastructure megaproject portfolio management under uncertainty and the demonstration of that framework via an indicative study of Australia's infrastructure megaproject portfolio.

Section one introduces the concept of uncertainty as it impacts megaprojects via a recent example and poses some questions about the management of a portfolio of national megaprojects. The second section provides the background theory and literature in relation to this study by outlining the theory with respect to framing and valuing infrastructure project options in the face of uncertainty. Section three establishes the design of this study, essentially an indicative analysis of megaproject real options using the Black Scholes option valuation model. It also details the sources of data used to develop and analyze project options and the assumptions made in carrying out the analysis. The fourth section presents the results and highlights findings with respect to the real options analysis at the project portfolio level. Section five discusses the results and section six recommends areas of future research. Section seven is the conclusion.

2 Background to This Study

Uncertainty in relation to infrastructure projects is to be expected and so there is a commensurate need for creating flexibility of choice and agility with respect to implementing the right solution in the prevailing circumstances. The managerial flexibility to enact the most applicable service delivery solution via infrastructure assets in a changing environment has considerable value [2, 3]. This section explores the theoretical framework for the valuation of different forms of flexibility in the delivery of services via infrastructure assets.

2.1 Asset Management and the Delivery of Services via Infrastructure Assets

The relatively recent discipline of asset management provides an appropriate context for the delivery of services via infrastructure. Societal requirements for various services, such as power, water, transport, telecommunications or waste removal are provided via various forms of infrastructure assets. Hence, the ISO 55000 series of International Standards on Asset Management places the emphasis on the provision of value via infrastructure assets to an organization and its stakeholders. Therefore, the infrastructure assets are necessary but secondary to the provision of value to stakeholders.

2.2 Real Options Theory (ROT) and Its Application to Infrastructure Assets

In 1973 Fischer Black and Myron Scholes developed an elegant mathematical solution for the value of a financial option that became the basis for financial option pricing. A financial option is the right but not the obligation to buy or sell a financial asset at an agreed price and date. Financial option valuation is firmly established as the basis for the design of financial assets and their management plans, that have the flexibility to cope with market uncertainty. Such was the value of this contribution that Scholes and one of his colleagues, Robert Merton, were awarded the Nobel prize in economics in 1997. It is most likely that Black would have been a co-recipient had he not passed away in the intervening period.

Real options, as opposed to financial options, were first described by Professor Stewart Myers in 1976 who defined them as "opportunities to purchase real assets on possibly favorable terms" [4]. Since then, the development of Real Options Theory (ROT), fostered by numerous theoreticians and practitioners, has grown to the point where it is used to assist with decisions related to design options, configurations, construction timing, alternate uses and even retirement or abandonment of infrastructure assets.

Real options, can be used as the basis for decisions related to the design of real infrastructure assets or their management and are directly analogous to financial options. The analogous relationship, as interpreted by several well-cited authors, Luerhman [5], Smit and Trigeorgis [6], Leslie and Michaels [7], Martins et al. [8], is shown in Table 1.

The evaluation of real asset options using ROT requires an estimate of the six parameters in Table 1. Parameter T, **the real option time to expiry**, is estimated from the time period for which contracts for the delivery of the asset remain valid. Technology life cycle [6, 7], contracts [6, 7] and competitive advantage [6] are mentioned as influencing factors for T. Clearly, though, other forms of business risk should be considered. For

Table 1. Analogy between real and financial options.

Symbol	Financial option variable	Real option variable Leslie and Michaels [7]	Real option variable Luerhman [5]	Real option variable Smit and Trigeorgis [6]	Real option variable Lint and Pennings cited in Martins et al. [8]
T	Time to expiry	Time to expiry	Length of time the decision may be deferred	Length of deferral time	Time until the investment opportunity disappears
X	Exercise price	Present value of fixed costs	Expenditure required to acquire the project assets	Present value of investment outlays	Costs of irreversible follow-on investment
σ	Uncertainty of stock price movements	Uncertainty of expected cash flows	Riskiness of the project assets	Volatility of project's returns	Variability of growth in project value
S	Stock price	Present value of expected cash flows	Present value of a projects operating assets to be acquired	Present value of expected cash flows	Present value of expected cash flows
r	Risk-free interest rate	Risk-free interest rate	Time value of money	Time value of money	Risk-free rate of return
δ	Dividends	Value lost over duration of option	–	–	Value lost by waiting to invest

example, the fallout from COVID-19 has severely impacted the commercial viability of many Public Private Partnership infrastructure projects [8] resulting in an effective and premature expiry of option for those projects.

Parameter X is the **cost estimate for the real underlying asset** in question. Parameter r, or **risk-free yield** is typically estimated using the yield of government bonds because they are considered as almost risk free. There appears to be widespread agreement on the analogical representation of parameters X and r.

Parameter S is the **expected present value of cash flows** for the real underlying asset in question. The distinction between our definition and "present value of expected cash flows" [6–8] is that expected present value takes correlation between contributing cash flow streams into account. Further, we specifically advocate building in business risks to cash flow estimates such as the risk of pandemic and note a lack of attention to those issues by others.

Parameter σ or the **volatility of expected cash flows** is, in many cases, estimated from the volatility of proxies such as relevant commodity or service prices and the demand for those commodities or services. Again, σ is severely impacted by existential business risks such as pandemic risk as demonstrated in the movement of share market volatility associated with news of the COVID-19 pandemic. The estimation approach for σ is highly dependent on the project in question. We will discuss σ in detail as we estimate relevant parameters for the examples in this study in Sect. 3.2.

Parameter δ or the **forgone rate of return** is equivalent to a dividend yield being paid on a share for which the option is yet to be converted. For real options, it must be estimated for each project on a case by case basis. We will examine this in detail for the projects under study in Sect. 3.2. Overall, we note agreement on this factor except that two [5, 6] have omitted δ as part of their analogy. In our opinion, ROT should consider δ because a dividend-like return is present in many projects.

A complete discourse on ROT is unnecessary in this forum, however a review of its key principles is worthwhile. ROT is a branch of decision theory applied to investments in real assets under conditions of uncertainty. Techniques derived from ROT such as the family of real option valuation (ROV) techniques take into consideration the value of flexibility in the face of uncertainty, unlike net present valuation (NPV) and cost benefit analysis (CBA) methods calculated using discounted cash flow (DCF). Another downfall of NPV and CBA is that they are simply expected values. They fail to consider the probabilistic distribution of DCF into the future whereas ROV does. ROV is therefore considered superior to NPV and CBA.

ROVs can be classified as European, for which the option is exercised at the expiry date, or American, for which the option can be exercised at any time up to and including the expiry date. The Black Scholes option valuation model (BSOVM) for European options [10], call option value C shown in Eq. (1) and put option value P in Eq. (2), can be used for both real and financial options. It has the advantages of being easy to calculate and precise but isn't applicable in all instances (e.g. where value is lost over the duration of the option). $N(d)$ is the cumulative normal distribution function.

$$C = SN(d_1) - Xe^{-rT}N(d_2) \tag{1}$$

$$P = Xe^{-rT}N(-d_2) - SN(-d_1) \tag{2}$$

where

$$d_1 = \frac{\ln\left(\frac{S}{X}\right) + \left(r + \frac{\sigma^2}{2}\right)T}{\sigma\sqrt{T}}$$

and

$$d_2 = d_1 - \sigma\sqrt{T}$$

The assumptions associated with applying the Black Scholes option valuation model BSOVM are clearly articulated in their seminal work [10] and many proponents of ROV have analyzed these assumptions in relation to the application of financial options valuation to ROV. A full discussion of the assumptions is not relevant here, however, some aspects of these assumptions are worthy of highlighting at this point.

The first of these is the assumption that the risk-free interest rate remains constant over the life of the option. We know this to be unrealistic from the behavior of bond markets. In the context of option valuation, the risk-free interest rate is usually approximated by the yield values of Government bonds. For this study, being about Australian projects for which the option life is expected to be no more than five years, it is appropriate to choose the Australian Government three-year bond yield as an approximation of the risk-free interest rate.

Another assumption is that the present value of expected cash flows for the underlying asset, over time, is in the form:

$$S_t = S_0 \cdot exp\left[\left(r - \frac{\sigma^2}{2}\right)t + Z_t\sigma\right] \quad (3)$$

Where r is a drift constant, which is actually assumed to be the same as the risk-free interest rate, σ is a volatility constant and the function Z_t is a standard geometric Brownian motion function. Whilst many real options researchers have implicitly agreed with this assumption through their steadfast application of option valuation models, several have highlighted the inconsistencies with the real world [11]. In summary the main issues are that:

- the value of S_t can never be negative yet in reality some projects have liabilities that cause the present value of expected cash flows to be negative.
- volatility σ is assumed to be constant and we know from experience this is almost never true.
- drift r is assumed to be constant and this is not consistent with the real world.

Davis [11] has addressed some of these issues via an approach that uses simple production capped business models under conditions of operating flexibility and no operating flexibility.

2.3 European and American Options

Recall the only difference between European and American options is that American options have the ability to be exercised at any time prior to maturity. Now, Merton [12] provided proof that the value of a European call option is higher than the value it would otherwise assume if it could be exercised immediately. Hence, rational investors never exercise American call options prior to maturity. An exception to this is possible under circumstances where a dividend is paid, remembering that dividends are payable only after an option has been exercised. For example, a toll road has a dividend yield, in this case the toll revenue, that is only realized when the road is operational, that is to say, only after the toll road design option has been "exercised". It is, of course, essential that the BSOVM is only applied to cases where no dividend is paid, however BSOVM is efficient and accurate and is therefore preferred where applicable.

Clearly, models used for American options reflect a higher level of flexibility than European options and should be employed, where possible, when a dividend or income stream is a feature. Bjerksund and Stensland [13] have developed a convenient closed form approximation of the value of call and put American options.

2.4 Options to Defer, Stage, Scale, Grow, Abandon, Shut Down and Restart

ROT extends well beyond the calculation of a number to represent the expected value of options. It provides the theory to support the identification of options to create, expand, switch use or abandon an infrastructure asset and guides ongoing data collection that supports value creating decisions for the asset stakeholders. Value creation is the end result of a process that takes advantage of opportunities whilst simultaneously limiting downside risks created by uncertainty. Trigeorgis and Reuer [2] have categorized four stages to this process, namely:

(i) identification of the option, during which hidden or "shadow" options are discovered;
(ii) the acquisition or creation phase, during which the information, third party agreements and organization of necessary resources are completed.
(iii) management of the option during which the option is maintained, kept relevant during periods of change and even further enhanced if opportunities are presented, and
(iv) exercising of the option during which plans for enactment of the option are implemented and benefits are realized.

Trigeorgis and Reuer [2] have reported a deficit of research in the real options identification stage. Moreover, it appears that the application of ROV to infrastructure asset management is still in its early stages [8, 14–16].

Lander and Pinches [17] identified six types of real options: defer, stage, scale, switch, grow and abandon. Trigeorgis [18], identified a seventh option to shut down and restart. All have the potential to be applied to major infrastructure assets, although some apply to the operational phase whereas others apply to the design and construction phase. A review of the applicability to infrastructure projects as well as the components of value and cost for each option is given here.

The option to defer a project is postponement of commencement until more favorable conditions prevail. It can apply to commencement of design, construction or operation. The value of deferral is in the flexibility to choose a start date that aligns with advantageous conditions. An infrastructure project start date may be deferred if there is a significant reduction in expected cash flow such that the completed project is predicted to be cash flow negative. The potential cost of deferral is escalation of labor and materials. Another potential cost is obsolescence of the project design or part of the design whereby the investment already made is either entirely or partly lost.

The option to stage a project implies a plan for possible deferral at the completion of logical parts or stages of a project. The value of staging is the flexibility to restart each subsequent stage at a time that suits. Alternatively, staging can be viewed as a series of options each on the next stage of the project. This is known as a compound option. Project staging is often planned in such a way that there is no need for additional demobilization and remobilization, hence the cost of staging is often limited to the escalation of subsequent stages if there has been a delay between stages large enough to result in escalation.

The option to scale up or down is associated with flexibility of design such that the capacity of the infrastructure can be increased or decreased at any future time. The value of an option to scale down is the flexibility to deliver infrastructure to match a lesser demand and to potentially save construction and operating costs commensurate with that smaller scale. Alternatively, the value of an option to scale up is the flexibility to deliver infrastructure to match a greater demand if needed.

The option to switch is associated with an ability to change between two or more different modes of operating, two or more designs, two or more asset configurations. An example of two different operating modes is weekday and weekend public transport schedules. In this example, we can create an option to switch between two different schedules. Another form of this option is a switch between inputs. This type of switching option is often associated with production lines that have a capability to manufacture different but similar products using different raw materials, however, it could be applied to a power station with multi fuel capability. The value of the switching option is found in the flexibility of choice in operating mode enabling reduction in operating costs or risks or increased revenue from an additional product line or revenue stream. The cost of this option is the extra capital cost associated with the switchable design and the losses associated with switching.

The option to grow is associated with the early identification of an additional type of revenue. Hence the growth is in new revenue streams rather than the scale of a given revenue stream. The value of an option to grow is in the possible improvement of net present value of the entire project. The cost typically begins with research including experiments and is followed by design, construction, commissioning and operating costs associated with the output of that research.

Another type is the option to abandon. This option is not taken lightly but is sometimes exercised when it is clear that the project will not be successful and that long term and unabated financial loss is apparent. The value of abandonment is the recovery of unspent project funds and possibly the salvaging of resources dedicated to the project. The cost of abandonment is the cost of demobilization of resources deployed to the project site.

Finally, the option to stop and restart may be of interest in situations where a period of substantial decline in the revenue associated with a project manifests unexpectedly. The value of such an option is in the ability to mitigate or even completely prevent significant financial loss for stakeholders. The cost of stopping and restarting is the cost of demobilization and the subsequent remobilization of resources when the project starts again.

3 Design of This Study

This study applies an indicative ROA to project stage management for national infrastructure assets. The infrastructure assets analyzed under this study are Australian based megaprojects that are in the option definition, design or construction stage.

Whilst considering the design of this study, there is an important distinction made between:

- analysis for the purpose of infrastructure asset investment, that essentially treats the infrastructure asset project as a black box by evaluating opportunities to invest in options relating to that infrastructure asset as a whole; and
- analysis for the purpose of project and portfolio managers seeking to evaluate managerial flexibility created via options **within** the infrastructure asset projects.

It is essentially the difference between real options "on" projects as opposed to those "in" projects [19]. This study is exclusively centered on the latter. Hence the study converges on the intersection between ROA, Asset Management and Project Management.

In addition to this, the perspective offered by this study of real options "in" projects is intended for those working at a national level on the portfolio of national infrastructure projects. Hence the study is designed to extract data from Australia's megaprojects and to perform ROA as strategic guidance for those responsible for national megaproject portfolio management.

3.1 Framing and Estimating the Cost of Megaproject Options

Framing a real option involves determining ways in which managerial flexibility can be introduced. One type of option that applies to all engineered projects is an option for flexibility of timing. Flexibility in project timing is gained by carrying out the design and locking in contractual arrangement with those suppliers and contractors that deliver a functioning asset.

An indicative cost of preparing the megaproject design plus the preparation and execution of the necessary contracts for megaproject procurement, construction and commissioning is estimated using the following formula. This is designated "the acquisition cost for the right to construct the project" (C_a) and it is derived from the estimated P50 project capital cost (C_{P50}), a project specific engineering complexity factor (μ) and a project specific contracts complexity factor (ρ) as follows:

$$C_a = (\mu + \rho)C_{P50} \tag{4}$$

The engineering complexity factor (μ) varies considerably from project to project and is largely a function of the design effort required in each project which in turn depends on the number of individual design elements, the number of standard design elements within the overall project and the level of design interdependency between individual design elements. Likewise, the contract complexity factor depends on the existence of standing agreements with suppliers, the existence of industry accepted standardized contract forms and the level of interdependency between individual contracts.

The nature of the project is also a factor, for example, the engineering complexity and contract complexity factors in relation to the design and procurement of nuclear power generation plants are quite different from those for highway projects. For the purpose of this study μ is estimated at 0.15 and ρ is estimated at 0.05.

3.2 Estimating the Value of Real Options

The first step is to correctly frame the option of interest. The option of interest is the option to carry out engineering design and execute procurement contracts required to deliver the project. We will subsequently call this the front-end engineering option (FEE option). The FEE option value is calculated using the BSOVM closed form Eq. (1). This concept was previously explored and utilized in 2010 in the context of the renewal of water utility assets [20].

The approach to estimating each of the six parameters used to value the real option defined above is examined individually here.

Parameter T: Time to real option expiry is the time period for which contracts for the delivery of the project remain valid. It is within the control of the project management team up to a point. Suppliers and contractors may be inclined to resist longer term contracts without an escalation clause and other types of risk protection. The project design may become obsolete or partially obsolete after several years. We have assumed a value of 5 years for all infrastructure projects in this study.

Parameter X: Cost estimate for the real underlying asset is derived directly from the P50 project cost estimates in the business plans of each respective project. It is subject to uncertainty to the extent not covered by contracts for the delivery of the project.

Parameter r: Risk-free yield. In the model, the current yield of three-year Australian government bonds is used as a proxy. The return for three-year Australian government bonds as at the 8th of September 2020 is 0.273%.

Parameter S: Expected present value of cash flows is derived from the business case, being the expected net present value for each project. It is subject to a significant level of uncertainty.

Parameter σ: Volatility of expected cash flows. This is a particularly difficult parameter to estimate although there are some proxies that can be used. The current share market volatility index provides some guidance because infrastructure companies such as Sydney Airport, a proxy for the proposed Western Sydney airport, and Transurban, a proxy for some toll roads are traded on the stock exchange. The current volatility index as of the 8th of September 2020 is 30.75% so a value of 30% is used.

Parameter δ: Forgone rate of return. This, too, is difficult to estimate. The forgone rate of return is equivalent to a dividend yield being paid on a share for which the option is yet to be converted. Shareholders are paid a dividend; option holders are not. Hence the value lost by an option holder is the dividend not received. It is almost impossible to determine the expected dividend on an asset that has yet to be built. However, we can say with surety that many public transport assets, a substantial proportion of the megaprojects of this study, only continue to operate under government subsidy and hence a dividend is not appropriate. Many of the highway and motorway projects in this study have no toll and will therefore have no associated dividend. Toll road projects expecting to charge a toll will likely have significantly reduced traffic volumes because of pandemic conditions. The planned Western Sydney airport is also likely to have significantly reduced usage and is not likely to be in a situation to pay a dividend for many years. In the understanding that this is study is indicative rather than definitive, an estimate of zero lost value would seem fitting.

The BSOPVM shown earlier in Eq. (1) is used to calculate the option values associated with engineering design and contract execution. As previously stated, the BSOVM is a European option valuation model and applies to projects for which no dividend is payable.

3.3 Analysis

As the prime step of the analysis, the FEE option value is compared to the FEE option cost for all of the megaprojects. Value for money is evident when the FEE option value exceeds the FEE option cost. Hence a decision to carry out the engineering for such a project is in the interests of the project customers because they would be receiving value for money. The option ratio is used as a numerical indicator of projects that are "in the money". The option ratio formula (R_o) is shown in Eq. (5).

$$R_o = C/C_a \tag{5}$$

The final part of the analysis is a comparison of option ratios across the entire portfolio of Australian infrastructure megaprojects to determine the portfolio of projects for which front end engineering is recommended.

4 Results

This section presents the results obtained from the option ratio analysis using data for current Australian infrastructure megaprojects extracted from Infrastructure Australia's Infrastructure Priority List "High Priority Projects" and "Priority Projects". The results also include an analysis of the Brisbane CRR project (never included on the infrastructure priority list but promoted by the Queensland Government) but exclude the Perth Metronet Morley to Ellenbrook project for which the capital cost estimate was unavailable.

The analysis showed that projects 1 to 13, on Table 2, had an option ratio significantly greater than unity. These are the projects on Australia's infrastructure list for which the FEE represents value for money, in that the cost of the FEE is significantly less than the theoretical value represented by the BSOVM. The P50 capital cost estimate for these projects is AUD 19.6 billion. Hence the analysis indicated that the FEE for just over thirty percent of estimated capital expenditure on Australia's infrastructure projects represented value for money.

Similarly, projects 16 to 28 inclusive had an option ratio significantly less than unity. The analysis indicated that the projects on Australia's infrastructure list representing poor value for money, in that the cost of the FEE is significantly higher than the theoretical value represented by the BSOVM, was an estimated AUD 38.8 billion worth of projects. Hence the FEE for just under sixty percent of the planned expenditure on Australia's infrastructure projects represented poor value for money by this analysis.

The remainder, that is projects 14 and 15, had an option ratio of close to unity. Given the uncertainties involved, the analysis showed that these projects may or may not represent value for money. This is approximately ten percent of planned infrastructure expenditure.

Overall, the analysis has shown that despite the fact that all projects on Table 2 had a positive net present value and a benefit to cost ratio greater than one, almost sixty percent of the planned expenditure was for projects that had option ratios significantly less than unity and therefore indicated the FEE represented poor value for money.

Note, the analysis undertaken here was indicative in nature, mainly because of difficulties in reliably estimating the unpredictability of expected cash flows (σ) but also

Table 2. Australian Infrastructure Projects Key Data [21] and ROA Results

ID	Project	Capital cost (P50) $m	NPV $m	Benefit cost ratio	Option ratio
1	M4 Motorway upgrade Paramatta to Lapstone: New South Wales (NSW)	792[1]	2640	5.30	16.20
2	Bruce Highway: Maroochydore Interchange	286[1]	529	3.21	7.32
3	University of Tasmania: North Transformation	279[1]	483	2.41	6.59
4	Bindoon Bypass: Western Australia	261[1]	462	3.40	6.26
5	Bruce Highway: Deception Bay Interchange Queensland	145[1]	234	3.03	5.89
6	Brisbane Metro: Inner City Public Transport	877[1]	1235	2.40	4.62
7	Inland Rail: Melbourne to Brisbane	9,890	13928	2.62	4.62
8	Nowra Bridge: Princes Highway NSW	221	268	2.20	3.50
9	Port Botany Rail Line: Sydney	379	430	2.68	3.06
10	Western Sydney Airport: Badgerys Creek	5,000	5441	1.90	2.82
11	Myalup-Wellington water project: Western Australia	394	388	1.60	2.28
12	University of Tasmania: Technology precinct	400	364	1.95	1.92
13	M80 Ring Road Upgrade: Melbourne	687	553	2.00	1.45
14	Bruce Highway: Caboolture-Bribie Island Road to Steve Irwin Way Queensland (Qld)	584[1]	422	1.91	1.11
15	Eyre Infrastructure project: South Australia	6,000	3800	1.30	0.79
16	Perth Metronet: High Capacity Signaling	1,157	688	2.60	0.66
17	M12 Motorway: Western Sydney	1,888	1171	1.80	0.57
18	Beerburrum to Nambour Rail Upgrade Qld	541	262	1.50	0.36
19	Peak Downs Highway Realignment Qld	145	67	1.50	0.32
20	Sydney Rail: More Trains, More Services: Stage 2	2,023	890	1.38	0.27
21	M1 Pacific Motorway, Varsity Lakes to Tugun Qld	966[1]	419	1.67	0.26
22	Brisbane Cross River Rail	5,400[1]	1877	1.41	0.12
23	Sydney Metro: City and southwest	8,680	2775	1.30	0.09
24	M1 Motorway: Eight Mile Plains to Daisy Hill Qld	710[1]	213	1.40	0.07
25	Bruce Highway: Cooroy to Curra Qld	1,005	274	1.36	0.05
26	North East Link: Melbourne	14,664[1]	2187	1.30	0.003
27	Gold Coast Light Rail: Stage 3A Qld	658[1]	56	1.10	0.0001
28	Bruce Highway: Cairns southern access Qld	1,005	49	1.14	0.000003
	Total	65,037			

Notes: (1) P50 Capital cost estimated from the P90 capital cost or other sources

because the value lost over the duration of the option (δ) has been assumed to be zero for all cases. However, it is worth noting that many ROA researchers have reported the same difficulty [11]. A more rigorous study would include confidence levels for the option ratios calculated.

5 Discussion

The option ratio analysis methodology employed here reveals a new way to analyze megaprojects and megaproject portfolios. It shows whether or not the FEE is worth doing from a rational "value for money" perspective and in full cognizance of uncertainty. For those megaprojects with option ratios much less than one, projects which could be labelled "out of the money" in options trading parlance, there is no rational argument for carrying out the FEE. If the FEE options for the out of the money projects were offered on the options exchange, it is likely that they would not trade. Interestingly, all of the projects in Table 2 have positive Net Present Value and a Benefit to Cost Ratio greater than 1. This would indicate to a rational investor that they are worth doing. However, since these measures don't take uncertainty into account, there is a reasonable argument to be skeptical about the inclusion of any of the projects on the list.

Should the option ratio be used to screen new project options? If so, should the projects that pass the option ratio test be scheduled to carry out FEE? This idea holds significant merit when viewed from the perspective of impending recessions. Economic recessions happen on a regular basis in most countries and the catch cry is usually for the government of the day to immediately implement "shovel ready" projects in order to generate jobs. Wouldn't it be good to have a portfolio of worthy projects that are ready to go?

So, Australia's portfolio of infrastructure projects appears to be lacking in value for money. This is not to say, with the advent of further uncertainty, that those projects will remain so. However, there appears to be a real need for a larger and more diverse portfolio of valuable projects, possibly smaller scale and distributed, that offer a literal smorgasbord of possibilities.

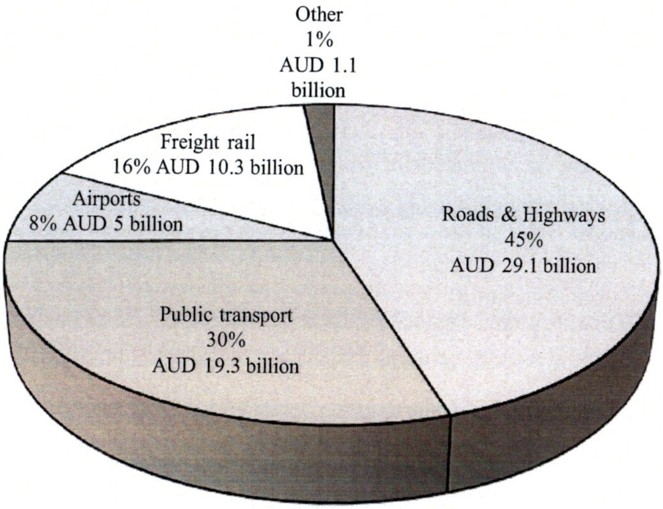

Fig. 1. Australias Infrastructure Priority List 2020: project categories [15]

Examining the current portfolio leads us to note another aspect, that the current infrastructure project list appears to be heavily weighted towards transport sector projects (see Fig. 1). One argument that could be posited is that such a one-sided portfolio is not ideal as an insurance against uncertainty especially if it were to materialize in the form of a disruptive technology. For example, disruptive drone transport systems are possibly not far over the horizon. How inappropriate would that render our proposed infrastructure projects? For that matter, how many of the projects now appear premature or inappropriate now that the pandemic has arrived? We have already discussed the Brisbane CRR project, however, there is a total of AUD 19.3 billion in public transport projects spread across Australia that is similarly placed. The AUD 5 billion Western Sydney Airport possibly should be postponed due to the longer-term outlook with respect to air travel. Many of the roads and highways were intended to service city or city fringe traffic needs during peak hour. The M12 Motorway project in Western Sydney was intended to service yet to be developed suburbs that will be left undeveloped for many years due to the now expected population decline.

Meanwhile, a broad and diverse array of infrastructure project types such as health and aged care facilities, justice and emergency services such as fire services infrastructure, hospitals, affordable housing, telecommunications, energy, water, wastewater, sea port infrastructure and coastal protection infrastructure should be included as options available for development and early deployment in the face of uncertain need. Project options should not overlook carefully designed demand management projects such as those needed for rationing of power, water, food, transport, hospital infrastructure and other forms of health facilities if necessary.

Of course, evaluation of infrastructure megaproject options can and should extend well beyond a financial analysis. Such projects have numerous stakeholders and multiple objectives. Often the objectives of one stakeholder oppose that of another. This is the complex realm of Multi Criteria Decision Analysis (MCDA) which is highly applicable to megaproject decisions, including those relating to megaproject options. Typically, the impact of a range of project options is considered in relation to socio-economic, environmental and financial objectives [20, 23] framed from the perspective of key stakeholders and their values.

In the case of the Brisbane CRR project the key stakeholders are Queensland Rail which is the owner/operator of the proposed CRR service, the passengers using the service, individual residents near to the railway service, the local community, the state and the nation. Some of the objectives important to stakeholders include establishing high levels of health and safety, jobs for community members, minimization of greenhouse gas emissions, the minimizing the cost of train fares (in the case of passengers) and maximizing revenue (in the case of the service operator).

Some of the options with the Brisbane CRR project are to:

(1) stage the project by simply doing the engineering and then completing the project as a second stage remembering that before the commencement of each stage there is the option to delay.
(2) Abandon the CRR project altogether.
(3) Switch between different operation modes such as a peak hour demand management model that provides fare price incentives for off peak services.

It is possible to construct an overall utility model through the results of a value trade-off survey involving members of each key stakeholder group. Hence, the option that maximizes utility is preferred. An approach that combines real option problem structuring and MCDA has been successfully attempted before for smaller scale asset renewal in the water utility industry [20, 23] but it appears that none have applied this approach to megaproject option decisions [24].

Now for some answers to questions posed in the beginning of this paper. Should the level of uncertainty in demand for services be a significant, or even the dominant factor, in relation to our design of solutions in response to demand for those services? It seems that uncertainty in the demand for services should be considered and this is supported by the results shown in Table 2.

It is interesting to note that the Brisbane CRR project is one of the projects for which the option ratio is very much less than 1. The questions in relation to this project were: Would another option be more appropriate at this juncture? Should the CRR project be delayed or even abandoned? It is pretty clear, given the option ratio is 0.12, that another option could better serve the travelling public. Perhaps a demand management option is more appropriate. And, yes, the Brisbane CRR project, which is currently underway, possibly should be delayed immediately or even abandoned. A hybrid MCDA ROA study in relation to these options could potentially save significant public money.

6 Recommendations

Several recommendations are offered for future research consideration:

- Apply a stochastic approach to the FEE option ratio calculation.
- Determine the confidence level for the option ratio figures. Indeed, develop a methodology for so doing.
- Collect data on the success, or otherwise, of projects to determine if the option ratio is a predictor of success. Indeed, carry out a longitudinal study based around the predictors of success for megaprojects.
- Carry out some further studies into megaprojects around options for abandonment, delay, switching and other pertinent options. Indeed, establish an international collaborative group for the expansion of a knowledge base in this area.
- Apply hybrid ROA MCDA decision processes to managing infrastructure megaprojects and their associated options.

7 Conclusion

There is serious inadequacy in Australia's preparedness with respect to its pipeline of proposed infrastructure projects. A significant proportion, sixty percent, of the proposed projects are not robust in the face of uncertainty when we use the Black Scholes option valuation model to test the value of engineering for that infrastructure against its estimated cost. Further to this, the proposed infrastructure project list also appears overly biased towards transport projects. There is not enough flexibility to cope with change.

We believe the COVID-19 pandemic has, through its unique pervasiveness, revealed weaknesses in the way Australia manages its portfolio of infrastructure projects. Traditional infrastructure projects in which a project passes through a long planning period before being delivered in exact accordance with that plan may no longer apply, given our uncertain and complex world. In fact, there appears to be a gap in the body of knowledge with respect to the management of a portfolio of infrastructure projects at a national level under conditions of uncertainty.

We advise a pipeline of carefully chosen pre-engineered options, some of which are constructible projects and others being non-asset, demand management solutions as the preferred portfolio of infrastructure projects. A flexible and diverse portfolio of "shovel ready" nationally distributed, project plans and solutions are necessitated as is a readiness to defer, stage, scale, grow, abandon, shut down and restart infrastructure projects as required.

References

1. Wang T, Chan A, He Q, Xu J (2020) Identifying the gaps in construction megaproject management research: a bibliographic analysis. Int J Constr Manage 1–12
2. Trigeorgis L, Reuer J (2017) Real options theory in strategic management. Strateg Manag J 38(1):42–63
3. Machiels T, Compernolle T, Coppens T (2020) Real option applications in megaproject planning: trends, relevance and research gaps: a literature review. Eur Plan Stud 1–22
4. Myers S (1977) Determinants of corporate borrowing. J Financ Econ 5(2):147–175
5. Luerhman T (1998) Investment opportunities as real options: getting started on the numbers. Harvard Business Review, Brighton
6. Smit H, Trigeorgis L (2004) Strategic investment: real options and games. Princeton University Press, Princeton
7. Leslie K, Michaels M (2000) The real power of real options. McKinsey Quar 3:1–23
8. Martins J, Cunha Marques R, Oliveira Cruz C (2013) Real options in infrastructure: revisiting the literature. J Infrastruct Syst 21(1):04014026
9. Casady CB, Baxter D (2020) Pandemics, public-private partnerships (PPPs), and force majeure | COVID-19 expectations and implications. Constr Manage Econ. https://doi.org/10.1080/01446193.2020.181751
10. Black F, Scholes M (1973) The pricing of options and corporate liabilities. J Polit Econ 81(3):637–654
11. Davis G (1998) Estimating volatility and dividend yield when valuing real options to invest or abandon. Quart Rev Econ Financ 38(special issue):725–754
12. Merton R (1973) Theory of rational option pricing. Bell J Econ Manage Sci 4(4):141–183
13. Bjerksund P, Stensland G (1993) Closed form approximation of American options. Scand J Manage 9:87–99
14. Maier S, Polak J, Gann D (2015) Appraising a portfolio of interdependent physical and digital urban infrastructure investments: a real options approach. In: 19th Annual International Real Options Conference
15. Martins J, Cunha Marques R, Oliveira Cruz C (2014) Maximizing the value for money of PPP arrangements through flexibility: an application to airports. J Air Transp Manage 39:72–80
16. Garvin M, Ford D (2012) Real options in infrastructure projects: theory, practice and prospects. Eng Proj Organ J 2(1–2):97–108
17. Lander D, Pinches G (1998) Challenges to the practical implementation of modeling and valuing real options. Quart Rev Econ Financ 38(Special Issue):537–567
18. Trigeorgis L (1996) Real options. Managerial flexibility and strategy in resource allocation. MIT Press, Cambridge
19. Wang T, de Neufville R (2004) Real options "in" projects. In: Trigeorgis L, Pereira P, Sousa M (eds) 9th real options annual international conference, pp 1–19. Real Options Group, Cyprus

20. Pudney S (2010) Asset renewal decision modelling with application to the water utility industry. PhD thesis, Queensland University of Technology, Queensland, Australia
21. Infrastructure Australia (2020) Australian infrastructure priority list, August 2020. https://www.infrastructureaustralia.gov.au/infrastructure-priority-list
22. Keeney R, McDaniels T (1999) Identifying and structuring values to guide integrated resource planning at BC gas. OR Practice 47(5):651–662
23. Lawrence J, Bell R, Stroombergen A (2019) A hybrid process to address uncertainty and changing climate risk in coastal areas using dynamic adaptive pathways planning, multi-criteria decision analysis & real options analysis: a New Zealand application. Sustainability 11(406):1–18
24. Marttunen M, Lienert J, Belton V (2017) Structuring problems for multi-criteria decision analysis: a literature review of method combinations. Eur J Oper Res 263:1–17

Strategies for COVID-19 Pandemic Recovery: Application of Engineering Asset Management Principles

Adolfo Crespo Márquez(✉)

SIM Research Group, Department of Industrial Management, School of Engineering, University of Seville, Seville 41092, Spain
adolfo@us.es

Abstract. In this Chapter, several efforts that the SIM Research Group, from the University of Seville, accomplished during the Covid-19 pandemic lockdown in March-June 2020, are introduced. These efforts had an important impact on media and were mainly concern with the pandemic recovery phase.

When confinement ends, recovery phase must be accurate planned at a local level. Health System (HS) capacity, specially ICUs and plants capacity and availability, would remain the key stone in this pandemic life cycle phase.

This Chapter describes: First the important of the action plans design by local level, while a unique decision-making center is considered by country; Second, a management tool based on a ICUs and plants capacity model to estimate ICUs and plants saturation, and with these results, set new local and temporal confinement measures. The tool allows a dynamic analysis to estimate maximum Ro saturation scenarios; Third, a practical management tools to tackle the deconfinement strategy design problem. A proper control system to follow the course of action, especially in a scenario with unprecedent uncertainty, is developed.

In all cases, it is remarked the importance of R (the pandemic basic and effective reproductive number), first as a variable to monitor and control the pandemic, to ensure its decline; second as a target to score risks associated to start certain activities over, after confinement. Reducing the potential increase in the value of R, when any type of activity is re-opening, guides the strategy.

One common objective in these initiatives: Applying asset management principles to accelerate as much as possible socioeconomic normalization with a strict control over HS relapses risk.

Keywords: Asset management principles · Indenture level · Health system capacity planning · Covid-19 recovery strategies · Basic reproductive number

1 Introduction

1.1 The Proper Pandemic Management Indenture Level

The relative position that an item occupies in a system it's called its level of indenture [1, 2]. Engineering systems usually have many such levels, and achieving the required

A. Crespo Márquez et al. (Eds.): WCEAM 2019, LNME, pp. 288–305, 2021.
https://doi.org/10.1007/978-3-030-64228-0_24

system's performance at a certain expense requires the identification of the most suitable indenture level for management [3]. In complex systems: reaching desired system performance is more expensive when the data, and management of their failures, are handled at higher indenture levels; Cost of strategies to ensure dependability of complex systems increases linearly when the cost of restoring the low-level item increases and the higher the indenture level the higher the increase rate; Cost of the strategy grows inversely proportional when dealing with high frequency of these systems' failure, triggering the cost when managing at the higher levels. In this Section of this Chapter, it is proposed:

- A local pandemic management indenture level: In Spain, current decision- making (until may 5th) applies to the entire country, simultaneously. We analyzed how applying adapted measures by lower geographical management units (management Indenture Levels) allow optimizing local and global behavior and socio-economic impact faster. Considering the political distribution of Spanish territory, this level could be assimilated to provinces (provincias). It is very important to note that sanitary resources allocation in Spain has, historically, a fundamental distribution by provinces. Nevertheless, decision making should be keeping centralized in a unique decision center.
- Quarantine time local determination: In accordance with the lower indenture level identification, it is possible to assume and predict different quarantine time levels by provinces. For instance, the prediction model used established a recommended quarantine of only 40 days for Huelva, while considering the global a unique decision for Spain (same measure simultaneously) would impose 70 days of quarantine using the same law, which means 40% more confinement than what Huelva really needs. This conclusion can be extended to other provinces where the "hard control measures" were taken before local pandemic free expansion point
- Early relapses detection: During the recovery phase, once quarantine ends, it is critical to detect potential relapses as soon as possible. The development of a unique method to monitor most important descriptors (not only test but also new infections rates, mobile geolocation data, etc.) detecting local relapses is a main concern to benchmark evolution of different provinces, which is crucial for getting a faster learning curve for decision making.

1.2 Modelling Health System Capacity

Citizen behavior after opening the confinement is key in order new infections do not grow exponentially again, leading back the country to a new quarantine with numerous deaths and a massive economic loss.

Saturation of hospitals plants and ICUs have been a fact during phase 1 (spread or expansion). This will also be one of the most important risk in the recovery phase. To deal with new massive demands of HS capacity, governments should find quick solutions (ASAP) and supported by contingencies plans and studies in order to guide as best as possible these quick investments [4]. IN this research The Health System (HS) capacity is modeled in order to control the optimal HS response. In this way, it is possible to use and take advantage of the great research community effort in infection prediction models for:

- Optimal decision making of sanitary resources allocation;
- Early detection of the force of infection the system can face under several circumstances; and
- Support citizen and organizations self-management.

The study is focused at local level (Indenture level definition). Relapses will emerge at local level also new HS saturation events will happen at local level. Inhabitants of a population will go to local hospitals, which will cause saturation in those hospitals in the provinces where the number of new infections is high. Through this work, it will also be possible to determine those provinces that, being the health system less saturated, will be able to receive people infected from other provinces that are in worse conditions, thus achieving an overall improvement in the health system.

1.3 A Control System to Follow the Pandemic Behavior

Reaching the end of the confinement phase, it is essential to have a strategy to face the recovery phase with the best possible guarantees. Previous studies demonstrated that saturation in hospitals could be modelled based on parameter R [5]. This is considered as the key parameter for designing recovery or de-escalation phase strategies. To propose an effective strategy, the parameter R will have to be known, monitored and controlled [6]. But uncertainty and lack of accurate knowledge are the principal handicaps of any attend of Covid 19 pandemic control. This crisis is not an IT problem or a mathematical modeling challenge. It is a catastrophe which impacts should be minimize with management reinforcement. Simplest management principles could be summarized and related to recovery management problem as follow:

- *Risk-benefit assessment* is the principal element of any decision. For pandemic recovery management the risk is composed by the high likelihood of relapses and their consequences severity. Consequences including not only infected and casualties but also the HS saturation. R parameter is emerging day by day as best risk indicator (short, mid and long term);
- Management can always be improved, even more, should always be improved. This is the *continuous improvement principle* that is universally accepted. In this moment, Covid19 crisis management has a tremendous improvement margin, because there is not any previous experience in the world;
- A *tool for planning* (measure and activities) and *a tool for checking* (control parameters and method) are needed. Both things are intimately related because you cannot manage what you cannot measure (Lord Kelvin *dixit*).
- *Hierarchization*. Not all possible options have the same relevance in terms of recovery optimization.

To start the recovery phase and to return the country to its "new" normal activity, a tool to monitor and control how the pandemic is evolving is needed. Otherwise, relapses could occur without realizing it and an essential time to stop the relapse could be lost. To that end, a control system allowing the detection, diagnosis and prognosis of pandemic abnormal behavior was proposed. Once there is a tool proposed to control the evolution of the pandemic, the work proceeds to design the post-confinement strategy. The methodology selected to design a strategy for the sequential opening of activities will minimize the risk of the potential impact of these activities on R. In order to do so, the number of contacts, the intensity of the contacts and the modification potential of each one of the activities will be assessed. Based on this evaluation, strategies are established for each level of risk, minimizing the possibility of contagion [7]. Logically, those activities with a higher risk of contagion will take longer to start up than those with a minimal risk. Companies whose activity is in a high-risk area will have to take measures to reduce the probability of contagion to an acceptable level and thus re-enter into economic activity.

2 Research Development

2.1 Selecting a Model to Calculate New Indenture Level Confinement Periods

In this Chapter we propose to follow Gañán-Calvo et al. [8] approach, which allows to model the pandemic "Confirmed" C(t) and "Deaths" D(t) using only time as independent variable, and supports the existence of a power-law in this variable. In their proposal, they concentrate in two non-dimensional parameters both: (i) the fundamental properties that the medium exposes to the action of the virus, and (ii) a simple model for the early behavior of the system prior to the asymptotic regime.

More precisely, a self-similar simple universal time-power law of the type $\varphi = \tau^{\alpha}$ is used to predict the behavior of COVID-19 pandemic before containment measures are enacted, where τ is the appropriate non-dimensional time since the onset of the free expansion, α is a fitting parameter with a value $\alpha = 3.75$ and φ the non-dimensional time descriptor of the infected population and mortality.

The predictors are defined as $\varphi_C = C/C_C$ or $\varphi_D = D/(m \cdot C_C)$ for confirmed and deaths respectively, where $C_c = 1.2 \times 10^4$ is a characteristic size of the pandemic infectious population, and m is an average early mortality descriptor observed with respect to confirmed cases, which may depend on the population structure and health system (showing homogeneity around the value m = 0:15, regardless the country).

Their more relevant result to this paper is the total confinement quarantine that they recommend after the first 100 cases of any unknown infection produced by a coronavirus are reported. They suggest to take a quarantine of the order of two times the period given by

$$T_Q = (t_{90\%} - t_c) = T_L \times (1 + (t_m - t_c)) \tag{1}$$

Where t_m is the day when measures are enacted, t_c the time where the expansion takes place and $t_{90\%}$ is the time when the 90% of the total maximum expected people infected is reached, specific for each geographic region. COVID-19 exhibits a characteristic infection time TL = 20.1 days according to measurements from the evolution of the pandemic in China. In the case that confinement is not yet in place, and the pandemic is in free expansion, quarantine measures must be taken immediately, and the death toll and quarantine period can be estimated using $\varphi_D(t)$ results once t_m is fixed. The model was applied to do the economic impact analysis of COVID-19 pandemic in Spain, Andalusia and its provinces and results obtained for these three government indenture levels as presented in Table 1.

Table 1. Table with model results for the three main government levels within Spain

	Date	t_c (days)	$t_m - t_c$ (days)	T_Q (days)
Spain	25-feb	0	18	76
Andalusia	4-mar	8	6	52
Almería	8-mar	12	2	44
Cádiz	10-mar	14	1	42
Córdoba	12-mar	16	2	44
Granada	12-mar	16	2	44
Huelva	16-mar	20	0	40
Jaén	11-mar	15	3	46
Málaga	5-mar	9	5	50
Seville	12-mar	16	2	44

2.2 Approaches and Variables for the Health System (HS) Modelling

The capacity planning problem, or sometimes the capacity expansion problem, is a classical problem in operations management literature. The problem the health systems face has important short-term dynamics, in fact, one can find some similarities to the problem that some organizations face with products rollouts to limit their short-term exposure while positioning themselves to capture the maximum long-term upside.

In this study, for instance, the magnitude of possible relapses will be uncertain and for our strategy development it is extremely important to know how quickly assumptions about can be converted to knowledge, and what to do when any assumption is invalidated, so managers must develop a kind of "discovery-driven planning" (as explained by McGrath et al. [9]). In these cases, the use of a disciplined process to uncover, test, and revise the assumptions behind the health system's response to pandemic, systematically, is required. By doing so, there is exposure to uncertainties common to pandemics, but uncertainties can be addressed at the lowest possible cost in public and economic health.

The problem the paper faces in this paper is the one of reaching a balance between the levels of the restrictive measures to take at the province level vs. the province HS's required capacity to deal with potential relapses. There are two common approaches to deal with this problem:

1. *Analytical models.* At this point, queue theory (QT) models (also known as compartmental models) are the most commonly used ones. The utilization of these models requires the knowledge of the rates of patient's arrival, and the distribution of the time for patients' treatments, in hospitals plants and intensive care units (ICUs). Other analytical models to deal with this problem such as Linear Programming models can also be used [10], although many authors of these models recognize that it is very complex to treat the general capacity planning problem in a single optimization model including all aspects of the problem.
2. *Monte-Carlo Simulation models.* A more general approach is based in stochastic simulation [11]. The simulation will be carried out in the computer, and estimates will be made for the desired measures of performance [12]. The simulation will be then treated as a series of real experiments, and statistical inference will then be used to estimate confidence intervals for the desired performance metrics.

In this Chapter both simulation modelling approaches are used. First a QT model is built to understand HS constraints under stationary conditions. Second a simulation model helps in the process of forecasting HS conditions under certain dynamic scenarios. Both approaches are found to be complementary, and QT model results offer a good information to validate patterns in simulation models results.

High number of variables allowed to define the system accurately, but on the other hand, this will increase the complexity of the model and therefore it will be more difficult to solve. Modelling optimization entail reaching an intermediate point where the system is well defined without making its resolution very difficult. The variables chosen to define the queuing model and the simulation model are shown in Table 2. The variables are divided into three groups.

Although one of the queuing problem hypotheses is to use a FIFO policy (First Inside, First Outside), circumstance that will not happen in real life, since the severity of the patient in the queue will finally prevail over the arrival order, due to the small percentage of the cases that occur daily, it is assumed that calculated averages will not be altered and therefore the use of this model continues to be successful. The formulation of the stationary queuing problem (in Table 3 and Fig. 1) is applied to the HS.

Once the steady state problem was analyzed with the model in Fig. 1. The health system capacity planning problem was modeled (see Fig. 2) using continuous time stochastic simulation (other examples can be found in the literature in discrete event simulation too, like un Günal et al. [13]). Chronological issues are considered by simulating the number of patients to be treated at any time in the different units. The model is built using VENSIM [14] as simulation tool, which has special features to facilitate Monte-Carlo type of simulation experiments, and to provide confidence interval estimations. Thus, for the dynamic modelling of the problem dealing with hospital demand determination, we selected a robust, understandable SEIR model similar to others in literature [15], that has been then re-formulated in a particular way. The general formulation of the SEIR model is a set of four first order differential equations as follows:

$$\frac{dS}{dt} = -\beta \times \frac{(S \times I)}{P} \tag{2}$$

Table 2. The HS model variables.

Variables group	Variables definition	Notation	Units
System states related variables	Number of infected that goes to the hospital	*C*	People/day
	Treated at the plant (Queue or bed)	*Ntp*	People
	Treated in Bed at the Plant	*Np*	People
	Waiting for plant	*Nwp*	People
	Treated at the ICU (Queue or bed)	*Nticu*	People
	Treated in bed at the ICU	*Nicu*	People
	Waiting for ICU	*Nwicu*	People
	Death toll	*Dtoll*	People
	Confined at home	*Nch*	People
	Immunized	*Ni*	People
Flows related variables	Entering at the plant	λep	People/day
	With bed assigned at the plant	μba	People/day
	Directed to the ICU from triage	λicu	People/day
	Entering the ICU	$\lambda eicu$	People/day
	Entering home confinement	λehc	People/day
	Entering plant from ICU	$\mu epfi$	People/day
	Entering the ICU from Plant	$\mu eifp$	People/day
	Dying at the ICU	μid	People/day
	Dying at the plant	μpd	People/day
	Released for home confinement	μrhc	People/day
	Cured at home	μch	People/day
	Leaving plant	μp	People/day
	Leaving ICU	μicu	People/day
Ratios, times and capacities variables	Ratio of patients derived to plant	*Rp*	Ratio
	Ratio of patients derived to ICU	*Ricu*	Ratio
	Plant to home ratio	*Rph*	Ratio
	Plant Death ratio	*Rpd*	Ratio
	ICU Death ratio	*Rid*	Ratio
	Average total time spent in plant per patient	*Ttp*	Days

(*continued*)

Table 2. (*continued*)

Variables group	Variables definition	Notation	Units
	Average total time spent in ICU per patient	*Tticu*	Days
	Time spent in bed at the plant	*Tp*	Days
	Time spent in bed at the ICU	*Ticu*	Days
	Average time at home	*Th*	Days
	Time waiting in Plant	*Twp*	Days
	Time waiting in ICU	*Twicu*	Days
	Plant capacity	*Cpb*	Beds
	ICU capacity	*Cicu*	Beds

Table 3. Input data to solve the queuing problem applied to HS.

Variables	Notation	Units	Value
Daily number of infected arriving to hospitals	*C*	People/day	Province based
Ratio of patients derived to Plant	*Rp*	Ratio	Province based
Ratio of patients derived to ICU	*Ricu*	Ratio	Province based
Average total time spent in plant per patient	*Ttp*	Days	11
Average total time spent in ICU per patient	*Tticu*	Days	14
Plant capacity	*Cpb*	Beds	Province based
ICU capacity	*Cicu*	Beds	Province based
Plant to home ratio	*Rph*	Ratio	80%
Plant death ratio	*Rpd*	Ratio	15%
ICU death ratio	*Rid*	Ratio	13%

$$\frac{dE}{dt} = \beta \times \frac{(S \times I)}{P} - a \times E \tag{3}$$

$$\frac{dI}{dt} = a \times E - \gamma \times I \tag{4}$$

$$\frac{dR}{dt} = \gamma \times I \tag{5}$$

Where *S, E, I* and *R* are for *Susceptible, Exposed, Infectious* and *Recovered* populations, respectively. *P* is the complete population of the region under study (e.g., the population of a province), β is the force of infection or the disease transmission rate, *a* is the inverse of the latent infection period and γ is the inverse of the infection duration time. For this model Ro, the disease basic reproduction number, is defined as $R0 = \beta/\gamma$.

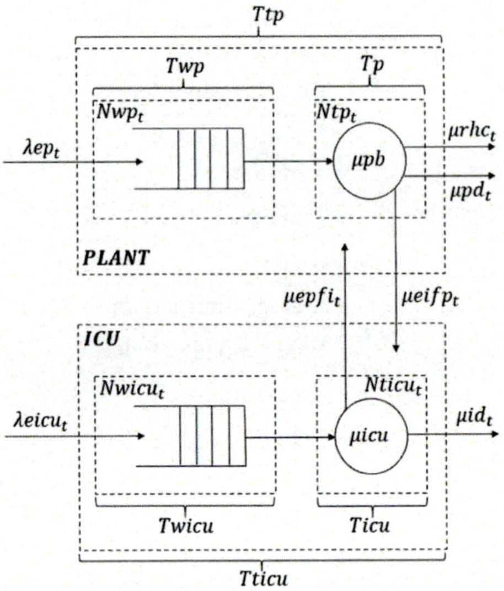

Fig. 1. Queueing problem applied to Plant and ICU

In Fig. 3, the model design selected for the empirical resolution of the above differential equations is presented.

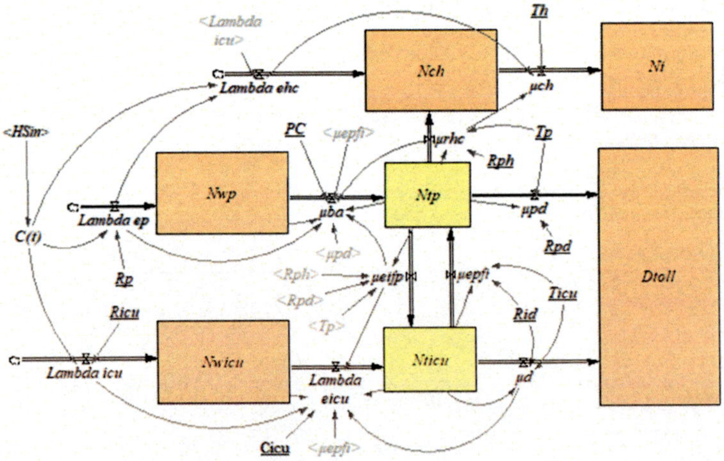

Fig. 2. Stock and flow diagram of the health system

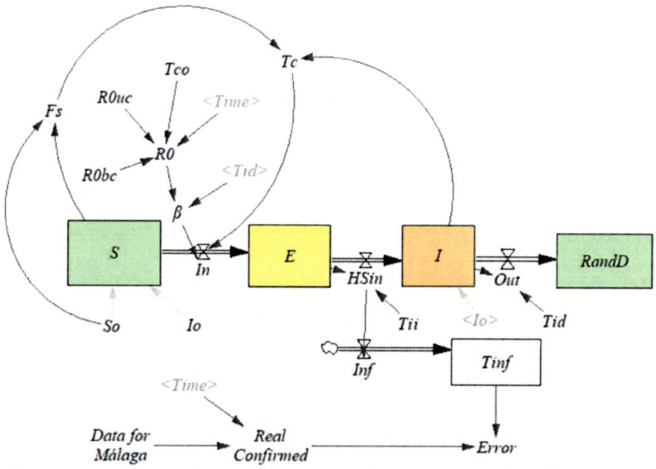

Fig. 3. Stock and flow diagram of the pandemic SEIR&D model.

2.3 Checking Tool: A Monitoring and Control System Proposal, Based on R

R parameter is the key element to pay attention to. It can be demonstrated that to control the pandemic rate of infection, $R < 1$, that means

$$R = \frac{N_C}{(\gamma \times I)} < 1 \tag{6}$$

Where N_C is the number of new cases that are diagnosed every day. Therefore, to monitor the number R, the three variables: N_C, γ and I, must be monitored. N_C and I would be taken from the information provided by the corresponding NHS and the estimation of γ must be improved as much as possible over time, incorporating not only accurate data for more infected cases, but also the evolution of knowledge from medical research about time duration and propagation potential of infected people, which is being improving significantly by week. Short-time R rising (days) will become mid-term (months) health system high strain scenarios. In order to implement a plan for a post-confinement de-escalation phase, tools for monitoring the status of the pandemic must be in place, and models for predicting its future potential evolution and consequences for the health system must be ready too. Beside the fact that the number R represents the pandemic status and its potential evolution at a given indenture level [16, 17], a very important point is that all re-opening measures can be evaluated in terms of their "R contribution" [7]. Also, any measure non-compliance and/or measure malfunction will increase the expected value of R. In summary, it is possible to employ R to monitor pandemic and to design, and control, measures that will be included in different recovery phase scenarios.

When R increases over one (principal warning threshold), this can be a clear symptom of a possible pandemic relapse. Most accurate alarm thresholds are supposed to be established in relation to HS capacity, as suggested in previous papers [18]. For instance, assuming a 1.5 R would generate a serious strain on the HS in less that 3 months, establishing an alarm threshold of 1.2 R could result in a reasonable conservative level

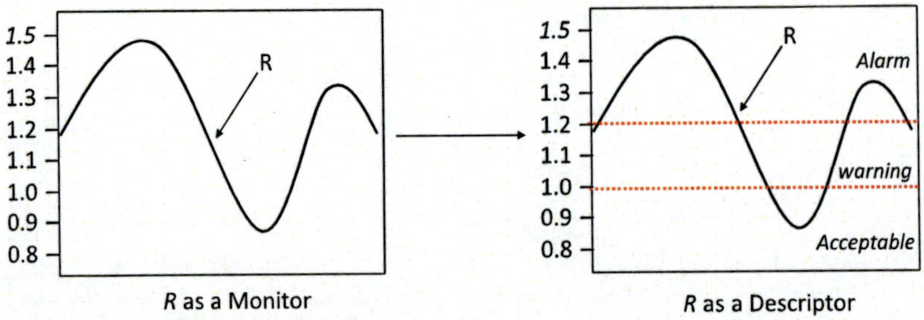

Fig. 4. R as a monitor (value) and as a descriptor (value, interpretation) of the problem

for the threshold. Therefore, *R* is not only a good monitor but also what is defined a good descriptor of the post-confinement plan situation (See Fig. 4). Consequently, there must be a Control System focused on *R*, and on the implementation of mechanisms for monitoring and integrating variables about the activities and resources that contributes to it. This Control System has the responsibility as a unique repository of information and governance coordination.

A pre-existing methodology, that is being applied successfully to monitor and control complex engineering assets [19, 20], was used to deal with this problem. The fundamentals of this proposal are:

- Providing a clear data/information structure for monitoring related to detection, diagnosis and prognosis issues;
- Offering a clear methodology for the accurate definition of the descriptors and interpretation rules, which are linked with different decisions and temporal horizons, formalizing expert knowledge within the monitoring process;

In order to describe, practical implementation of this Control System for *R* monitoring, Fig. 5 is a UML proposal about how structure data and integrate information flows. For this purpose, five different blocks are proposed (see Table 1) according to International standards.

Each one of the five blocks is considered a level of information, first two are sources of resources and activities variables, which are processed into monitors in block 3, block 4 behavior is centered on R and other predictors definitions, and the last one is oriented to standardize analysis and guide the decision making, see an example in next Table. The term "measure" is used here to name concrete activities of the pandemic recovery plan. Obviously, data quality is crucial for an adequate control, and control mechanisms have to evaluate continuously this issue, as it is explained in the next section.

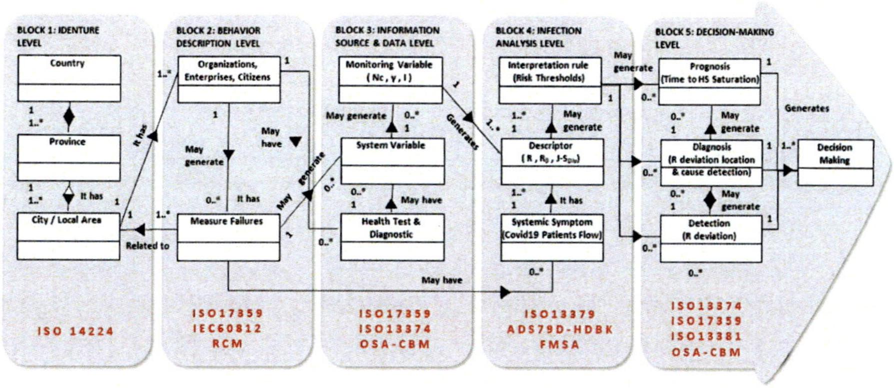

Fig. 5. UML simple diagram of the monitoring and control system proposed

3 Results

3.1 Economic Impact of Confinement When Changing the Management Level

The model has been applied to do the economic impact analysis of COVID-19 pandemic in Andalusia, defining the following three scenarios:

- Scenario 1. The management Indenture Level is Spain. Confinement period is calculated with data aggregated at a national level and confinement applies to the entire country, regions and provinces, simultaneously.
- Scenario 2. The management Indenture Level is Andalusia. Quarantine time is obtained with data aggregated at regional level and confinement applies to all region's provinces, simultaneously.
- Scenario 3. The management Indenture Level is each province. Quarantine time is obtained with data of the province and confinement applies to each province, separately.

In Table 4, results of the tree different scenarios are presented by province. In the last two columns we show potential reductions of the GDP loss toll relative to the first scenario. Considering an estimated value for the Andalusia's GDP around 160,222 M€ (For precise data see [21, 22]), the absolute savings in GDP loss toll for would be around 160,222 *M€* × (4.579% − 2.712%) = 2,991.34 M€ result of using the estimated quarantine times by province (which supposes an earlier return to economic activity) instead of the estimated quarantine time for the whole country.

3.2 HS Capacity Modelling and Implications on R

In the recovery phase of the Covid-19 pandemic, what is the force of infection that our HS can bear? What is the basic information that citizens should perfectly know in order take good decisions to collaborate in relapses risk reduction? How long will this next phase last? In order to consider this important aspect, the model is projecting results until

Table 4. Expected impact on current GDP per region, provinces and scenario

	GDP Loss (Scenario 1)	GDP Loss (Scenario 2)	GDP Loss (Scenario 3)	Reduction of GDP loss toll SC3 vs. SC1	Reduction of GDP loss toll SC3 vs. SC2
Andalusia	4,579%	3,133%	2,712%	40,776%	13,441%
Almería	4,745%	3,247%	2,747%	42,105%	15,385%
Cádiz	4,682%	3,203%	2,587%	44,737%	19,231%
Córdoba	4,239%	2,900%	2,454%	42,105%	15,385%
Granada	4,521%	3,093%	2,617%	42,105%	15,385%
Huelva	4,490%	3,072%	2,363%	47,368%	23,077%
Jaen	4,214%	2,883%	2,550%	39,474%	11,538%
Málaga	5,207%	3,563%	3,426%	34,211%	3,846%
Seville	4,176%	2,857%	2,418%	42,105%	15,385%

day 200, that is to say, 125 days after the release of the confinement, assuming this will take place the day 75, 56 days after the confinement (May 10th for Málaga). Multivariate sensitivity analysis has been done. It is considered the following hypothesis:

- [$Tinf_t \pm$ 30%] variability added in prediction for infected (random uniform),
- [$R\&T \pm$ 20%] variability added in flow rates and times in HS (random uniform),
- $R0ac$, after confinement, within the interval [0.85 – 1, 5], also (random uniform).

An example of results for the sensitivity analysis 400 simulations is presented in Fig. 6, where ICU occupations for 125 days since the release of the confinement is analyzed. These results show:

- The risk of ICU saturation is very low, in only 5% of the simulations.
- No saturation would take place in Plant, never reaching the 2000 patients regardless the possible queue of patients in plant waiting for ICU.
- Although this would be the maximum level of risk to bear within the period analyzed (R0 = 1.5), ensuring control to limit R0 to 1.1 persons maximum would be advisable. It would be attainable by monitoring the status of the variable, preparing for eventual confinement that could take place if needed due to an important relapse.
- According to the data and calibration of the model in Málaga, in case of sudden relapses (monitored over a certain period, for instance 5 days) a 2 weeks quarantine would be enough to lower R0 to reach levels below 1 person (see strategy in Qun Li et al., 2020).

3.3 Planning Tool: A Plan for De-escalation of Confinement Based on R

Once understood the important of the effective reproductive number R, and established in advance a surveillance, monitor and control system for R, a plan for the de-escalation

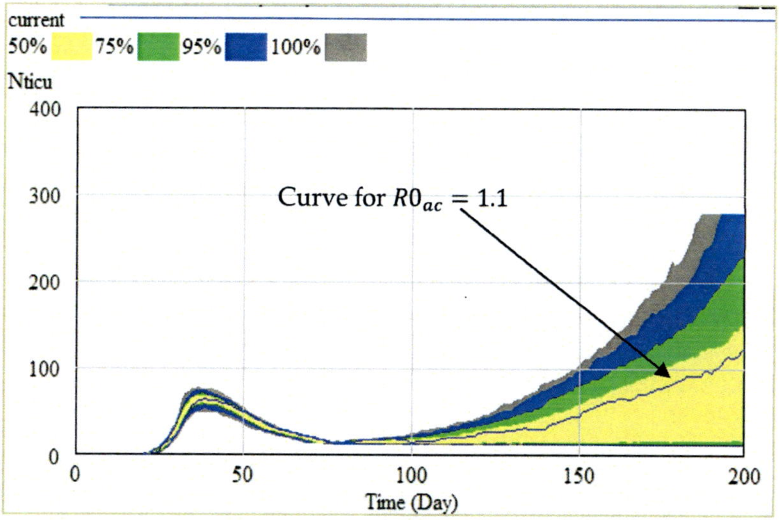

Fig. 6. Sensitivity results for ICU occupation.

of confinement will now be elaborated with the intention to reduce the risk of sudden R increase. At this point, and before proceeding with any possible escalation of activities, the following advice should be a golden rule [23]: (i) to keep physical distancing measures, and to ask people to remain at home; (ii) To increase access to diagnostic testing (focusing testing and resources on individuals with disease who may be infectious and their close contacts); (iii) Sectors to start reopening when the following 4 criteria are met: (1) two weeks decline in the number of new cases; (2) rapid diagnostic testing capacity sufficient to test people with symptoms, close contacts and those in essential roles; (3) the HS has appropriate PPE for healthcare workers; and (4) there is sufficient public health capacity to conduct contact tracing for all new cases and their close contacts.

Considering that all previous paragraphs requirements are fulfilled, this Section provides a strategy, a manner to proceed with a staggered reopening of activities, that is based on modelling their risk, and the expected time to the establishment of effective risk mitigation measures preventing the spread of the virus.

Two features of an activity, with a direct impact on R, will drive this course of action [7]: The activity contact rate (R_C) and the likelihood of infection per contact (R_i) — between a susceptible person and an infectious person or vector—during the time the activity takes place. In addition, the contact rate will be characterized by: the contact intensity (C_I) and the expected number of contacts (N_C) that the infected is supposed to have during that activity. The activity contact intensity (C_I) is now rated as either low, medium, or high. It depends on the contact type (close-to-distant) and its duration (brief-to-prolonged). Low contact intensity activities are brief and fairly distant interactions while high contact intensity activities involve prolonged close contact (like sharing a dormitory). Sharing a meal in seats separated by several feet can be considered a medium contact intensity activity. The number of contacts (N_C) will also be rated as either low, medium, or high. Defined as the approximate number of people in the setting at the same time, on average. A higher N_C is presumed to be riskier. Finally, each activity will have a *modification potential* (the degree to which mitigation measures can reduce those risks, R_i). For instance, businesses that can effectively incorporate physical distancing and engineering controls are considered to have a higher modification potential than those relying on administrative controls or personal protective equipment, to reduce risk. A hierarchy of COVID-19 mitigation measures can look like:

- Physical Distancing: wherever possible having people work from home; including restructuring efforts to minimize physical presence.
- Engineering controls: creating physical barriers between people.
- Administrative controls: restructuring responsibilities to reduce contacts and using technology to easy communication.
- PPE: having people wear gloves and masks.

Depending on the location in the matrix, the risk of the activity changes and so changes the likelihood of that risk to be mitigated in the near term (Fig. 7). Thus, the rules for the reopening strategy can be formulated.

R_c			
25	**Last to go !! with possible measures.** High risk of contact and high potential for modification.	**Cannot go!!.** High risk of contact and with potential means of modification.	**Cannot go !!** Very high risk of contact and low potential for modification.
15	**Could go with high safety measures.** High risk of contact and high potential for modification.	**Last to go !! with all safety measures.** High risk of contact and potential means of modification.	**Cannot go!!** High risk of contact and low potential for modification.
9	**Can go with protective measures.** Medium-high contact risk and high modification potential.	**Could go with high safety measures.** Medium-high contact risk and high modification potential.	**Last go!!** Medium-high contact risk and low modification potential.
5	**Can go with caution.** Activities with average contact risk and high modification potential.	**Can go with protective measures.** Medium risk of contact and with potential means of modification.	**To avoid as much as possible.** Medium risk of contact and low potential for modification.
3	**First to go.** Low-medium contact risk and with high modification potential.	**Can go with caution.** Low-medium contact risk and with potential means of modification.	**Can go with extreme caution.** Low-medium contact risk and potential means of modification.
1	**First to go.** Low contact risk and high modification potential.	**First to go.** Low contact risk and potential means of modification.	**Can go with caution.** Low contact risk and low modification potential.
	1	3	5 R_i

Fig. 7. Risk matrix definition and action plan for the activities.

Of course, the de-escalation of activities must be applied to their non-operative portion only, thus allowing to recover the "normal" sectors' conditions. Figure 8 shows the recovery of six relevant national economic sectors over a one-year horizon. Note that sectors such as agriculture and industry will be the first to achieve full recovery, while the entertainment sector will not return to full normality until May 2021 (where we may expect a vaccine could be released). It is very important the fact that, although the sector reaches100% of activity, that does not mean the production will be at the same level since demand may drop. For example, in the case of hostelry, it could be working 100%

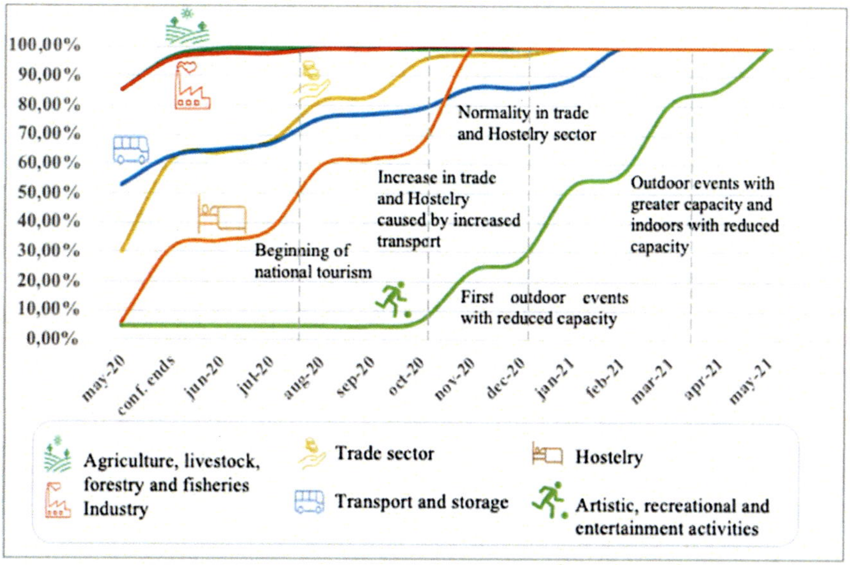

Fig. 8. Recovery by sector (in%)

after the end of 2020, but transportation activities would not reach 100% until February 2021, which would cause a drop in hostelry sector's revenues.

4 Conclusions

In this Chapter, the efforts done by the SIM research group in Spain concerning COVID19 pandemic management analysis are presented. SIM group aim was always to put the best skills and experience serving the society facing this unprecedent situation. The first impression when reading, in mid-march, very serious papers on COVID 19 emergency was that the generation of practical management tools to tackle the crisis was being forgotten. A great volume of data, mathematical models, etc. could be found, but the question was: how could government, companies and citizen take advantage of these results to reduce the pandemic impact?

Since the very beginning the SIM group highlighted great similarities and analogy points between pandemic recovery management and maintenance management of a complex engineering asset. Our research discipline provides specific verified tools and powerful analysis methodologies that we really think can be used as reference for this brand-new scenario management. At this point, and as a conclusion of this work, we would like to highlight the main SIM group contributions:

- Analysis of the importance of correct determination of the indenture level to manage the COVID recovery. This level was linked with geographical areas where management should focus on. We proposed and justify for Spain to use at least the province level as a reference, from local different quarantine time estimation according with real local pandemic behavior and GDP province expect impact. Three weeks after paper 1 publication date, Spain Government has followed a similar strategy, either also countries as Germany. Thus confirms, at least by the current knowledge, the goodness of this approach.

- A tool based on a ICUs and plants capacity model. Principal outputs: (i) ICUs and plants saturation estimation data (according to incoming rate of patients), (ii) with this results new local and temporal confinement measure can be planned and also a dynamic analysis can be done to estimate maximum Ro saturation scenarios, and finally (iii) provide citizen with clear and accurate data allow them adapting their behavior to authorities' previous recommendations. Just releasing this research to be published we could see Chancellor Angela Merkel presenting a similar approach in Germany.
- A complete process (end-to-end) describing with great detail a tool to control the performance of proposed recovery measures and their impact. Three fundamental pillars are considered: Definition of activities and their hierarchy in terms of risk in R contribution, consideration of uncertainty of data and R model calculation and monitoring and control system proposal.

Acknowledgements. We dedicate this Chapter and our research to "the people in the front line", those valuable individuals giving out their best to our community at the risk of their own lives. To all of them... thanks!!!

SIM group members participating in this research were the following, by alphabetic order: Eduardo Candón Fernández, Adolfo Crespo Márquez, María Crespo Márquez, Antonio de la Fuente Carmona, Juan Francisco Gómez Fernández, Vicente González-Prida Díaz, Antonio Guillén López, Pablo Martínez-Galán Fernández, Pedro Moreu de León, Enrique Paloma Castro y Antonio Sola Rosique. Also participating, Alfonso M. Gañán-Calvo is an external collaborator of our group. We also thank Beatriz Crespo for her kind collaboration with data ETL.

References

1. CEN (2010) EN 13306:2010. Maintenance Terminology. Brussels
2. British Standards Institute (1991) BS 5760–5. Reliability of Systems, Equipment and Components Part 5: Guide to Failure Modes, Effects and Criticality Analysis (FMEA and FMECA). London
3. U.S. Department of Defense (1981) Reliability modeling and prediction. MIL-STD-756B. Military Standard
4. Mcclellan MB, Silvis L, Gottlieb S, Rivers C, Watson C (2020) National coronavirus response, American Enterprise Institute
5. Fraser C et al (2009) Pandemic potential of a strain of influenza A (H1N1): early findings. Sci NIH Publ Access 324(5934):1557–1561. https://doi.org/10.1126/science.1176062
6. Dietz K (1993) The estimation of the basic reproduction number for infectious diseases. Stat Methods Med Res 2(1):23–41. https://doi.org/10.1177/096228029300200103
7. Rivers C, et al (2020) Public health principles for a phased reopening during COVID-19: guidance for governors 2 authors
8. Gañán-Calvo AM, Hernández Ramos JA (2020) The fractal time growth of COVID-19 pandemic: an accurate self-similar model, and urgent conclusions, 27 March 2020. arXiv:submit/3104086

9. McGrath RG, MacMillan IC (2000) The entrepreneurial mindset: strategies for continuously creating opportunity in an age of uncertainty, vol. 284. Harvard Business Press, Boston
10. Duffuaa SO (2000) Mathematical models in maintenance planning and scheduling. In: Ben-Daya M, Duffuaa SO, Raouf A (eds) Maintenance, modelling and optimization. Kluwer Academic Publishers, Boston
11. Dekker R, Groenendijk WP (1995) Availability assessment methods and their application in practice. Microelectron Reliab 35(9–10):1257–1274
12. Hoyland A, Rausand M (1995) System reliability theory. Models and statistical methods. Wiley, New York
13. Günal MM, Pidd M (2010) Discrete event simulation for performance modelling in health care: a review of the literature. J Simul 4(1):42–51
14. Vensim, Version 5.4. (2004) Ventana Systems Inc. Harvard, Massachusetts
15. Iwata K, Miyakoshi C (2020) A simulation on potential secondary spread of novel coronavirus in an exported country using a stochastic epidemic SEIR model 1(6). https://doi.org/10.20944/preprints202002.0179
16. Flahault A et al (2006) Strategies for containing a global influenza pandemic. Vaccine 24(44–46):6751–6755. https://doi.org/10.1016/j.vaccine.2006.05.079
17. Crespo Márquez A, SIM Research Group (2020a) A COVID-19 recovery strategy based on the health system capacity modeling. Implications citizen self-management. https://idus.us.es/handle/11441/95407. Accessed 30 April 2020
18. Crespo Márquez A, SIM Research Group (2020b) Strategies for COVID-19 pandemic recovery. Applying engineering asset management principles'. https://idus.us.es/handle/11441/94965. Accessed 30 April 2020
19. Guillén AJ, et al (2016) A framework for effective management of condition based maintenance programs in the context of industrial development of E-maintenance strategies. Comput Ind 82:170–185. https://doi.org/10.1016/j.compind.2016.07.003
20. Ferrero Bermejo J, et al (2019) A review of the use of artificial neural network models for energy and reliability prediction. A study of the solar PV, hydraulic and wind energy sources'. Appl Sci (Switzerland) 9(9). https://doi.org/10.3390/app9091844
21. Contabilidad Regional de España, Revisión Estadística 2019. Instituto Nacional de Estadística. www.ine.es
22. Contabilidad Provincial (2019) Anual de Andalucía. Instituto de estadística y cartografía de Andalucía. Consejería de Economía, Conocimiento, Empresas y universidad. www.juntadeandalucia.es
23. National coronavirus response: A road map to reopening | American Enterprise Institute – AEI (no date). https://www.aei.org/research-products/report/national-coronavirus-response-a-road-map-to-reopening/. Accessed 30 April 2020

Editors Bio Briefs

Adolfo Crespo Márquez (PhD FISEAM) is currently Full Professor at the School of Engineering of the University of Seville, and recently Head of the Department of Industrial Management. He recently received the Spanish Maintenance Award 2020 from the Spanish Maintenance Association (AEM). Prof. Crespo is the author of nine books and many research publications in journals conference proceedings. He leads the Spanish Research Network on Asset management, the SIM (Sistemas Inteligentes de Mantenimiento) research group and has extensively participated in many engineering and consulting projects for different companies, for the Spanish Departments of Defense, Science and Education as well as for the European Commission (IPTS). He is the President of INGEMAN (a National Association for the Development of Maintenance Engineering in Spain) since 2002.

A. Crespo Márquez et al. (Eds.): WCEAM 2019, LNME, pp. 307–308, 2021.
https://doi.org/10.1007/978-3-030-64228-0

Dragan Komljenovic (P. Eng PhD FISEAM) works as a Senior Research Scientist at the Hydro-Quebec's Research Institute (IREQ) in the field of reliability, asset management, risk analysis and maintenance optimization. Dragan worked almost 12 years as a Reliability and Nuclear Safety Engineer at the Gentilly-2 Nuclear Power Plant, Hydro-Quebec. He collaborates with several universities in Canada and abroad. Dragan has published more than 80 refereed journal and conference papers. He is a Fellow of the International Society of Engineering Asset Management (ISEAM), and Vice-President of the Montreal Chapter of Society of Reliability Engineers (SRE).

Joe Amadi-Echendu (BSc MSc DPhil CEng PrEng FISEAM) is a Professor of Engineering and Technology Management at the Graduate School of Technology Management, University of Pretoria, South Africa. Joe is a registered professional engineer in both UK and South Africa and has more than 38 years of work experience in industry and academia. Joe is a founding Fellow of the International Society of Engineering Asset Management (ISEAM), the current chair of the board of directors of ISEAM; as well as a current board member and treasurer of International Association for the Management of Technology (IAMOT). Professor Amadi-Echendu is the founding Editor-in-Chief of *Engineering Asset Management Review* series publication by Springer.

Author Index

A. Crespo Márquez et al. (Eds.): WCEAM 2019, LNME, pp. 309–310, 2021.
https://doi.org/10.1007/978-3-030-64228-0